In trop
crops in a small area for subsistence with
limited resources a good understanding of how environmental conditions affect the characteristics and performance of these crops is essential. This book considers the response of tropical food crops to environmental factors such as climate, soil and farming system. Three types of
crop are considered – cereals, legumes and non-cereal energy crops –
with individual chapters on the four most important crops in each
group. This material is set in context by introductory chapters on tropical farming systems, tropical climates and tropical soils.

This new, updated edition retains the successful formula of the
first edition, and will serve the needs of advanced students of tropical
agriculture, as well as professionals engaged in research and extension
work in tropical crop production.

The ecology of tropical food crops

The ecology of tropical food crops

Second Edition

M. J. T. NORMAN

Emeritus Professor of Agronomy, University of Sydney, Australia

C. J. PEARSON

Professor of Agronomy, Department of Crop Sciences, University of Sydney, Australia

&

P. G. E. SEARLE

Program Manager, ANUTECH Pty Ltd, Australian National University, Canberra, Australia

CAMBRIDGE
UNIVERSITY PRESS

Published by the Press Syndicate of the University of Cambridge
The Pitt Building, Trumpington Street, Cambridge CB2 1RP
40 West 20th Street, New York, NY 10011-4211, USA
10 Stamford Road, Oakleigh, Melbourne 3166, Australia

First published 1984
Second edition 1995

Printed in Great Britain at the University Press, Cambridge

A catalogue record for this book is available from the British Library

Library of Congress cataloguing in publication data

Norman, M. J. T. (Michael John Thornley)
The ecology of tropical food crops/M. J. T. Norman, C. J. Pearson,
P. G. E. Searle. – 2nd ed.
p.   cm.
Includes bibliographical references (p.   ) and index.
ISBN 0 521 41062 2 (hardback). – ISBN 0 521 42264 7 (paperback)
1. Food crops – Ecology – Tropics.   2. Tropical crops – Ecology.
I. Pearson, C. J.   II. Searle, P. G. E.   III. Title.
SB176.T76N67   1995
630′.2′5745264–dc20   94-31156 CIP

ISBN 0 521 41062 2 hardback
ISBN 0 521 42264 7 paperback

TAG

CONTENTS

PREFACE

The first edition of *The Ecology of Tropical Food Crops* was published at the end of a grand era of research on the physiology, ecology, agronomy, soil science and micrometeorology which underlie crop production. This research was driven by international concern about the need to increase food production to meet demands from increasing population, particularly in tropical countries, and by developments in scientific methodology and funding. As we began the preface of that edition: 'No-one in the 1980s needs to be reminded of the necessity to increase food production in tropical countries...the number of people in the tropics ...will rise by AD 2000 to over 3 billion.'

In the decade since, some things have changed markedly and others have remained unchanged. Agricultural science has changed: research in the disciplines which immediately underpin crop production has waned while more attention has been given to two complementary areas. More attention is now directed to farming systems and particularly to environmental research and resource management, and to more basic fields such as molecular biology. There is thus a broader attack on problems and opportunities of tropical food crops than ten years ago.

While predictions made in 1984 about population growth proved true, few people foresaw the consequences of that growth, the rural poverty which would persist and deepen in parts of the tropics, and the widespread degradation of soil and water which would result from poor management of tropical cropping systems. World population is now predicted to reach 8 billion by 2025 and most models suggest recurrent serious food shortages in tropical countries from 2000. Food shortages will be exacerbated in tropical countries by reduced yield potential in some areas due to global warming and to soil degradation resulting from widespread removal of tropical vegetation and over-cropping.

Further, in developed temperate countries, affluence, concern with 'quality of life' and the environment, and relatively poor economic rewards for farmers, all suggest that tropical countries are going to have to rely on their own crop production to meet their food shortages early in the twenty-first century.

The deepening need for increasing productivity of tropical crops will be met through creating opportunities based on better understanding of the crop environment. This approach remains unchanging. Also unchanging is our definition of the environment: its components are the atmosphere, the soil, the biotic environment, particularly pests and diseases (which we do not report), and the cropping systems of which the individual crop forms a part. Thus, after a decade it is timely to update this book while retaining its original format, as a source of information on the components of the ecology of tropical crops. As in the first edition, we have not reported on deliberate modifications to the environment through agronomic or other practices, nor on unplanned degradation or improvement of the environment as outcomes of cropping. Instead we have chosen to keep this volume of limited size and suggest complementary reading of crop management practices and regional resource assessments.

The book is in four parts. The first is a general account of the three environmental components which we cover, namely cropping systems, climate and soil. This is followed by parts devoted to the most important crops within each of the three broad groups of food crops: cereals, legumes and non-cereal food energy crops.

We are grateful to those who helped us with the first edition and whom we listed in that Preface. In addition, C.J. Asher, R.M. Bourke, E.T. Craswell, D.G. Edwards, M. Hutchinson, R.F. Isbell, A. Koppi, P. New and P.A. Sanchez have helped with this second edition. However, we are responsible for errors of fact, interpretation and judgement.

June 1994   M.J.T.N.
                                                             C.J.P.
                                                             P.G.E.S.

I

GENERAL

1

Tropical cropping systems

1.1 Introduction

Food crop systems are communities of plants which are managed to obtain food, profit, satisfaction or, most commonly, a combination of these goals. Conway (1987) encapsulated the goals in the term 'increased social value'. Such systems are purposeful in that farmers (who are part of the system) can set goals and change them even when their atmospheric, soil, technical, economic or social environment may not be changing (Bawden & Ison, 1992). Purposefulness also implies that farmers may pursue the same goals by following different behaviour patterns, either in the same or in different environments. Thus, although the market and physical environments are major constraints on the type of food crops grown, and cause farmers in a specific locality to grow similar mixes of crops, the management of any particular crop — its environment as modified by the farmer — will depend on the goals of that farmer. Figure 1.1 describes a root-crop-based farming system to illustrate the interrelationships between the purpose for which a system is managed and its human, market and physical components.

The individual crop is influenced by the system of which the crop is a component. Thus, the ecological conditions under which an individual crop is grown are determined not only by its atmospheric environment and soil, and the modifications made to this environment by the farmer — through ploughing, weeding, irrigating, applying fertiliser, cutting leaves for livestock, etc. — but also by the preceding crop, particularly if it was a legume. Farmers may elect to grow cassava during a low-fertility phase of their cropping sequence, not because such conditions are favourable to the crop, but because they know that it will yield moderately in the circumstances while other crops would fail. When the farmer plants two crops together in an intercropping pattern, the expectation is that neither will yield as well as it would if grown separately,

Fig. 1.1. A schema representing the components of a traditional mixed root-crop-based farming system in the wetter regions of sub-Saharan Africa. Source: Juo (1989).

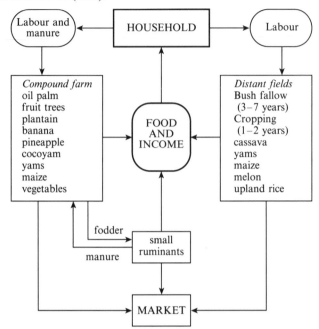

but that the combined yield will exceed that from two single crops collectively occupying the same area, and the diversity of product will give greater security of income.

This interdependence of crops within a cropping pattern provides the rationale for those sections in the chapters on individual crops that are concerned with that crop's place in tropical cropping systems. It is appropriate to precede these with a general account of the character of cropping systems in the tropics. Since the food crops covered in the book are all grown as annuals, with the exception of bananas and some cassava, this introductory chapter will be restricted to annual crop systems and to mixed annual/perennial systems in which annuals are a major component.

1.2 Classification of cropping systems

Each of the general texts concerned with tropical farming systems (Grigg, 1974; Manshard, 1974; Norman, 1979; Ruthenberg, 1980) uses different methods of classification and different nomenclature. The difficulties in arriving at a general-purpose farming systems typology which is at the same time rational and useful are discussed by Norman (1979). However, the restriction of this chapter to a particular group — tropical cropping systems of which annual crops (excluding vegetables) are a dominant or important component — makes the task of classification somewhat easier. The main categories here adopted, and given below, are based on those of Norman (1979):

1. Shifting cultivation systems.
2. Semi-intensive rainfed systems.
3. Intensive rainfed systems.
4. Irrigated and flooded systems.
5. Mixed annual/perennial systems.

The first two classes are rainfed cropping systems characterised by a fallow period during which no crop is grown and native or adventive species recolonise the cropped area. These and the third class form a series based on increasing *cultivation frequency*; that is, the duration of the cropping phase as a percentage of the total duration of the cultivation cycle (crop plus fallow). Thus if in a shifting cultivation system the land is on average cropped for 1 year and fallowed for 9 years, cultivation frequency is 10%. The boundary between shifting and semi-intensive cultivation is arbitrarily set at a cultivation frequency of 30%, and that between semi-intensive and intensive cultivation at 70%.

The fourth category has two subclasses: irrigated upland (i.e., non-flooded) systems and flooded or 'wet' rice systems. It is important to recognise that rice is grown under three sets of conditions: as a rainfed upland crop and as a flooded crop (wet rice) that may or may not be irrigated. The global area of non-irrigated wet rice substantially exceeds that of irrigated wet rice.

The fifth category, mixed annual/perennial systems, includes the following three subclasses: systems of annual crops in association with herbaceous perennials or semi-perennials, 'mixed garden' systems, and systems of annual crops in association with perennial tree crops.

1.3 Shifting cultivation systems

The term 'shifting cultivation' describes systems of rainfed cropping with annuals, biennials or short-lived perennials in which a cropping period alternates with a longer rest or fallow period, during which the abandoned crop area is recolonised by native herbaceous, shrub or tree species or by adventive species that find the ecological conditions favourable (Norman, 1979). The limit of cultivation frequency beyond which such cropping patterns are no longer termed 'shifting' is quite arbitrary: 30% (Ruthenberg, 1980) is the figure used here, while the limit for Sanchez (1976) is 50% and for Allan (1965) 10%.

One consistent feature of areas in which shifting cultivation is practised, and when cultivation frequency has become too high for the full re-establishment of forest before the next cropping break, is the development of savanna or scrub vegetation in place of forest. The term 'bush fallow' (Boserup, 1965; Morgan, 1969) is often applied to this type of fallow vegetation and to the cropping systems associated with it, as distinct from 'forest fallow' systems where the fallow break is long enough for trees to re-establish. Allan (1965) and Spencer (1966) carry the categorisation of shifting cultivation systems further. Regional accounts of shifting cultivation systems are given by Allan (1965), Vine (1968), Morgan (1969) and FAO (1974) for Africa; by Spencer (1966) and Kunstadter, Chapman & Sanga Sabhasri (1978) for southeast Asia and by Watters (1971) for Latin America. At this point it is worth noting that shifting cultivators, particularly in wet tropical forests, often gather a significant proportion of their total food supply from wild plants. Powell (1976) lists for New Guinea 251 species used for food: about 20% cultivated, 20% both cultivated and harvested wild, and 60% harvested wild.

The essential principles of shifting cultivation are that, during the fallow phase, nutrients are taken up by the recolonising vegetation and are returned to the surface soil as litter. Nutrients accumulated in the above-ground vegetation are made available to subsequent crops when the vegetation is cut down and burned. Cropping continues until available soil nutrients in the crop root zone decline, through crop uptake and perhaps immobilisation or leaching, to a level that gives an unacceptably low yield. 'Unacceptable' signifies the point at which the cultivator decides that the expected increment in crop yield from abandoning the current site and preparing a new one will repay the additional input of fresh clearing. The decision to abandon a crop area and clear elsewhere may also be influenced by the increased ingress of arable weeds as the cropping phase is extended.

Nutrient cycling in shifting cultivation systems is a complex of processes; major contributions to our understanding have been made by Nye & Greenland (1960) in relation to African conditions and by Sanchez (1976) in relation to Latin America.

During the fallow phase, nutrients are returned to the soil from the vegetation through litter fall, rainwash, timber fall and root decomposition. Litter fall is the most important. Below are given the average annual rates of nutrient return in litter from three African forest sites (Nye & Greenland, 1960):

Dry matter	12.5 t ha^{-1}	K	73 kg ha^{-1}
N	184 kg ha^{-1}	Ca	142 kg ha^{-1}
P	6 kg ha^{-1}	Mg	47 kg ha^{-1}

Under tropical forest, litter decomposition is rapid: 50–500% per year, according to McGinnis & Golley (1967), quoted by Sanchez (1976). Much of the resulting available nutrient store is, of course, subsequently taken up again and recycled through the forest vegetation; the net gain to the cropping system when the land is cleared is largely those nutrients taken up by trees from subsoil layers that would otherwise be unavailable to shallow-rooted annual crops.

When the fallow vegetation is cut down and burned, much of the nitrogen and sulphur is lost to the atmosphere, but other nutrients remain in readily available form in the ash. The figures given below for nutrient addition to the topsoil after burning 17-year-old secondary forest in Peru are from Sanchez (1976):

N	67 kg ha^{-1}	Ca	75 kg ha^{-1}
P	6 kg ha^{-1}	Mg	16 kg ha^{-1}
K	38 kg ha^{-1}		

The magnitude of the available nutrient increment depends not only on the amount of nutrients accumulated in the vegetation before burning, which is broadly related to biomass and hence to the duration of the fallow, but also on the efficiency of the burn.

The rate of decline in crop yield as the cropping phase progresses is dependent upon the natural fertility of the soil, on the amount of nutrients available in the ash, and on the nutrient requirements of the crops grown in the sequence. Figure 1.2 summarises some of the more reliable data. The data from Peru illustrate the effect of soil type and those from Zaïre the effect of crop type.

Various cropping tactics have evolved to mitigate the effects on crop production of declining fertility, the recovery of woody vegetation from cut stumps, and the increased incidence of arable weeds as the cropping phase proceeds. It is usual for the most valued crops — cash crops or primary subsistence energy crops — to be grown in the early phases of

Fig. 1.2. Yield decline with continued cropping in shifting cultivation systems. Figures below columns represent years of cropping; figures above columns are yields in t ha^{-1}. Source: Sanchez (1976).

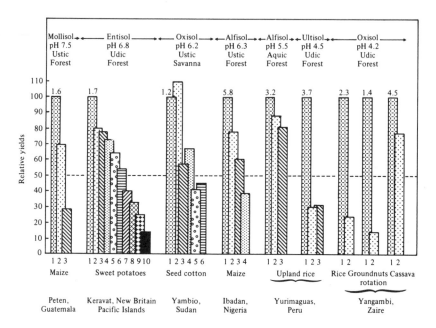

the cropping period, and for legumes to become more prominent in the later phases as available soil nitrogen declines. In the final year before the site is abandoned, it is common to plant crops with one or more of the following characteristics:

1. Tall robust crops able to compete with weeds.
2. Crops capable of yielding moderately well under low fertility.
3. Crops with an extended growing period, including biennials and perennials, that will continue to yield, with little or no attention from the cultivator, during the early part of the fallow phase.
4. Crops planted as a subsistence energy reserve in the case of failure of the primary crop (such crops may never be harvested).

Cassava, for example, has all these characteristics: in a survey of shifting cultivation systems in Zaïre, Miracle (1967) recorded cassava 5 times more frequently as the last crop than as the first crop in forest fallow systems and 11 times more frequently in bush fallow systems.

It is a common generalisation that cropping patterns in shifting cultivation are characterised by a very complex and irregular array of crop species grown in a single cleared area. This is only true of cropping in lowland forest in areas of favourable soil water regime. With increasing limitation to the environment, e.g., a shorter growing season, lower fertility, or lower temperature determined by altitude, diversity is reduced.

Where mixed cropping is practised, the exploitation of micro-environments within the cleared area by crops with specific requirements or tolerances is ingenious. Thus shade-tolerant species are planted on the shady margin of the plot, moisture-demanding species at the bottom of sloping sites, fertility-demanding species on localised ash concentrations or on hoed-up mounds of topsoil, and climbing species against unfelled tree trunks or beside rigid upright crops.

1.4 Semi-intensive and intensive rainfed systems

In Section 1.3, reference was made to the complex and irregular time and space pattern of crops in a single clearing that is characteristic of shifting cultivation in favourable environments. Such patterns may be described by the general phrase 'mixed cropping', signifying any situation where more than one crop is grown on a given land area at any one time. As will be seen in Section 1.4.2, the transition from shifting cultivation to more intensive cropping is associated with the emergence of more regular cropping patterns. In order to be able to discuss these, some definitions are given below; for a more complete list of terms relating to cropping systems, see IRRI (1984).

> *Intercropping.* More than one crop on a given area at one time arranged in a geometric pattern. A typical pattern in India might be two rows of sorghum alternating with one row of pigeon pea.
>
> *Relay cropping.* A form of intercropping where not all the crops are planted at the same time. A typical Indonesian example is rice and maize planted together and cassava interplanted a month or so later.
>
> *Sequential cropping.* More than one crop (or intercrop) on a given area in the same year, the second crop being planted after the first is harvested.
>
> *Cropping index.* The number of successive crops grown on the same land each year.
>
> *Land Equivalent Ratio (LER).* The ratio of the area needed under sole cropping to the one under intercropping to give equal amounts of yield at the same management level.

Useful regional accounts of semi-intensive and intensive rainfed cropping systems are given by Harwood & Price (1976) for Asia, Okigbo &

Greenland (1976) for Africa, and Pinchinat, Soria & Bazan (1976) for Latin America. Francis (1986) provides a general account.

1.4.1 *Historical development*

Shifting cultivation is — and historical evidence suggests that it always has been — the rainfed cropping system practised by rural populations with only hand-tool technology where labour, rather than land, is the major factor limiting crop output. In the past this has been true of both temperate and tropical climates. With increasing population density and a decline in the land area available per head for exploitation, average cultivation frequency increases: the cropping phase is extended and the fallow phase curtailed and finally eliminated. The biological and socio-economic consequences of such an increase in rural population are many and varied: Table 1.1 shows the important changes in an African context (Protheroe, 1972).

The application of animal manure to rainfed field crops, as distinct from house compounds or gardens, is not a widespread general practice in the tropics, and the use of mineral fertilisers is very largely confined to irrigated annual cropping and plantation crop systems. Hence when population density has reached a level at which all the available arable land is under continuous cultivation and the fallow eliminated, and there is no accession of subsoil nutrients through fallow vegetation burning, the general level of crop output is broadly determined by the inherent annual nutrient-supplying capacity of the soil in the crop root zone. Thus on Alfisols in intensive rainfed cropping areas of India and Africa, average cereal yields are of the order of 500 kg ha^{-1}: in contrast, in intensively cropped areas of Java on Andisols they are of the order of 1500 kg ha^{-1}.

1.4.2 *Cropping system changes with increasing cultivation frequency*

Table 1.1 indicates the general socio-economic pattern of change with increasing population density and the consequent transition from shifting cultivation to semi-intensive and intensive cropping. Narrowing the focus to the actual changes in cropping systems, the most significant features are:
1. Change in the character of the fallow vegetation.
2. Localisation of cropfields.
3. Systematisation of the cropping sequence.

Table 1.1. *Characteristics and components for a model of African population/land relationships*

	LOW DENSITY →	increasing numbers →	HIGH DENSITY Possible surplus rural population, seasonal/permanent migration, urban drift
Population	LOW DENSITY →	increasing numbers →	HIGH DENSITY Possible surplus rural population, seasonal/permanent migration, urban drift
System	SHIFTING CULTIVATION Increasing length of cultivation period →	ROTATIONAL CULTIVATION FALLOW Decreasing length of fallow period → SUBSISTENCE FOOD CROPS Decreasing importance → CASH (FOOD AND EXPORT) CROPS Increasing importance →	SEMI-PERMANENT/PERMANENT CULTIVATION Manuring and fertilising
Tenure	COMMUNAL RIGHTS TO LAND → (individual usufructary rights) Land allocation by need → Fragmented/dispersed holdings → No permanent demarcation of holdings →	Communal rights decreasing, individual rights increasing → Land transfer by pledge, rent, lease and sale Consolidated holdings Permanent demarcation of holdings	INDIVIDUAL RIGHTS TO LAND
Settlement	IMPERMANENT/MIGRATORY → SMALL VILLAGES/DISPERSED	Increasing permanence and nucleation →	PERMANENT/FIXED NUCLEATED AND DISPERSED
Exchange	NON-EXISTENT/LOCAL →	Increasing involvement at local, regional, national and international levels	

After Protheroe (1972).

4. Simplification of the within-field cropping pattern.

5. The use of draft animals.

These changes are discussed in greater detail by Ruthenberg (1980), Sanchez (1976) and Norman (1979).

Fallow vegetation. The fallow vegetation characteristic of semi-intensive cropping systems, here defined as those with a cultivation frequency of between 30% and 70%, is normally savanna grassland. The duration of the fallow, rarely more than 5 years, does not permit the re-establishment of forest. Furthermore, the general frequency of burning in an area under semi-intensive cultivation, the demand for firewood and, in many instances, the presence of ruminant livestock, all tend to favour the development of grassland. Because of its lower biomass and shallower root system, the cycling of nutrients by grassland is much less than that by forest.

Localisation of cropfields. In shifting cultivation systems, the individual cropping areas of the farming unit are scattered as temporary islands in an ocean of fallow-phase or unutilised natural vegetation. As the ratio of fallow land to cropland declines, the boundaries of the cropping areas become contiguous and permanently defined. However, this does not imply that all the cropfields of one farming unit are necessarily consolidated in a single land area: a geographical dispersion of the fields of one holding is very common.

Systematisation of the cropping sequence. With increase in cultivation frequency the somewhat irregular time sequence of crop and fallow crystallises into a rotation which, in its most intensive phase, becomes a rotation of crops only. On the other hand, it must not be assumed that all the fields of one farming unit are necessarily cropped at the same cultivation frequency. In a dispersed pattern of settlement, i.e., one characterised by isolated domiciles not strongly nucleated into villages, a roughly concentric zonation of cropping frequency tends to develop: fields near the domicile are cropped more frequently than outlying fields, and may also receive some animal manure so that the higher frequency can be maintained. In a nucleated settlement pattern, a clear distinction tends to emerge between the outlying field crop area and the house compound or garden, where a range of fruits and vegetables, with some staple crops, is grown at a more intensive level of input (see Section 1.6.2).

Simplification of the within-field cropping pattern. The change from shifting cultivation to intensive cropping is marked by a reduction in the number of crops grown in each field at one time, by the development of a more regular pattern of cropping and by the adoption of geometric planting. The less orderly planting and harvesting procedures associated with a complex crop mix and the exploitation of micro-environments within the field are replaced by more orderly intercropping, relay cropping and sequential cropping practices appropriate to a smaller number of crops and a more uniform cropfield. There is no necessary connection between row planting and the use of draft animals, for in general they are used only for the pre-planting operations of ploughing and harrowing.

The use of draft animals. Draft animals are very uncommon in shifting cultivation systems: the cultivators are often too poor to own large ruminants; in wet forest areas the environment is unsuitable in respect of both forage supply and health; the degree of land clearance achieved is usually inadequate for ploughing; and cropping is often undertaken on slopes too steep for draft animals to work on. The use of cattle and buffalo for draft is, however, very characteristic of semi-intensive and intensive cropping in tropical Asia, though far less so in Africa or Latin America. It should be noted that in rainfed cropping systems forages for animal consumption are rarely grown: ruminant stock subsist on common grazing land and crop residues.

1.4.3 *Climate and rainfed cropping systems*

In the general classification given in Section 1.2, the distinguishing criterion between the three groups of rainfed cropping systems was cultivation frequency, with no reference to type of crop. It is also possible to group rainfed systems very approximately on the basis of rainfall and length of growing season, which in turn is related to the dominant food energy crops grown:
1. Systems based on cereals in wet-and-dry climate zones.
2. Systems based on non-cereal energy crops in wet tropical climate zones.

This is most clearly illustrated in tropical Africa (Fig. 1.3). At the rainfall limits of agriculture south of the Sahara, cropping systems are dominated by pearl millet; moving to regions of higher rainfall and

Fig. 1.3. Cropping systems of tropical Africa in relation to rainfall. Dashed lines represent approximate limits of tropical crop production. (*a*) Annual rainfall (mm), (*b*) dominant crops in cropping systems. Source: Okigbo & Greenland (1976).

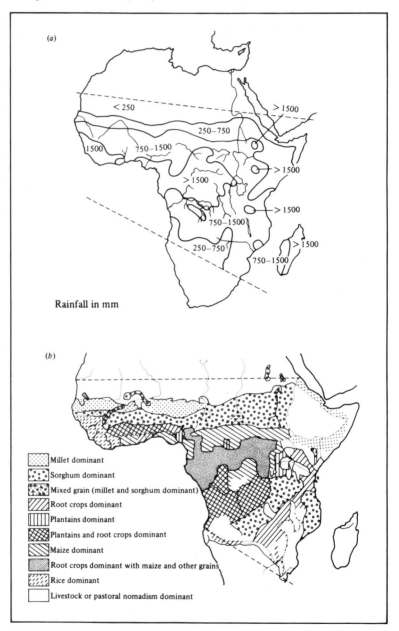

longer growing season, sorghum becomes important and then maize. In wet tropical zones with an extended growing season, root and tuber crops and bananas become dominant.

1.4.4 *Desertification*

Increasing human population, increasing livestock population, and increasing intensity and frequency of cropping may lead to a reduction in soil nutrients (e.g., Fig. 1.2) and degradation of soil structure. These processes make dryland cropping soils susceptible to erosion by natural agents (wind, water) and livestock. Population-induced over-cropping has exacerbated land degradation to the point of desertification in semi-arid areas such as parts of the African Sahel. Since land degradation is largely associated with misuse of soils (although it is accelerated by recent environmental change) it is considered in Chapter 3.

1.5 Irrigated and flooded systems

Ruthenberg's definition of irrigation as 'those practices that are adopted to supply water to an area where crops are grown, so as to reduce the length and frequency of the periods in which a lack of soil moisture is the limiting factor to plant growth' is adequate when applied to the irrigation of upland, i.e., non-flooded, crops. However, throughout the tropics 'wet' rice, as distinct from upland rice, is often grown under water supply and drainage conditions to which the above definition has no clear application.

Thus it is common for a slope or hillside to be terraced to form a series of fields for rice culture, through which excess surface water drains from the upper to the lower fields. The uppermost field may receive nothing but rainfall, and hence by no definition may be termed irrigated; the field below receives rainfall plus surplus water from the field above, and in turn its surplus drains into the field below it. Such a pattern is an impounding and drainage system, but not an irrigation system. On the other hand, where rainfall is augmented by water supplied in a controlled fashion (either by gravity flow from a higher storage or drawn up from wells, ponds or streams) and provision is made for controlled drainage, the system is 'irrigated'.

Thus a more inclusive definition of irrigation is 'the controlled supply of water to a crop area, additional to the rain that falls on it, and the controlled drainage of surplus water from the area' (Norman, 1979).

However, for the agricultural scientist, if not for the irrigation engineer, there is little essential difference between a flooded rice field with controlled water supply and one where the supply is uncontrolled; it is therefore appropriate to include them both in one category. Tropical irrigated annual cropping systems are described in detail in Ruthenberg (1980), Norman (1979), IRRI (1977a), and ASA (1976).

1.5.1 *Irrigated upland cropping systems*

The term irrigated upland cropping systems is here applied to irrigated cropping systems in which all the crops concerned are grown under non-flooded conditions. (The difficulties of farming systems typology become apparent when we attempt to classify systems in which irrigated upland crops are grown in fields in which, for the wetter part of the year, flooded rice is grown. These are considered in Section 1.5.2 under wet rice systems.)

Upland crops in the tropics are conventionally irrigated by gravity flow schemes, where water from a storage reservoir or stream is delivered through a reticulation system to fields laid out in ridge-and-furrow or bed-and-furrow patterns and surplus water is disposed of through a drainage channel system, usually into a river. (Sprinkler irrigation is scarcely used at all for tropical annual crops.) The largest of such systems is the Sudan Gezira, where annual rainfall is only 200–400 mm. Formerly, a fixed cropping pattern of cotton, fallow, sorghum and lubia bean (*Lablab purpureus*) was followed and only 40–50% of the land was cropped each year in order to conserve fertility. Cropping·is now more intensive and more diverse.

The Gezira is an example of a large-scale irrigation project, but throughout the tropics, particularly in the Indian subcontinent, there are smaller concentrations of irrigated upland cropping in which cotton, sorghum, pearl millet, wheat, maize, groundnuts, etc., are grown (see ICAR, 1972; Ruthenberg, 1980). It is worth noting that in India these crop systems are often associated with small-scale milk production and represent one type of tropical annual cropping system in which forages (e.g., sudan grass, berseem clover) are grown for ruminant consumption.

The above systems involve the periodic irrigation of crops during their growth period, but in some tropical areas crops are grown only on water stored in the profile. In riverine situations, land submerged during the period of high river flow may be planted as the water recedes to crops that do not receive any further water supply. Since the supply is not controlled, such operations can scarcely be termed irrigated cropping. However, in arid northern Sudan upland crops are grown under a system known as 'basin irrigation' where water supply and drainage are controlled. Channels from the Nile carry water into adjacent natural depressions, which may be subdivided by cross banks. When the river rises, the basins are flooded (to 70–100 cm), the water is impounded for about a month and then drained back into the river. The basins are then planted to crops such as sorghum and chickpea.

1.5.2 Wet rice systems

There is a wide range of cropping systems based on flooded rice, particularly in Asia (Harwood & Price, 1976; IRRI, 1975, 1977*a*; Vergara, 1992). Their pattern is governed by the period in the year for which water is available, from either rainfall or irrigation, and by the drainage characteristics of the cropping area. The hydrological aspects of rice land are discussed in detail by Moormann & van Breemen (1978).

However, before reviewing the types briefly, it is important to note two special soil nutrient features of wet rice cropping: nitrogen fixation under flooded conditions — particularly by autotrophic blue-green algae, *Azolla* spp. and heterotrophic bacteria — and the increased availability of soil phosphorus under flooding (see Section 5.5.3). These features help to explain the long-term stability of wet rice cropping systems without added fertiliser. On the other hand, with the advent of nitrogen-responsive, lodging-resistant and high-yielding cultivars, wet rice has become the main focus of increased fertiliser usage in the tropics.

Wet rice cropping systems, which are treated in more detail in Chapter 5, may be classified under three headings:

1. Shallow-water rice only.
2. Shallow-water rice with upland crops.
3. Long-standing floodwater rice

The first and second categories include cropping patterns in which rice is either the only crop or the wet-season crop and where it is flooded to a depth of less than 1 m. These areas comprise about 40 million (m) ha of rice in high altitude, semi-arid and arid environments and 111 m ha in rainfed and irrigated lowlands (Vergara, 1992). The degree of water control varies widely, and the pattern of rainfall and seasonal availability of irrigation water, if any, govern the number of crops that may be grown each year. Thus in some semi-arid regions where rainfall is limited and erratic and storages are small, only a single crop can be grown. In wet-and-dry climates with a 5–6 month wet season but with no irrigation, it is still possible to grow only one crop, but in twin-peak wet-and-dry and wet tropical climates, or in areas with ample irrigation during the no-rain period, two or even three crops may be grown. Fourteen climatic situations for rice-growing in south and southeast Asia are analysed in detail in IRRI (1975).

While the difficulty of drainage in low-lying tropical regions during periods of heavy rainfall may determine that only rice can be grown in the wet season, the water requirement of flooded rice generally exceeds that of upland crops of comparable duration by at least 50% (Oldeman & Suardi, 1977), and in many wet-and-dry climate zones where dry-season irrigation water is too limited for a second rice crop, upland crops are grown after wet-season rice (see Harwood & Price, 1976; Brammer, 1977). Intensive rice-upland crop systems with a cropping index of 3 or more are characteristic of subtropical Taiwan (Dalrymple, 1971).

The floodwater systems occupy about 13 million ha (Vergara, 1992). Here, floodwaters of more than 1 m remain for more than 1 month during the growth of the rice crop and, since the rise and fall of the water is rapid, special rice cultivars are planted. Throughout substantial areas rice is grown in water up to 6 m deep, and harvested by boat before the rapid lowering of the floodwaters. Floodwater or deep-water rice systems, which are all single-crop systems, are characteristic of deltas of the great rivers of south and southeast Asia.

1.6 Mixed annual/perennial cropping systems

Small-scale tropical farmers, even when they are producing cash crops, normally endeavour to be as self-sufficient for food as possible. In addition to annual subsistence crops and vegetables, a few perennial fruit or nut trees are usually also grown where the environment permits. Exceptions include rainfed cropping systems in marginal semi-arid climates where the dry season is too long, and shifting cultivation systems in forest where fruit and nuts are more likely to be gathered from wild than from planted trees. This section, however, deals only with cropping systems in which both annuals and perennials are major components of the cropping pattern.

1.6.1 *Annuals with herbaceous perennials or semi-perennials*

There are two main subtypes to consider: where the two types of crop are integrated into a cropping sequence and where they are not. An example of the first type is the intensive irrigated rice/sugarcane/upland crop pattern of southeast Asia, where the cane is treated as a long-season annual or biennial within a relatively short cropping cycle. A typical rotation in East Java is given in Table 1.2. Although the cropping index of this sequence is only 1.3, the level of input is high, since the land surface has to be modified from the flat wet-rice field to the bed-and-furrow configuration for sugarcane and upland crops, which require good drainage.

The second type is exemplified by a common cropping pattern in areas of East Africa where the dry season is relatively short, where sweet and non-sweet bananas are grown as the main source of food energy in association with annual subsistence crops such as finger millet (*Eleusine coracana*) (Parsons, 1970). A more widespread pattern in East Africa, which could also be classified as an annual crop/perennial tree crop system (see Section 1.6.3), is banana/coffee/annual crops. The time and space relations of the bananas and annual crops vary: the perennial may represent a semi-permanent component of a mixed cropping area with annuals growing between, or be separate from the annual crop area. In the latter case, however, it is normally intercropped with annuals during the establishment phase.

1.6.2 *Mixed gardens*

The term 'mixed garden' was coined in the early years of this century to describe the dense mixture of fruit, nut and vegetable crops characteristic of the house compound area in many parts of the wet tropics and high-rainfall zones of the wet-and-dry tropics, particularly southeast Asia (Terra, 1958), coastal south Asia and the West Indies. Strictly speaking, the term refers only to the house compound component of

Table 1.2. *Intensive rice/sugarcane/upland crop rotation, East Java*

Year	Months	Period (months)	Crop
1	January–May	5	Rice
1–2	June–December	up to 18	Sugarcane
2–3	January–May	5	Rice
3	June–December	6	Upland crops

cropping systems that also include separate areas of field crops (see Fig. 1.1) but it is convenient to apply it to the whole system.

In Indonesia the total area of house compounds exceeds 1 m ha, and their food energy output per unit area is comparable to that of wet rice. Terra (1954) lists 57 fruit trees, 7 nut trees, 6 spice trees, 35 'vegetable' trees yielding edible leaves or non-sweet fruit, and 45 species of annual vegetable or fruit crops grown in Indonesian mixed gardens. The wide range of crops grown provides a varied diet and often a source of cash from the sale of fruit or vegetables. The gardens also yield 'industrial' products: bamboo for construction, palm leaves for thatching and gourds for domestic vessels.

1.6.3 *Annual crops with perennial tree crops*

The term 'annual crops with perennial tree crops' is used to designate cropping systems where a single type of tree crop is a major component of the system and is grown as a cash crop in association with annual subsistence crops. Of the six major tropical perennials that are grown under plantation systems — tea, coffee, cocoa, rubber, coconut, oilpalm — all, with the general exception of tea, are also grown in small holdings by cultivators who grow annual crops for subsistence. Examples include the previously mentioned banana/coffee/annuals system of East Africa; cocoa (or oilpalm) with yams, maize, cassava in West Africa; rubber and rice in southeast Asia; and coconuts with annual crops in coastal areas of south and southeast Asia and Oceania. Ruthenberg (1980) gives a useful survey.

Further reading

Francis, C.A. (ed.) (1986). *Multiple Cropping Systems*. New York: Macmillan, 383 pp.

IRRI (1984). *Cropping Systems in Asia: On-farm Research and Management*. Los Baños, Philippines: International Rice Research Institute, 196 pp.

Ruthenberg, H. (1980). *Farming Systems in the Tropics*, 3rd edn. Oxford: Clarendon Press, 424 pp.

Tivy, J. (1990). *Agricultural Ecology*. Harlow, UK: Longman, 288 pp.

2

Tropical crop/climate relations

2.1 Environment and its broad influences on crop growth

The relation between crop and environment is a true interaction in that
it operates in both directions: the environment affects the crop and the
crop modifies its environment. The effects of the crop on its environment will generally not be dealt with in this book, but it is important to
appreciate that they are sometimes of major significance: for example,
root growth and exploration of increasing soil volume must be
described in order to predict realistic water use by crops.

2.1.1 *Radiation*

There are two atmospheric inputs which are used directly by a crop: radiation and carbon dioxide. The tropics have a year-round positive radiation balance and relatively small seasonal changes: monthly radiation during the period of highest solar angle is within 15% of that at the period of
lowest solar angle. Variation in total radiation has two effects on crop
growth. First, optimum plant population and potential grain yield per
hectare vary with season, being higher in high-radiation, cloudless seasons
than in wet or monsoon seasons. Naturally, this applies only when water
is non-limiting, e.g., to irrigated rice (Fig. 5.2). Second, crop yield components and grain yield per hectare vary according to radiation received, at
least during reproductive growth. Radiation influences yield components
in the order in which the component is determined. For rice, the number
of inflorescences per hectare is only weakly dependent on radiation
($r^2 \simeq 0.5$), whereas the number of grains per hectare is highly sensitive to
radiation until the late stages of crop growth (Evans & de Datta, 1979).

   The spatial variation in radiation in the tropics due to cloudiness is
related both to orographic rainfall (caused by altitude) and to seasonality of rainfall: e.g., monsoonal effects in India and southeast Asia

(Budyko, 1968). This variation is an important constraint on potential productivity; a compensation is that locations with less favourable incoming radiation usually have more favourable soil water status.

Net radiation (R_N) is the net amount of radiation absorbed by the crop and soil, i.e., short-wave (0.4–3 μm) radiation minus long-wave (5–300 μm) outgoing radiation. R_N is given by:

$$R_N = S(1 - A) - L,$$

where S is total radiation, A is the albedo of the crop/soil surface and L is the net loss of long-wave radiation. The albedo or reflection coefficient increases with increasing crop leaf area. It is approximately 0.2 after canopy closure; values of $(1–A)$ ranging from 0.68 to 0.86 are cited for wheat, maize, rice and cotton (Monteith, 1976). Likewise net long-wave radiation flux is about 0.2 of total incoming radiation, so that R_N/S ranges from 0.4 to 0.6 for tropical crops with closed canopies. Net radiation may, of course, be measured directly, but in practice acceptable accuracy to within 20% (Denmead, 1976) can be obtained by regressions of R_N on S, such as $R_N = 0.86S - 56$ for rice and $R_N = 0.82S - 90$ for maize, where R_N and S are in W m^{-2} (Denmead, 1976; Uchijima, 1976).

2.1.2 Daylength

Daylength, defined as the time from sunrise to sunset, is 12.1 h in all seasons at the equator and ranges from 10.6 to 13.7 h at latitude 25°. The range is much smaller than in temperate regions: e.g., 8.7–15.7 h at 45°. Nevertheless, the initiation of flowering in many tropical crops is sensitive to daylength. In West Africa, informal selection by farmers for daylength sensitivity in sorghum has resulted in crops that mature as available soil water is exhausted in the early part of the dry season; this ensures that the crop fully utilises the growing season yet avoids diseases associated with high humidity during grain maturation (Bunting & Curtis, 1968). In Asia, informal selection and more recent deliberate breeding have led to a range of rice cultivars from those with extreme daylength sensitivity to cultivars that are essentially daylength insensitive.

2.1.3 Temperature

Mean air temperatures at sea level in the tropics are usually between 20 and 27 °C, although in the wet-and-dry tropics means may reach 33 °C towards the end of the dry season and maxima may be as high as 45 °C.

Seasonal changes in temperature are small (e.g., Money, 1978). Most tropical crops display a broad temperature optimum from 25 to 35 °C. This results in optimum plant population and grain yield per hectare at a given tropical location not being detectably correlated with temperature, except where seasonal differences in temperature are substantial, as in continental wet-and-dry tropical climates of north India.

Between-year variation in temperatures in the tropics is generally small. At Hyderabad, India, the highest and lowest daily temperatures recorded since 1891 are within only ± 4 °C of the long-term mean highest or lowest daily temperature in the month. As a consequence, seasonal variation in yield of a given crop is most likely to reflect variable rainfall (and hence, possibly, radiation) and the incidence of pests and diseases.

Spatial variation in temperature is mostly brought about by altitude: temperatures fall by 0.65 °C for every 100 m increase in altitude to 1.5 km (Lockwood, 1974). This variation in temperature is confounded with greater cloudiness and lower water vapour pressure deficits, and hence lower evaporation, at high altitudes.

The rate of development (i.e., progression through the life cycle) of any particular crop is positively related to temperature (see Section 2.2). Development is slowed down when a crop is grown at higher altitude. The longer crop duration, coupled with a broad thermal optimum for growth, usually leads to grain yields at high altitude being equal to or greater than those at low altitude in the same climatic region. However, because of the slowing-down of development, the cropping system (particularly intensity of cropping) is often closely related to altitude. Thus, in lowland East Java rice takes 90–100 days from transplanting to maturity and three crops per year may be grown, whereas at 1600 m the same cultivar requires *c.* 220 days and is single-cropped. Ultimately the cultivation of any tropical crop is limited by altitude, e.g., coffee at 1500 m and maize and wheat at 3000 m (Trewartha, 1968), because of specific cardinal temperatures that must be met for crop developmental events to occur.

2.1.4 *Evaporation*

Potential evaporation from crop and soil surfaces crudely follows net radiation:

$$E_p = R_N/\lambda,$$

where E_p is potential evaporation (kg m^{-2}), R_N net radiation (MJ m^{-2})

and λ is latent heat of evaporation of water (2.453 MJ kg⁻¹ at 20 °C). Thus, if average daily net radiation is 9 MJ m⁻², average evaporation is 9/2.453 or 4 mm per day. Therefore, commonly occurring values in the tropics of 8–10 or even 15 mm per day are due to both direct radiant energy and advected thermal energy. Moreover, in contrast to temperate climates, high daily evaporation occurs over less than 8 h because of short daylength. This results in very high hourly water fluxes which cannot be met by crops even on moist soils; transient leaf water deficits are common.

Most tropical crops have evolved mechanisms of stomatal control in order to minimise or survive water deficits associated with such high evaporative demand: stomata close in response to increasingly negative leaf water potential and/or to increasing dryness of the air (saturation deficit) (Muchow *et al.*, 1980). Actual evaporation (E) is thus usually below that predicted from R_N/λ because transpiration is restricted by partial stomatal closure. Only one crop (pineapple) closes its stomata throughout the day, thereby reducing E/R_N to 0.08–0.28 (Ekern, 1965). Sunflower has evolved an alternative strategy whereby stomata remain open under conditions of high evaporative demand (Rawson & Constable, 1980), so that actual evaporation is reduced only by loss of leaf turgidity and consequent reduction in the interception of net radiation.

Papadakis (1975) recognised the sensitivity of crops to atmospheric water vapour (or more specifically to the gradient of water vapour from crop to atmosphere) and the impracticability of detailed environmental measurements in the tropics; he proposed an empirical estimation of E_p:

$$E_p = 5.625 \; \Delta q,$$

where Δq is the difference between saturated and average vapour content at the maximum daily temperature and the constant refers to E_p in millimetres and Δq in millibars.

Actual evaporation from a crop/soil system may be related to net absorption of radiation by the crop, R_{NA}, which may be measured as R_N above the crop canopy minus R_N at the soil surface. However, actual evaporation from a crop also depends on the gradients of latent heat (temperature ΔT, °C) and water vapour (Δq, kg water) from the crop to the atmosphere. Thus,

$$E = R_{NA}/(\lambda + C_p \Delta T/\Delta q),$$

where C_p is the heat capacity of moist air (1 kJ kg⁻¹) (Milthorpe & Moorby, 1979). Waggoner (1968) emphasised the dependence of actual

evaporation on radiation and water vapour content; he showed evaporation to be 5 times more sensitive to changes in radiation than to
changes of the same relative magnitude in windspeed.

It is appropriate here to note that the amount of water vapour which
may be held in air, generally referred to as saturated vapour pressure,
q_w(kPa), increases exponentially with temperature over the range
0–50 °C: $q_w = 0.611$ exp $(17.3T(T + 237))$. This contributes to high rates
of actual evaporation (Newell *et al.*, 1974). Total actual evaporation is
relatively constant from year to year in the tropics, having a coefficient
of variation of 5% according to Griffith (1972).

2.1.5 *Rainfall and erosivity*

Rainfall patterns in the tropics vary continuously from the wet equatorial zone to dry zones where rainfall is below the limit required for crop
growth and maturation (e.g., Newell *et al.*, 1974). The lower limit for
cropping is approximately 350 mm average annual rainfall (AAR),
although pearl millet is reputed to grow in areas of 125 mm AAR (Jain
& Mathur, 1961).

A higher total amount and a higher proportion of rain falls as high-
intensity storms (intensities of 1 mm min^{-1} for at least 5 min) in the
tropics than in temperate regions of comparable total rainfall. The high
intensity is due almost equally to the increased volume of each raindrop
(Fig. 2.1a) and the increased number of drops falling per unit time.

The kinetic energy of rain per volume or intensity of occurrence
increases curvilinearly with increasing intensity until it reaches a plateau
of about 28 J m^{-2} at intensities over 70 mm h^{-1}:

$$K = 210 + 89 \log I,$$

where K is kinetic energy (t ha^{-1}) per centimetre rain, and I is intensity
(cm h^{-1}) (Wischmeier & Smith, 1958). As a rough approximation, to
avoid measuring intensity, K may be related to total rainfall in any one
occurrence (R, mm) by the following equation (Kowal & Kassam, 1976):

$$K = 43R - 155.$$

High total rainfall and high kinetic energy of rain result in high kinetic
energy loads. Such loads may be distributed relatively evenly through
the year, as in the wet tropics, or they may be markedly seasonal, as in
wet-and-dry climates (Fig. 2.1b).

When rainfall intensity exceeds the infiltration capacity of soil and the rainfall has high kinetic energy, there is a high likelihood of erosive runoff. Generally runoff and soil erosion occur when rainfall exceeds 20–25 mm, either per hour (Hudson, 1971) or during the rainfall occurrence (Kowal & Kassam, 1976; Fig. 2.1b). By these criteria the proportion of tropical rainfall that is erosive is about 40% compared with 5% in temperate regions (Hudson, 1971); in northern Nigeria, 58% of the year's rainfall (635 mm) is erosive (Fig. 2.1b).

Soil loss, unrelated to slope, may be exacerbated by the low infiltration capacity of some tropical soils: cumulative infiltration (I_c, cm) is related directly to soil sorptivity (S, cm min$^{-1/2}$) and saturated hydraulic conductivity (K_0, cm min^{-1}):

$$I = St^{1/2} + K_0t/2.8,$$

where t is elapsed time (min) from the beginning of rainfall. Many tropical soils exhibit surface sealing (e.g., Alfisols under impact of large raindrops) or low long-term infiltration due to low terminal hydraulic conductivity (e.g., Vertisols). Infiltration data for four soils in Thailand (Tienseemuang & Ponsana, 1977) show eight-fold differences among soils, from 5.8 to 45 cm of water entry in 100 min. Furthermore, pedologically different soils having the same initial infiltration characteristics may, at the same location, have very different runoff (Table 2.1).

The erosive impact of rainfall is further confounded, and to an extent may be managed, by varying the density and architecture of the crop

Fig. 2.1. (a) Drop size distribution of rainstorms at Samaru, north Nigeria. A, Typical rainfall; B, composite (storm and drizzle); C, drizzle; D, heavy rainfall. (b) Kinetic energy load of rainstorms during the wet season at Samaru. Filled circles, total rainfall; open squares, erosive rainfall; open circles, non-erosive rainfall. Erosive rainfall defined as >20 mm per occurrence. Source: Kowal & Kassam (1976).

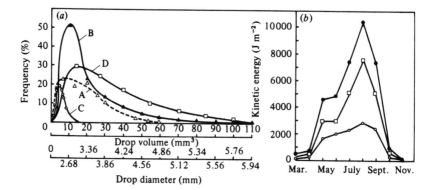

canopy: crop type markedly affects the amount and intensity of rainfall that reaches the soil surface, and conquently, soil loss (Fig. 2.2). The effect of vegetation type on erosivity of rainfall has led to the classification of crops on an empirical scale from 0 to 1, where 1 is the maximum loss that will occur from bare soil (Roose, 1977).

Table 2.1. *Rain infiltration characteristics and runoff under cropping during the monsoon season of a Vertisol and an Alfisol at Hyderabad, India (slope 0.6%)*

	Vertisol	Alfisol
Water infiltration characteristics (mm h^{-1})		
0–0.5 h	76	73
after 144 h	0.21	7.7
Runoff (mm per season)		
1976	73	141
1977	1.5	45
Runoff (% of rainfall)		
1976	15	28
1977	0.43	13

After Krantz, Kampen & Russell (1978).

Fig. 2.2. Rainfall interception and soil erosion under different cropping systems in West Africa. Source: Aina, Lal & Taylor (1979).

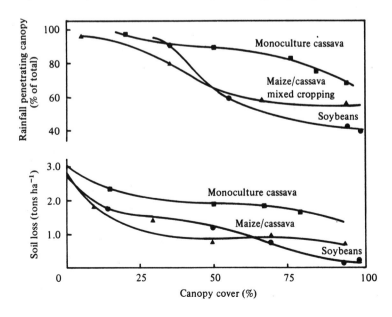

2.1.6 *Water budgets*

After infiltration, water is stored in the soil or drains from it. Changes in soil water are estimated from recognition of the crop/soil water balance equation (Fig. 2.3):

$$\Delta M = R + C + N - (O + D + E),$$

where M is soil water in the crop root zone, R is rainfall, C capillary rise from subsoil to root zone (usually ignored), N run-on, O runoff, D drainage below the root zone and E crop evaporation.

Soil water storage characteristics are most important in determining the crop growing season and the cropping system. These vary appreciably among soils: the amount available for crop growth (the moisture content at field capacity minus that at wilting point, say −1.5 MPa) is twice as great for an Andisol as for an Ultisol or Inceptisol, whereas in sandy soils most soil water is used and the crop wilts appreciably before −1.5 MPa (Fig. 2.4).

In wet tropical climates, where rainfall always exceeds evaporation, cropping is possible throughout the year. In the seasonally wet-and-dry tropics where there is a single rainy season, cropping is based on the

Fig. 2.3. Fate of rainfall: a schema to describe crops/soil water balance.

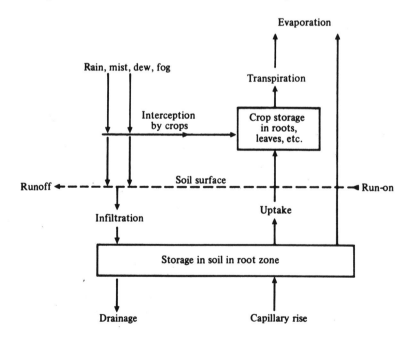

expectation of a period of soil water recharge followed by a time when crops may use stored water despite evaporation exceeding rainfall.

Agricultural events such as land preparation, planting or transplanting and harvesting depend on the probability of occurrence of rainfall that will permit or not permit the operation concerned. Hence it is desirable to assign probabilities to rainfall occurrence. This is particularly so in tropical wet-and-dry climates because of the risks associated with untimely planting and because rainfall is least reliable in these climates: at Hyderabad the coefficient of variation in rainfall in the monsoon months ranges between 45% and 57%. Probabilities of rainfall occurrence obviously depend on season (Fig. 2.5). When coupled with stochastic definitions of the start and end of the growing season, they permit the derivation of probabilities of experiencing a growing season

Fig. 2.4. Relationships between soil water potential and water content in the surface soil of a sandy Inceptisol, an oxidic Ultisol, a clayey Vertisol and an Andisol. FC, field capacity; PWP, permanent wilting point; water available for crop growth indicated by horizontal bars at top to figure. Sources: Inceptisol, Ultisol and Vertisol, N. Collis-George (unpublished data); Andisol, calculated from data of Sanchez (1976).

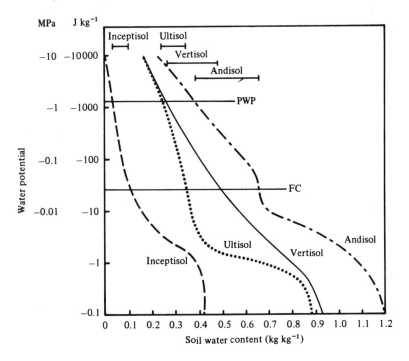

of any particular duration on various soil types. Virmani (1975) for India defines 'sowing rains' as more than 20 mm rain on not more than two consecutive days in a week having dependable rainfall (a more than 70% probability of receiving more than 10 mm rain per week). The growing season continues as long as there is 70% probability of receiving at least 10 mm rain per week. Cochemé & Franquin (1967) define, for Africa, the beginning of the 'sowing' or 'first intermediate' period as when rainfall exceeds half potential evaporation; the active growing season continues as long as the ratio remains higher than this; growth ends when the soil reserve is exhausted. Figure 2.5 illustrates the relations between rainfall probabilities, soil type and length of growing season at Hyderabad: it indicates that soil water-holding capacity may have a major effect — in this case a 2-month difference.

2.2 Crop development

For agricultural land management we aim to quantify the growth of a particular crop in a particular location. This requires knowledge of climate and of the effects of environment on both crop development (progression through a life cycle) and growth (accumulation of dry weight and, ultimately, harvestable yield).

Fig. 2.5. Probability of occurrence of (*a*) a given rainfall intensity (mm week^{-1}) and (*b*) of a given length of growing season (weeks) for three soil types at Hyderabad, India. Adapted from Virmani (1975).

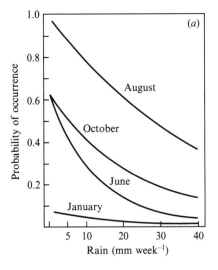 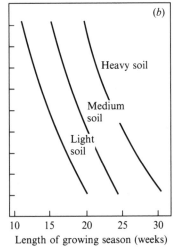

Insofar as all crops have a great deal in common, it is convenient to outline in general the effects of environment on each developmental stage. This physiological approach forms the framework for discussion of the ecology of individual crops. The life cycle of an annual seed crop is shown in Fig. 2.6. The developmental events are germination, emergence, floral initiation, anthesis and physiological maturity of the seed. The timing of these events, relative to physiological maturity, is most dependent on temperature and, in some crops, daylength; events vary between crops, but they usually occur at times within 20% of the values shown in Fig. 2.6. It is therefore useful to define a relative time scale or development index (DI), which will be used in later chapters to relate each stage of development to genotype and climate: DI ranges from 0 to 1, where seedling emergence is 0 and seed maturity is 1. Notice that the maxima for leaf area and above-ground dry weight do not coincide: they are usually reached before and some time after anthesis, at about DI = 0.5–0.6 and 0.8–0.9, respectively. Patterns of development of root and tuber crops do not fit within Fig. 2.6; they are discussed in Chapters 15–17.

The speed of crop development increases more or less linearly with temperature from a threshold, at which there is no development, to an

Fig. 2.6. Schematic life cycle of an annual crop. Developmental events are: G, germination; E, emergence; Fl, floral initiation; A, anthesis; PM, physiological maturity of the seed. The developmental index is set at 0 at E and 1.0 at PM.

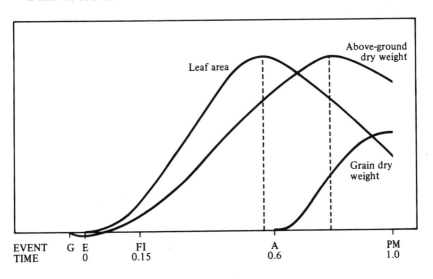

| EVENT | G E | FI | A | PM |
| TIME | 0 | 0.15 | 0.6 | 1.0 |

optimum. The threshold for tropical legumes and grasses is usually 8–10 and 8–13 °C respectively and the optimum ranges from 25–35 and 33–40 °C respectively. Beyond the optimum, development may be delayed, as has been documented in temperate grasses but is seldom reported with tropical crops because of their high temperature optimum. The upper limit for development (at least, for processes such as germination) lies between 40 and 45 °C for tropical crops.

Speed of development (S, units 1/time or d^{-1}) may, for tropical crops, be approximated by:

$$S = (T-T_b)/D$$

where T is daily mean temperature, T_b is the threshold temperature and D is the thermal time (day-degrees) required for a particular development process, e.g., emergence, to be completed. The threshold and optimum temperatures for various developmental processes in rice are given in Table 5.2; the thermal times required for developmental events to occur in pearl millet (*Pennisetum glaucum*) are in Table 8.1.

Daylength, water deficit and nutrition may also affect speed of development, either directly or indirectly through an effect on growth which, if slowed, may reduce the number of cells and size of organ primordia and hence affect cell differentiation and the onset of the next stage of development of the crop. Effects of daylength are specific to particular crops: rice cultivars are classified as having floral differentiation which is strongly dependent, weakly dependent, essentially independent or completely independent of daylength between 12 and 13.5 h (Chang & Oka, 1976). Effects of daylength will be discussed in crop-specific chapters.

Water deficit, i.e., increasingly negative soil water potential, delays development and, as water potential approaches wilting point, it is likely that some processes such as percentage germination and emergence and establishment will decline. It is, however, difficult to generalise about the sensitivity of germination to water because, as Milthorpe & Moorby (1979) point out, rate of water uptake depends on the size of soil aggregates relative to that of the seed. Perry (1976) regarded the ideal as aggregated soil with a mean particle size 0.1–0.2 of the diameter of the seed.

A major ecological problem associated with seedling establishment is crusting and slaking of the surface soil. Crusting occurs when soil particles form a highly bonded structure which offers strong mechanical impedance to penetration (upwards, by seedlings, or downwards, by rain). It is most likely to occur in soils having high levels of iron or aluminium, low divalent base status and low organic matter content; a thick crust is most likely to form if the soil is relatively wet before rain and dries rapidly. Arndt (1965), measuring upward impedance (i.e., as encountered by a seedling) in an Alfisol showed a two- to three-fold increase in impedance during wetting but a six- to nine-fold increase relative to that of dry soil as the soil dried from field capacity to 6% water content. Slaking — the breaking down of soil aggregates resulting in the levelling of the surface — reduces gas diffusion through the soil and may impose short-term anaerobic conditions upon seedlings. Thus, crusting and slaking, independently or in combination, may cause seedling mortality.

Genotypes that are less susceptible to water deficits or soil impedance and show faster and higher percentage establishment have some of the following attributes: large seeds (e.g., groundnut, maize), hypogeal germination (legumes) and strong hypocotyls (presumably related to cell wall characteristics). Agronomically, a more important aspect of seed type is quality, which is defined not on size or chemical criteria but on performance (International Seed Testing Association, 1976). In the tropics, seed is often produced and matured at high temperature. This may cause rapid maturation and small seed size, which results in lower rates of seedling development and greater susceptibility to environmental stress; limited field data (Fussell & Pearson, 1980) support this contention. Furthermore, humidity is often high during maturation and storage, which is likely to encourage fungal and insect damage. Because post-harvest respiration is correlated with seed moisture content (Fussell & Pearson, 1980), a significant loss of seed weight before planting may be expected in the humid tropics.

Fertilisation of grain primordia is widely believed to be the developmental event most sensitive to environmental stress. Low fertility may result from cloudiness (low radiation), from temperatures greater than about 12 °C above or below the optimum for whole plant growth, and from water deficit. Crops such as maize, with separately positioned male and female flowers, are particularly sensitive to environmental stress during fertilisation: water deficit at this time causes male and female flowering to be asynchronous so that embryos do not develop (Hall *et al.,* 1982). Experiments with wheat suggest that atmospheric stress, probably through elevation of crop temperature, is most serious in reducing spikelet fertility and grain set (Fischer, 1980). Fertilisation is followed by a short period during which cell division establishes the cell number, and thus the potential size, of embryo and endosperm.

Subsequent growth of the grain is predominantly through importation of carbon and minerals until growth ceases and the transport pathway collapses or is blocked at physiological maturity.

The effects of nutritional deficiencies on development are not as well documented as are those of temperature or water. Deficiencies slow the speed of development, increasing the duration of the life cycle, and reduce the fraction of the life cycle which is spent in the reproductive phase. The most recognised effect of nutrient deficiencies is on speed of leaf and tiller appearance; for example, thermal time for the appearance of each leaf of pearl millet increases from 60 to almost 100 day-degrees as nutrient treatments become more deficient (Coaldrake & Pearson, 1985a).

2.3 Crop growth

Accumulation of dry matter, and ultimately an edible product, depends on climate and soil factors. Each growth process, e.g., photosynthesis, increases with increasing radiation, has a temperature optimum (which is lower than that for development), an optimal requirement for water and nutrients, and sometimes a sensitivity to daylength. General responses of physiological processes are described elsewhere (e.g., Goldsworthy & Fischer, 1984) and details for particular crops are given in later crop-specific chapters.

Tropical crops may be classified into two groups: those having a C4 photosynthetic pathway, no net photorespiration, a 'dorsiventral' leaf anatomy and membrane phase changes at 10–14 °C; and those with a C3 or Calvin photosynthetic pathway, photorespiration and membrane impairment at about 4 °C. The former group, called cold-sensitive, includes maize, sorghum and pearl millet; their seeds fail to germinate below about 12 °C. The latter group, cold-resistant, includes wheat and rice; their seed may germinate at 4 °C. Angus *et al.* (1980), listing minimum temperatures required for at least 50% emergence for 30 crops, illustrate the discrete difference in tolerance between cold-sensitive and cold-resistant species. However, despite this difference, C4 and C3 species overlap in productivity (Snaydon, 1991). Climatic effects on growth and development and partitioning of dry weight among organs may be important in explaining differences in growth rates such as the six-fold difference reported between *Saccharum officinarum* and *Pennisetum purpureum*, both C4 grasses with superficially similar canopy structure (CIAT, 1980c) .

In the tropics, high radiation (and hence high transpiration) may result in high rates of net uptake of carbon dioxide and nitrogen. Furthermore, day temperatures are usually at about the optimum for net carbon dioxide exchange for canopies, which is a broad plateau from 20 to 35 °C for C3 cereals (e.g., rice) and legumes (e.g., soybean) and from 30 to 45 °C for tropical C4 cereals, although single leaves have a more distinct optimum at 5–8 °C above the midpoint of the plateau exhibited by canopies. Above-ground growth rates exceeding 0.5 t ha^{-1} d^{-1} have been recorded for most C4 cereals, and rates of 0.2–0.4 t ha^{-1} d^{-1} are common over short periods for legumes (e.g., Cooper, 1975).

However, it is uncommon for these high growth rates to be maintained for more than the 2 or 3 weeks towards the end of vegetative growth, between canopy closure and first flowering. In the first place, radiation, temperature and soil water availability are not likely all to remain at optimum or near-optimum levels for extended periods. Furthermore, high growth rates are possible only when the canopy reaches a critical developmental stage: high temperature accelerates the attainment of this stage, but by accelerating flowering and senescence, it restricts its duration. Increasing temperature, or increasing leaf area beyond the optimum, increases dry matter loss via respiration and leaf and root detachment.

The relationship between growth rate and intercepted radiation is described (Warren-Wilson, 1971; Charles-Edwards, 1982) by:

$$G = EI[1-\exp(-KL)]$$

where G is net crop growth rate, E is the efficiency of conversion of intercepted radiation into dry matter, I is daily PAR (photosynthetically active radiation of 400–700 nm wavelength) and $1-\exp(-KL)$ is the proportion of radiation intercepted by a crop with leaf area index L and extinction coefficient K. Thus, if water and nutrients are non-limiting, there is a linear relation between seedling crop growth rate and intercepted radiation, or leaf area index, before canopy closure (ICRISAT, 1978). Branching or tillering is initiated early in the life of a crop (at about DI = 0.1) and leads to full radiation interception between a leaf area index of 3.5 (dicotyledons; e.g., groundnut) and 8 (tall grasses; e.g., maize) . Values of E range from 1.3 to 4.2 μg dry weight per J PAR (Charles-Edwards, 1982). The proportion of radiation which is intercepted by tropical crops reaches a maximum of 0.3–0.8 when leaf area reaches its maximum, at about DI=0.5, and, with leaf area, declines thereafter (Fig. 2.7). Note that from Fig. 2.7 it can be inferred that the efficiency of conversion of radiation to dry matter for each crop (E) falls as canopies age, in association with intra-crop shading, reduced photosynthetic efficiency as leaves age, and the shift to partitioning of dry matter to grain.

While it can be argued that radiation is the primary driving force for crop growth, temperature and water (or more specifically, plant water status) and nutrition will also markedly affect rate of growth. Sub-optimal temperature or water deficits depress growth rate through reducing stomatal aperture and physiological activity. When crop growth and yield during the season is the focus individual species may be grouped into four on the basis of relationships between growth and temperature. The grouping is modified from Fitzpatrick & Nix (1970) and more recent experience: temperate grasses and legumes, subtropical grasses which originated in the high altitude tropics and subtropics, tropical legumes, and tropical grasses. Their cardinal temperatures for growth are summarised in Table 2.2.

Fig. 2.7. Relationship between crop dry matter and intercepted solar radiation for four crops of pearl millet at Hyderabad, India, and Niamey, Niger. A is the line of dry matter accumulation corresponding to a crop which converts radiation into dry matter with an efficiency E of 2 g MJ^{-1}; B is a monsoon-season crop at Hyderabad with average (from sowing to harvest) E = 1.5; C a dry-season crop at Hyderabad with average E = 1.4; D is a dry-season crop grown on drying soil at Hyderabad with average E = 1.1; and E is a dry-season crop at Niamey with average E = 0.4. Source: various authors, reproduced in Squire (1990).

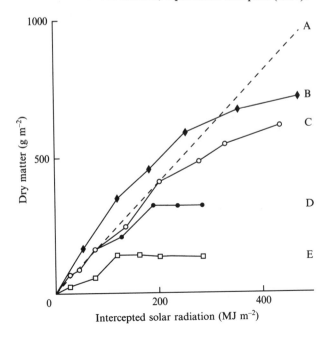

With respect to crop-water relations, there is a linear correlation between crop growth and water use:

$$G = E_w E$$

where E_w is water use efficiency (sometimes abbreviated WUE, dry matter per kilogram transpired water) and E is evaporation from the crop. Crop evaporation can be calculated from meteorological data (e.g., Section 2.1.4) or is often assumed to be 0.8 E_p when the crop canopy is intercepting essentially all radiation and available soil water does not restrict evaporation. This is not always appropriate for tall tropical grasses, for which E may exceed E_p, or for wet rice, in which seasonal evaporation is in the range 4–7 mm d^{-1} and E is 1.2 E_p (Tomar & O'Toole, 1980). Ritchie (1973), working with maize, cotton, sorghum and alfalfa, showed that the transpiration coefficient declines linearly when available soil water falls below 20% of maximum available water. de Wit (1978) provides an example of this computation of E_p for crops under various radiation conditions. Water use efficiency declines as air becomes drier — the saturation deficit increases — so that the product of E_w and the saturation deficit may be reasonably stable or 'conservative' (Squire, 1990) for a particular crop, irrespective of whether it is growing rapidly in a humid environment or slowly and subject to water stress. The physiological bases for observed effects of water deficit on plant growth have been reviewed by Turner (1986*a*, *b*), and osmoregulation, one way in which plants adjust to stress to maintain physiological activity, by Morgan (1984).

Nutrients, particularly nitrogen, influence growth in three ways: directly, e.g., by affecting the activity of photosynthetic enzymes; secon-

Table 2.2. *Approximate cardinal temperatures for dry weight accumulation of various broad classes of food crops*

	T_b	T_o	T_H
Tropical grasses	10–12	28–40	45–50
Tropical legumes	8–10	25–30	40–45
Subtropical grasses	8	23–28	40
Temperate grasses and legumes	4	18–20	35

T_b threshold or base temperature below which growth does not occur; T_o optimum range; T_H high or critical temperature above which growth does not occur.

darily, by affecting the rate of expansion of the leaf surface and thus subsequent interception of radiation; and indirectly through development, e.g., by affecting the speed of leaf and tiller production. Thus, in the wet tropics nitrogen, after radiation, has the greatest impact on the rate of plant growth (e.g., Murata, 1975). Figure 2.8 shows, for rice, that growth per unit of nitrogen declines at high levels of nitrogen availability in a way which is analogous to the decline in efficiency of growth per unit of radiation at high levels of radiation interception. Naturally, where water deficits are imposed, as in the wet-and-dry tropics, the role of nitrogen is less crucial and relationships such as Fig. 2.8 will be masked by the over-riding effects of water deficit.

Grain growth may be described by a logarithmic function with time but it is an accepted convention to divide it into three periods: the lag, linear and maturation phases. The actual grain-filling period (AGFP) is then defined as the length of time from the end of the lag phase to the attainment of maximum grain dry weight. Since each phase has a specific response to the environment, the most useful measure of grain

Fig. 2.8. Relationships between photosynthetic rates (*a*, per unit leaf area; *b*, per unit leaf nitrogen) and nitrogen concentration for IR72 rice at Nueva Ecija, Philippines. Source: Dingkuhn *et al.* (1992).

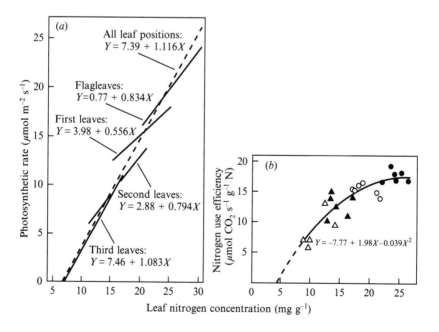

growth is probably that calculated over the AGFP, and this will be used in subsequent chapters. Rate of grain growth over the AGFP increases with increasing temperature in wheat, maize and sorghum, whereas Fussell & Pearson (1980) found it to be constant in pearl millet from 21 to 33 °C. The rate or the AGFP, or both, are reduced under drought conditions (Fischer, 1980) .

Ultimate grain size is constrained by cell division, rate of grain filling, duration of filling and post-maturity loss of weight, that is, the decline in grain weight following physiological maturity often observed in field crops. The little experimental evidence that is available and regional yield data suggest that grain size is low when high temperature or water deficit reduces the AGFP, when climatic extremes abruptly terminate grain-filling, or when high temperature or high humidity induce loss of grain weight after maturity.

2.4 Classification of climate

Climates have traditionally been described and classified on the basis of meteorological data and type of vegetation. Knowledge of plant growth and computer technology has now developed to the point where zones may be defined in terms of seasonal patterns in potential crop growth. A crop-oriented approach prompted Papadakis (1970) and Fitzpatrick & Nix (1970) to develop scalars which broadly relate crop growth to seasonal radiation, temperature and water availability. Papadakis's (1975) monthly growth index (A, ranging from 0 to about 100) is positively related to average daily maximum temperature, water availability and daylength and negatively to minimum temperature:

$$A = 1/[(1/10^{0.1T} + 10^{0.1t}/10^5 + 0.5/10^{2.5H} + 1/10^2) \times (12/D)^{0.75}],$$

where T and t are average daily maxima and minima (°C), H is a humidity index [(rainfall + stored soil water)/potential evaporation] and D is daylength (h). Index values unfortunately do not permit identification of the climatic variables with most effect on A, except in the obvious situation of unirrigated and irrigated wet-and-dry tropical regions.

Fitzpatrick & Nix (1970) define a growth index (from 0 to 1) as the product of temperature, radiation and moisture indices, each of which ranges from 0 to 1, the maximum value representing no limitation to growth. Their temperature index (TI) is specific for each of the broad classes of temperate crops, tropical legumes and tropical grasses, being set at 0 at the threshold or base temperature for growth, 1 at optimal temperatures, and falling to 0 again at the high temperature above which no growth occurs (Table 2.2).

Index values or auxograms (Papadakis's term) give an insight into the relative growth (or conversely the relative stress) that a particular crop type might, in an average year, experience in a given location. These descriptions of overall crop response can thus be used as a basis for calculating actual crop growth rates, by multiplying the growth index by a maximum absolute growth rate determined experimentally at an appropriate level of soil fertility, and by a further index to account for the developmental pattern before canopy closure. They are also used for the classification of climates according to crop potential rather than on the traditional basis of meteorological data.

Figure 2.9 classifies world climates on the basis of crop growth, using Fitzpatrick & Nix's (1970) growth indices. Its authors, Hutchinson, Nix & McMahon (1992), delineate 34 zones, half of which are represented in the tropics. The tropical zones fall into three broad categories: wet [subcategories hot (code J), and warm (code F)]; seasonally wet-and-dry [subcategories hot (I) and warm (E)]; and dry (category H; and very dry (G)]. The tropics also include small areas which have an alpine climate in which temperature is too low for cropping: cropping is discontinued at about 3000 m at the equator. The seasonal patterns of crop growth within the tropical climates are illustrated in Fig. 2.10.

2.3.1 *The wet tropics*

The wet tropics may be defined as regions where rainfall infiltration exceeds crop/soil evaporation for at least 10 months per year; Hutchinson *et al.* (1992) quantify this in a comparable way, as having an average annual moisture index (MI) of 0.66–0.97. The hot wet tropical climate classes (J, in Fig. 2.9) coincides with parts of Köppen's Af, Am, Aw classes, while the warm wet classes (F in Fig. 2.9) overlap Köppen's Cfa, Cwa and Aw (Trewartha, 1968). Collectively, the hot and warm wet tropics are roughly equivalent to A and B climates from Schmidt & Ferguson's (1951) Q system, where Q, the percentage of dry months (less than 60 mm) relative to wet months (above 100 mm), is less than 33. They coincide with Papadakis's (1975) humid equatorial

---

Fig. 2.9. An agro-climatic classification of world climates, in which 17 classes are represented in the tropics within broad zones of E–J. Source: Hutchinson *et al.* (1992).

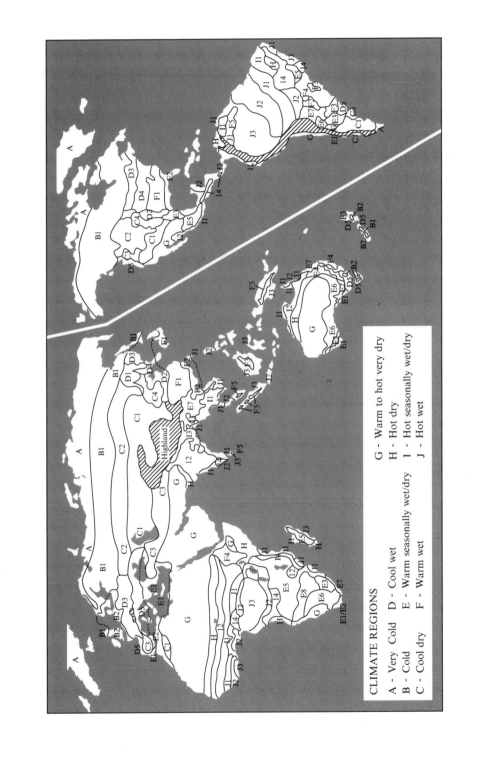

CLIMATE REGIONS

A - Very Cold D - Cool wet
B - Cold   E - Warm seasonally wet/dry
C - Cool dry   F - Warm wet

G - Warm to hot very dry
H - Hot dry
I - Hot seasonally wet/dry
J - Hot wet

Fig. 2.10. Growth index curves which estimate weekly growth of vegetation for a representative station for each of the six broad tropical agro-climatic zones: wet (*a*), seasonally wet-and-dry (*b*) and dry (*c*). The horizontal scale is weeks beginning in January. Continuous line, growth index; short dashes, temperature index; long dashes, moisture index; dots, light index. Source: Hutchinson *et al.* (1992).

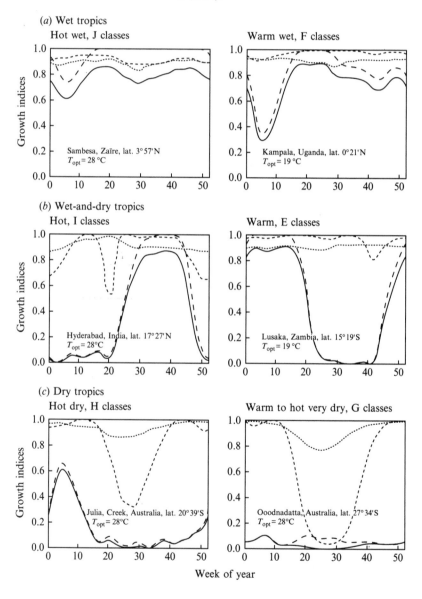

(*a*) Wet tropics

Hot wet, J classes — Sambesa, Zaïre, lat. 3°57'N, $T_{opt} = 28\ °C$

Warm wet, F classes — Kampala, Uganda, lat. 0°21'N, $T_{opt} = 19\ °C$

(*b*) Wet-and-dry tropics

Hot, I classes — Hyderabad, India, lat. 17°27'N, $T_{opt} = 28°C$

Warm, E classes — Lusaka, Zambia, lat. 15°19'S, $T_{opt} = 19\ °C$

(*c*) Dry tropics

Hot dry, H classes — Julia, Creek, Australia, lat. 20°39'S, $T_{opt} = 28°C$

Warm to hot very dry, G classes — Ooodnadatta, Australia, lat. 27°34'S, $T_{opt} = 28°C$

Week of year

class 1.1 and to some areas of Papadakis's 1.4, 1.7, 4.1, 4.7 and 7.1. Interestingly, other research and classification of 850 m ha in South America suggests that the wet tropics are best delineated not according to rainfall or rainfall/evaporation balance, but by total potential evaporation during the wet season, TWPE (CIAT, 1980*c*) . Under this classification, the wet tropics have TWPE in excess of 1061 mm.

The wet tropics are represented by the Amazon and Congo basins and the lowlands of southeast Asia. Much of the area receives an AAR in excess of 2500 mm.

2.4.2 *The wet-and-dry tropics*

The wet-and-dry tropics have on average at least one period of more than 2 months in which rainfall infiltration and stored soil water are not adequate to maintain crop growth. The wet-and-dry tropics have either one or two wet and dry seasons. Bimodal rainfall regimes are characteristically found in latitudes between the wet tropics and the unimodal wet-and-dry rainfall zones; however, their distribution is governed more by pressure systems and land masses. For example, bimodal patterns are rarely found in east Asia. The wet season(s) coincide with high solar angle except for a relatively small area in east Brazil.

The wet-and-dry tropics are commonly called monsoon (in Asia) or savanna (Africa) although there is confusion over these terms; Köppen uses 'monsoon' in a restricted sense, as a climate with a dry season but able to support rain forest. The wet-and-dry tropics are called I and E by Hutchinson *et al.* (1992) on the basis of temperature. These climate zones have mean annual moisture indices as low as 0.29, but a growth index during the highest quarter in excess of 0.3 and commonly 0.5–0.8, i.e., sufficient to support good crop growth. Hutchinson *et al.*'s I zone is coincident with parts of the Aw, Bsh and Cwa types of Köppen and Trewartha. The zone corresponds with *Q* percentages from 33 to 300 or more (Schmidt & Ferguson, 1951) and with types 1.3, 1.4, 1.5, 1.7 and 1.9 in Papadakis's (1975) terminology. Hutchinson *et al.*'s E zone, which is not as hot, coincides with parts of Cfa, Cwa and Bsh and Papadakis zones 1.7, 2, 4.2 and 5 in the tropics.

Within the wet-and-dry tropics, length of growing season and cropping system are determined not only by rainfall distribution but by soil water-holding characteristics (Fig. 2.4) and topographic position, that is, by run-off to and run-on from neighbouring regions. On soils with favourable water-holding characteristics where rainfall is bimodal or unimodal, with Q 100% and some run-on water, year-round cropping is feasible (e.g., Surabaya, Indonesia, AAR 1800 mm, Q = 75%). Q values in excess of 150, without run-on water, are associated with single cropping or sometimes relay cropping, where the crop selected must reach maturity within the constraint of declining soil water. The length of time available for the cropping sequence at Hyderabad, India (AAR 760 mm, Q = 175), as affected by soil water-holding characteristics, is shown in Fig. 2.5.

Because the wet-and-dry tropics cover a broad spectrum of water availability and cropping pattern, they are subdivided into several zones by Hutchinson *et al.* (1992) (Fig. 2.9). Similarly, in the Bolivian lowlands (below 1000 m) Cochrane (1975) distinguishes four subclasses: very humid, humid, semi-humid and dry, where soil moisture is adequate for crop growth (at a 75% rainfall probability) for 8, 7, 5 and less than 5 months per year respectively. In West Africa, Kowal & Kassam (1978) distinguish seven bioclimatic regions within savanna, of which six, supporting arable cropping, would fit within our definition of wet-and-dry tropics and one (the southern Saharan fringe), having no distinct wet season and no permanent cropping, would clearly be classed within the dry tropics. Kowal & Kassam (1978) provide climatic boundaries which, although strictly applicable only to West Africa, illustrate the gradation in wet-and-dry climates (Table 2.3) . The dry margin of the wet-and-dry tropics in West Africa has a rainy season of 55 days, providing soil water adequate for crop growth for 75 days, which corresponds to a minimum AAR isohyet of about 350 mm.

Table 2.3. *Subclasses within the seasonally wet-and-dry tropical climates of West Africa*

	Increasing dryness→			
	South Guinea	North Guinea	Sudan	Southern Sahel
Rain period (d)	190–240	140–190	95–140	70–95
AAR (mm)	1200–1600	880–1200	500–880	350–500
Solar radiation during rain period $(MJ\,m^{-2}\,d^{-1})$	16–18	17–18	18–19	19–20

After Kowal & Kassam (1978).

2.4.3 *The dry tropics*

Dry tropical climates are here defined as those in which rainfall infiltration exceeds crop/soil evaporation for such a short period that even the most rapidly maturing crops cannot be grown without irrigation; Hutchinson *et al.* (1992) consider their hot, dry climate (H) 'too dry or very marginal for rain-fed crops' while the very dry climate (G) is unquestionably unable to support cropping. These areas range from supporting dry savanna vegetation to hot deserts. They coincide with Köppen's arid BWh, BWk and BSh climates, but not with his broader B group, in which annual evaporation need only exceed rainfall, and with parts of Papadakis's 1.5, 1.8, 3.1, 3.3 and 4.3. Broadly, the dry non-arable tropics have an AAR below 250 or 300 mm. They stretch across Africa, the Arabian peninsula and Pakistan between 15° N and the Tropic of Cancer, and across Australia between 15° S and the Tropic of Capricorn.

2.4.4 *Predicting crop productivity and land potential*

There is now a range of computer-based aids to predict the growth of individual crops in relation to environment. These include databases, deterministic simulation models, stochastic models and rule-based decision support systems. Such aids are used to predict which crops may be grown in untried situations (e.g., Hackett, 1991) as well as to obtain information on spatial variation in climate and soils: van Diepen *et al.* (1991) review these aspects of land evaluation. Knowledge of spatial variation in environment, and of plant attributes, allows prediction of spatial variation in plant performance (e.g., rooting depth) and of crop requirements (e.g., water) (Madsen & Holst, 1990). Where background data are available, it is now feasible to couple climate and soil information with crop growth models to predict spatial variation in crop yields throughout a country (e.g., Zambia, van Diepen *et al.*, 1989).

Further reading

Bunting, A.H. (1975). Time, phenology and the yields of crops. *Weather*, **30**, 312–25.

Hutchinson, M., Nix, H.A. & McMahon, J.P. (1992). Climate constraints on cropping systems. In *Field Crop Ecosystems*, ed. C.J. Pearson, pp. 37–58. Amsterdam: Elsevier.

Squire, G.R. (1990). *The Physiology of Tropical Crop Production*. Wallingford, UK: CAB International, 236 pp.

3

Tropical crop/soil relations

3.1 Broad features of soils in the tropics

Important applications for the study of tropical soils include the relation of soils to the productivity of farming and forestry systems, and the fate of areas under native vegetation which are being converted to new forms of farming and forestry. For instance, recent estimates indicate that over the period 1981–90 the rate of deforestation has increased from 11.3 to 17 m ha yr^{-1}, about 8.5 million ha being deforested annually in the moist tropical zone (FAO, 1990). Some reports suggest that deforestation rates in South America increased nearly two-fold between 1980 and 1989 (Houghton, 1990). The decline in soil fertility after clearing can be rapid, particularly in high-rainfall areas on soils susceptible to erosion, and if cleared in inappropriate ways (Lal, 1981a, 1984). The properties and management of soils cleared of natural vegetation in the tropics need to be fully comprehended if fertility is to be conserved in the transition to new forms of land use. There is an urgent need to develop sustainable systems in deforested and cropped areas and to reduce the deforesting of further areas; clearing rainforests is considered a significant cause of global warming (EPA, 1990) climate change (Shukla, Nobre & Sellers, 1990) and loss of biological diversity.

Figure 3.1 provides globally collated estimates of the deterioration of soils through deforestation and inappropriate soil and crop management. Degradation by water is widespread throughout the wet tropics and wet-and-dry tropics whereas soil degradation associated with wind erosion is supposedly confined to the wet-and-dry tropics and tropical areas too arid for cropping. In this survey (Middleton & Thomas, 1992),

Fig. 3.1. Estimates of the degree of soil degradation from wind, water and structural deterioration. Source: Middleton & Thomas (1992).

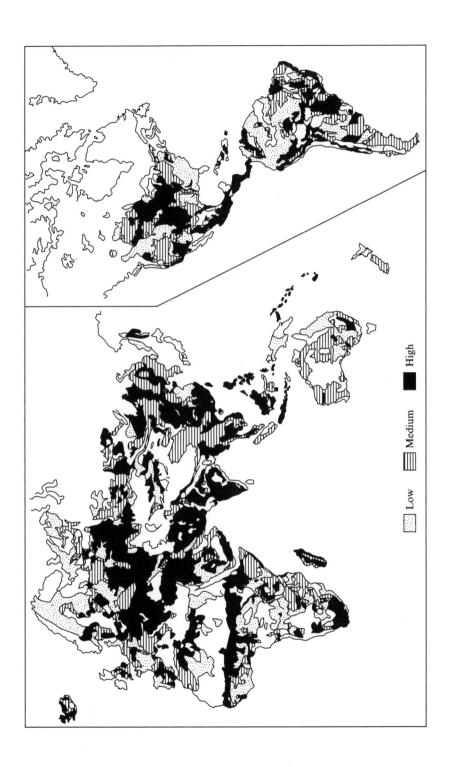

Low
Medium
High

physical soil deterioration (not shown) arising from compaction by tillage etc., was identified mostly in Europe, although its apparent absence from tropical cropping areas may be related to lack of awareness and to survey techniques.

The extent of deforestation and soil degradation may be reduced by converting shifting cultivation to more productive forms of sustainable agriculture: Sanchez, Palm & Smyth (1990) show, for instance, that one hectare of sustainable wet rice production can save clearing 11 hectares of tropical forest annually. Similarly, soil productivity may be improved by developing more intensive cropping systems that include legumes, in place of extensive systems which include unmanaged, heavily grazed 'fallows' (Pearson, 1994). If we do not develop appropriate soil management technologies, Sanchez (1987) warns that development of land in the humid tropics will fail economically and ecologically.

3.1.1 *Soil classification*

In this book the nomenclature in *Keys to Soil Taxonomy* (4th edition) of the United States Department of Agriculture (Soil Survey Staff, 1990) is used. Soil names from other classification systems are used only where the *Keys to Soil Taxonomy* (4th edition) equivalent is not known.

In *Keys to Soil Taxonomy* as published in 1990 there are 11 soil orders. This is an increase of one over the earlier US classification system, *Soil Taxonomy* (Soil Survey Staff, 1975), and is based on the reclassification of Andepts (volcanic ash soils) and some andic subgroups within the order Inceptisols in a new order, Andisols (Parfitt & Clayden, 1991). The soil orders are differentiated by the presence or absence of diagnostic horizons or features that indicate differences in the degree and kind of the dominant sets of soil-forming processes (Soil Survey Staff, 1990). A comparison of units with the FAO/UNESCO legend is given in Table 3.1.

The US classification system separates soils within the tropics from others at the great group and family level on the basis of temperature. Tropical soils are defined as those in which the mean soil temperature of the three warmest months is less than 5 °C different from that of the three coolest months, when measured at 50 cm or at the bottom of the profile if it is shallower. All soils within this 5 °C amplitude occur in what are called isotemperature regimes. The correlation of boundaries of the isotemperature regimes, as presently defined, with the Tropics of Cancer and Capricorn could be improved. Data from tropical Australia show that only in the extreme north of the continent do soils have an isotemperature regime, and that clearing a rainforest to produce a bare soil can change the temperature regime from 'iso' to 'non-iso' (Murtha, 1986).

3.1.2 *Formation and distribution*

There are no uniquely tropical soils (Isbell, 1983). All soil orders are represented in the tropics: tropical soils differ from temperate soils not in kind but in relative extent. All soil-forming factors that are present in temperate areas and have consequences for soil fertility are also present in the tropics, but the relative magnitude of their effects is influenced by latitude. Mean temperature and rainfall are higher in tropical areas, so that mean rates of profile development resulting from processes such as leaching and clay formation are higher. For this reason, on stable surfaces at least, tropical soils in lowland humid areas tend to be less fertile, redder and deeper than their counterparts in temperate areas, owing to greater leaching, desilication and relative accumulation of oxides of iron and aluminium.

Table 3.1. *Approximate correlation of the present US soil taxonomy with the FAO/UNESCO legend*

Present US soil orders[a]	FAO/UNESCO soil units[b]
Oxisols	*Ferralsols*,[c] Gleysols
Ultisols	*Acrisols*, Nitosols
Entisols	*Fluvisols, Lithosols, Regosols*, Gleysols, Arenosols
Alfisols	*Luvisols*, Planosols, Nitosols, Podzoluvisols, Solonetz
Inceptisols	*Cambisols, Gleysols*, Solonchaks, Rankers
Vertisols	*Vertisols*
Aridisols	*Yermosols, Xerosols*, Solonetz, Solonchaks
Mollisols	*Chernozems, Phaeozems, Kastanozems, Greyzems, Rendzinas, Gleysols*, Planosols, Solonchaks, Solonetz
Andisols	*Andosols*
Histosols	*Histosols*
Spodosols	*Podzols*

After Beinroth (1975).
[a]According to Soil Survey Staff (1990).
[b]FAO (1970).
[c]*Italics* indicate the predominant FAO/UNESCO correlatives of the respective orders of soil taxonomy.

Parent material has also influenced the relative extent of soil types in the tropics and in temperate regions. Extensive areas of quartz sands supporting coniferous woodland in Europe have favoured the development of infertile Spodosols (podzols). Such soils in the tropics are limited mainly to narrow strands of coastal sands so that Spodosols are only a minor soil in the tropics. Glaciation, which occurred in high latitudes, particularly in the northern hemisphere, has resulted in a suite of soils developing on moraine, whereas glaciation occurred in the tropics only at high altitude (e.g., Papua New Guinea: Bleeker, 1983). Loess is also more commonly found in temperate areas, e.g., China and the USA. On the other hand the majority of the large river systems of the world are found in tropical areas, contributing to the extensive plains of fertile alluvial soils such as those of the Brahmaputra, Irrawaddy, Chao Phraya and Mekong rivers of southeast Asia. The areal extent of Andisols, resulting from volcanic ash showers, is about the same in tropical and temperate areas. Sometimes quite 'tropical' soils may be found far from the equator. Oxisols, which represent one end-point of intensive weathering and leaching, and are therefore at their greatest extent in the humid tropics, can also be found at intervals down the Australian coast into Tasmania (more than 40° S). The occurrence of Oxisols in these places is related both to the presence of basaltic parent material and to their formation under an earlier warm pluvial period during the Tertiary.

Figure 3.2 shows the distribution of soils in the tropics at the order level. The most widespread soils are Oxisols, Ultisols, Entisols, Alfisols and Inceptisols (Table 3.2). The remaining orders make up only 23% of the tropical land area. However, precise information on the distribution of tropical soils is lacking and there remains a continuing need for more accurate survey and documentation of their properties.

The Oxisols, Ultisols, Entisols and Alfisols, constituting 66% of the total area, are basically infertile, and many Inceptisols, Aridisols, Histosols and Spodosols are also infertile and unproductive because of inherent or climatic constraints. Most of the infertile soils are acid and have low base saturation of the exchange complex. On the other hand, there are also highly fertile and productive soils in the tropics, and improvements in social and economic conditions could lead to inputs which would vastly improve the productivity of soils which are presently so infertile.

The greatest potential for agriculture in the tropics is in the 1496 m ha of humid tropics where there is nearly constant temperature, a dry

Table 3.2. *Generalised areal distribution of soils in the tropics*

Soil Associations dominated by	Tropical America (m ha)	Tropical Africa (m ha)	Tropical Asia (m ha)	Tropical Australia (m ha)	Total (m ha)	Proportion of tropics (%)
Oxisols	502	550	15	—	1067	21
Ultisols	320	135	286	8	749	15
Entisols	124	300	75	93	592	12
Alfisols	183	550	123	55	911	18
Inceptisols	204	156	169	3	532	11
Vertisols	20	46	66	31	163	3
Aridisols	30	704[a]	23	33	790	16
Mollisols	65	—	9	0	74	<2
Andisols	31	1	11	0	43	<1
Histosols	4	5	27	—	36	<1
Spodosols	10	3	6	1	20	—
Total	1493	2450	810	224	4977	100

After Sanchez & Salinas (1981) and other sources.
[a] Of this area, approximately only 1 m ha have more than 150 days growing season.

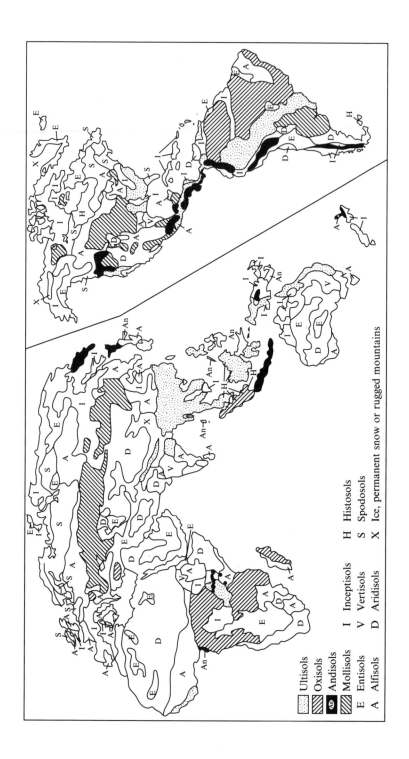

Ultisols	I Inceptisols
Oxisols	V Vertisols
Andisols	D Aridisols
Mollisols	H Histosols
E Entisols	S Spodosols
A Alfisols	X Ice, permanent snow or rugged mountains

season of less than four consecutive months, and a tropical rainforest or seasonal tropical forest as native vegetation (Sanchez, 1989). However, this area, comprising some 30% of the 4977 m ha within the tropics (Table 3.2), has extensive soil constraints (Table 3.3).

3.1.3 *Properties of the clay-size fraction*

Crop growth and soil management to a large extent are influenced by the most active particle size fraction in the soil: the clay-size fraction. The proportion of this fraction and its mineralogy determine a number of soil properties which affect the kind of crop that may be grown, its growth performance and the potential duration of the seasonal cropping period. The clay-size fraction influences cation exchange capacity (CEC), availability of nutrients and retention against leaching, and also modifies many physical properties such as soil structure, water movement, available water, plasticity and soil temperature.

In mineral soils most colloids in the clay-size fraction are clay minerals and iron and aluminium oxides and hydroxides, the remainder being organic. Our knowledge of the properties of the clay minerals and their relationship to crop behaviour is most advanced for soils in temperate regions. While the same kinds of clay minerals are also found to varying extent in many tropical soils, where they would be expected to behave as in temperate areas under similar conditions, their properties are not emphasised in this chapter.

In temperate areas, where soils are not as highly weathered as in much of the tropics, the bulk of soil colloids are crystalline silicate clays: smectites, illites and vermiculites with a 2 : 1 silica : alumina molar ratio and chlorites with a 2 : 2 silica : alumina ratio. Most of the surface charge is negative and permanent owing to isomorphous substitution of Al^{3+} for Si^{4+} in silica tetrahedra and Fe^{2+} and Mg^{2+} for Al^{3+} in alumina octahedra of the clay lattice (Brady, 1974). The proportion of pH-dependent charge and the proportion of positive charge are small. They are sometimes referred to as constant surface charge clays or high activity clays.

The more intensive weathering of soils in the humid tropics has resulted in more extreme loss of bases and silica than in their temperate

Fig. 3.2. Distribution of soils in the world. Adapted from Sanchez & Uehara (1980) and other sources.

Table 3.3. *Areal distribution of soils constraints in humid tropical regions*

Soil constraint	Tropical America		Tropical Africa		Tropical Asia		Humid tropics	
	m ha	(%)	m ha	(%)	m ha	(%)	m ha	(%)
Low nutrient reserves	543	(66)	285	(67)	101	(45)	929	(64)
Aluminium toxicity	490	(61)	226	(53)	92	(41)	808	(56)
High phosophorus fixation	379	(47)	84	(20)	74	(33)	537	(37)
Acid, not Al toxic	88	(11)	92	(22)	74	(33)	255	(18)
Slopes steeper than 30%	145	(18)	22	(5)	73	(33)	241	(17)
Poor drainage	90	(11)	59	(14)	42	(19)	191	(13)
Low ECEC	68	(8)	87	(20)	10	(5)	165	(11)
Shallow depth	54	(7)	17	(4)	27	(12)	98	(7)
No major limitations	28	(3)	7	(2)	5	(2)	40	(3)
Shrink swell	11	(1)	2	(1)	3	(2)	17	(1)
Allophane	8	(1)	1	(−)	4	(2)	13	(1)
Acid sulphate soils	2	(−)[a]	5	(1)	6	(3)	13	(1)
Gravel	2	(−)	6	(1)	3	(1)	10	(1)
Salinity	3	(−)	1	(−)	4	(2)	8	(−)
Sodic soils	5	(1)	3	(1)	1	(−)	9	(−)

After Sanchez, Couto & Buol (1982).
ECEC, effective cation exchange capacity.
[a] <1%.

region counterparts. This has further resulted in a clay-size fraction which is dominated by iron and aluminium oxides and hydroxides which may be crystalline (e.g., haematite, goethite and gibbsite), paracrystalline or poorly ordered, and 1 : 1 silica : alumina crystalline silicate clays. Iron and aluminium oxides (including the hydroxides) are found in greatest abundance in highly weathered soils such as Oxisols and Ultisols, but are also found in Alfisols and Spodosols. Soils high in oxides are often described as oxidic, and while they are also to be found in temperate areas, the proportion is considerably lower than in the tropics. The main aluminium silicate clays in oxidic soils are usually kandites.

Oxides and kandites confer properties on soils which are quite different from those of high activity clays. Isomorphous substitution is not the source of negative charge, except perhaps to a small extent in the kandites (e.g., kaolinite: Smith & Emerson, 1976). Instead the charge arises because of the nature of the surface, and the magnitude of the charge depends principally on the pH and ionic strength (as determined by electrolyte concentration and valence of counter-ions) in the ambient solution, and is therefore variable. Thus they are called variable charge minerals.

The pH-dependent charge of these colloids can vary from net negative through zero to net positive (Uehara, 1977). The effect that the potential-determining ions (Bowden, Posner & Quirk, 1981) H^+ and OH^- have on the oxide surface is shown in Fig. 3.3. The surface charge density (σ), expressed in mol (e^-) m^{-2} is equal to $e(\Gamma_{H^+}-\Gamma_{OH^-})$ where e is the charge on the electron and Γ_{H^+} and Γ_{OH^-} represent the adsorption densities of the potential-determining ions, in mol (p^+) m^{-2} for Γ_{H^+} and mol (e^-) m^{-2} for Γ_{OH^-}. From this it follows that when $\Gamma_{H^+} = \Gamma_{OH^-}$, the surface charge density (σ) is zero. This is the point of zero net charge. The pH at which this occurs can be used to characterise oxidic soils (Stoop, 1980). It is usually lower than the pH of soils in which plants grow. At any given pH there may be positive, negative and neutral sites. The usual situation is for the soil to have a net negative charge, i.e., CEC exceeds anion exchange capacity. For instance, in ten soils within the orders Andisols, Ultisols, Alfisols and Inceptisols examined by Marsh, Tillman & Syers (1987), positive charge varied from 0 to 4.3 cmol (p^+) kg^{-1} and surface negative charge varied from 5.0 to 16.3 cmol (e^-) kg^{-1}. The positive charge attracts anions such as sulphate and phosphate. As soil pH is increased, as occurs on liming, the net negative charge increases in the soil. Conversely, as soil pH is reduced the net negative charge decreases. The influence of pH on surface positive and negative charge in two soils is illustrated in Fig. 3.4.

The CEC (measured in mol (p^+) kg^{-1}) of a soil is related to the surface charge density (σ) (measured in mol (e^-) m^{-2}) through the expression

$$CEC = \sigma S,$$

Fig. 3.3. Development of surface charge in the presence of potential-determining ions (H^+ and OH^-) on an oxide surface.

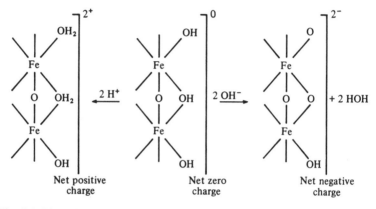

Fig. 3.4. The relationship between surface charge and pH in two contrasting soils, which had been incubated with varying rates of lime. (*a*) Andisol subsoil; (*b*) Alfisol surface soil. Source: Marsh, Tillman & Syers (1987).

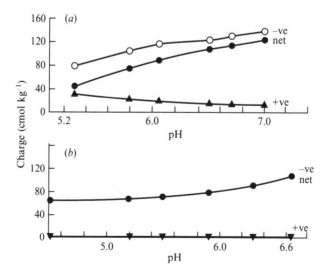

where S is the specific surface ($m^2 \, kg^{-1}$). It is therefore apparent that, since S remains constant for a given soil sample, the CEC of a variable charge soil is affected by the same factors as affect σ, viz., pH and ionic strength.

Measurement of CEC using, say, 1 mol l^{-1} ammonium acetate buffered at pH 7.0, as is common for soils of constant surface charge, will give higher values for variable charge soils than are actually present in the field where the pH is, say, 5.2. Measurement of CEC using an unbuffered neutral salt solution (e.g., potassium chloride) allows the CEC to be measured near the actual pH of the soil, and it is therefore called the effective cation exchange capacity (ECEC). Differences in cation exchange capacity and base saturation values obtained using the two methods are shown in Table 3.4. CEC values typically exceed ECEC values in variable charge soils. Their base saturation values show the reverse trend. The preferred salt and concentration to use in these determinations have been debated, but the concentrations should be low because ionic strengths in such soils are also low: <0.005 in the surface and <0.002 in the subsurface have been reported by Bell & Gillman (1978). Increasing ionic strength from 0.002 to 0.05 can significantly increase ECEC values in these soils (Gillman, 1981).

Variable charge oxidic soils generally have low ECEC values (e.g., Table 3.4) and are therefore also referred to as low activity soils. The constant surface charge kandites which are found associated with the oxidic soils also have low CEC values, usually less than 10 cmol (p^+) kg^{-1} of clay. This contrasts with less weathered soils, e.g., Vertisols, containing constant surface charge smectites whose CEC values may reach 150 cmol (p^+) kg^{-1} of clay.

There are also non-crystalline minerals (e.g., allophane), crystalline (e.g., imogolite) and poorly crystalline soil constituents, which are called variable charge colloids because they have pH-dependent charge characteristics, but which behave somewhat differently from variable charge low activity oxides (Uehara, 1978). Allophane, a hydrated aluminium silicate with silica : alumina ratios between 1.0 and 2.0, is found in volcanic ash soils (Andisols; Wada, 1981). Although it is X-ray amorphous, high-resolution electron microscopy has indicated that it consists of 'hollow spherules' with diameters of 3.5–5 nm (Henmi & Wada, 1976). This structure gives it a high specific surface which, combined with its pH-dependent surface charge properties, results in very high negative charge being induced at high pH values. This also results in greater contrasts between ECEC and CEC values compared with oxidic

Table 3.4. *Comparison of the cation exchange capacity and base saturation determined by two methods on north Queensland soils*

Soil[a]	Depth (cm)	pH[b]	Total exchangeable bases[c] (cmol (p$^+$) kg^{-1})	Exchange capacity (cmol (p$^+$) kg^{-1}) ECEC (unbuffered, 0.01 mol l^{-1} KCl)	CEC (1 mol l^{-1} NH$_4$OAc, pH 7)	Base saturation based on ECEC (%)	CEC (%)
Haplustox	0–10	5.2	0.3	0.6	4.2	50	7
	120–135	5.3	0.3	1.0	1.7	30	18
Eutrustox	0–10	6.6	4.5	4.3	6.2	100	72
	150–165	6.1	3.0	3.5	5.0	86	60
Acrohumox	0–10	5.2	3.9	4.2	21.4	93	18
	210–240	4.8	0.3	0.9	8.2	33	4

After Bell & Gillman (1978).

CEC, cation exchange capacity; ECEC, effective cation exchange capacity.

[a]Alternative classifications for the Eutrostox and Haplustox are Paleustalf and Paleustult, respectively.

[b]1:5 soil:water.

[c]Extracted with 1 mol l^{-1} NH$_4$Cl.

soils; e.g., 8 cmol (p$^+$) kg^{-1} ECEC at pH 5.2 and 65 cmol (p$^+$) kg^{-1} CEC at pH 7.0.

Though we have been considering distinctly different categories of soil colloids, they rarely occur alone in a particular soil. Instead, it is usual to find clays of constant surface charge, such as illite, mixed with colloids of variable charge. The presence of mixtures means that the properties of the dominant category are blurred by the presence of the others in the mix. In surface soil, organic matter, which also has a pH-dependent variable charge (Bache, 1979), is usually also present, much of it bonded to inorganic clay.

Understanding the properties and behaviour of variable charge colloids is important to tropical agriculture, since it has been estimated that at least half the land surface in this region has soils with variable charge properties. The figure for the temperate zone is only about one-tenth (Theng, 1981). The relevance of this point will be better appreciated when the relationship between variable charge soils and soil factors of agronomic significance is considered.

3.1.4 *Nutrient availability*

The fertility of oxidic soils is usually very low because of the intensive weathering conditions that led to their formation (Fox, 1981). Leaching of nutrients and desilication is accompanied by removal of primary minerals, which are important reservoirs of nutrients. This results in soils that are typically acid and low in exchangeable bases. Where soil pH values are low, high levels of exchangeable aluminium, and sometimes of exchangeable manganese, may be present, both of which can be toxic to plants. Furthermore, the low ECEC of oxidic soils results in a low capacity to retain fertilisers against leaching. Among Oxisols, the great groups Eutrorthox and Eutrustox are exceptional in that they are more fertile, having >35% and >50% base saturation respectively. Their extent is not large. Andisols, which contain allophane, are of variable fertility. Those which have been least leached, owing to recent formation or dry climate, may be quite fertile. Their generally higher ECEC values give them a greater capacity to retain added nutrients against leaching compared with the variable charge soils with low specific surface.

Phosphorus. As variable charge soils are extensive in the tropics, so too is phosphorus deficiency. For example, of 1.04×10^9 ha of acid, infertile soils in tropical America within which variable charge clays predominate,

96% are phosphorus deficient (Sanchez & Salinas, 1981). Phosphate in applied superphosphate fertilisers becomes fixed in these soils (i.e., less soluble and less readily available) owing to sorption of phosphorus on iron and aluminium oxide surfaces, and to precipitation by exchangeable aluminium if present (Sanchez & Uehara, 1980). The influence of mineralogy on sorption of added phosphorus, a measure of phosphorus fixation capacity, is shown in Table 3.5. In Andisols the capacity of soil to adsorb phosphate can be extremely high (Fox & Searle, 1978; Table 3.5), and is related directly to the degree of desilication (Fox, 1974).

Large additions of superphosphate to variable charge soils, to increase phosphorus availability, may also result in agronomically significant increases in ECEC, which retard the leaching of calcium, magnesium and potassium (Gillman & Fox, 1980). Phosphorus fertilisers may also increase adsorption of applied zinc (Saeed & Fox, 1979) and precipitate toxic levels of exchangeable aluminium. In some variable charge soils the net charge may be positive, particularly at depth, which will retard the leaching of anions such as nitrate nitrogen (Black & Waring, 1979; Wong, Hughes & Rowell, 1990).

Most plant species are able to form associations with fungi on the surface of, and within, their roots. The most important of these symbiotic fungi are the vesicular-arbuscular mycorrhizal (VAM) fungi which are widespread except in members of the families Brassicaceae, Cyperaceae and Chenopodiaceae (Mosse, 1981). VAM fungi provide considerable advantage to tropical crops growing in low phosphorus

Table 3.5. *Relationship between soil mineralogy and sorption of added phosphate*

Standard P sorption[a] (mol P kg^{-1} soil)	P sorption group	Usual mineralogy encountered
< 0.0003[b]	Very low	Quartz, organic materials
0.0003–0.0032	Low	2:1 clays, quartz and 1:1 clays
0.0032–0.0161	Medium	1:1 clays with oxides
0.0161–0.0323	High	Oxides, moderately weathered ash
> 0.0323	Very high	Desilicated amorphous materials

After Juo & Fox (1977).

[a] The standard P sorption value for a particular soil is that concentration of added P required to support a concentration of 6.4 μmol l^{-1} P (0.2 ppm P) in the soil solution of the Fox & Kamprath (1970) soil test.

[b] 0.0003 mol P kg^{-1} soil is equivalent to 10 μg P g^{-1} soil.

soils as the mycorrhizae assist in the uptake of phosphorus (and some other nutrients) (Dodd *et al.*, 1990*a,b*), the degree of dependence varying with the species of plant and VAM fungus, and with soil phosphorus level (Howeler, Sieverding & Saif, 1987). Species with thick fleshy roots and few root hairs, such as cassava, are more dependent than species which have a finer root system, such as soybean (Yost & Fox, 1979). The variable dependence of selected tropical plants on root infection by VAM fungi is seen in Table 3.6.

Nitrogen. Nitrogen deficiency is widespread: in tropical America for instance, 89% of cropland is considered deficient (Sanchez & Salinas, 1981). Because of the prohibitive cost of nitrogen fertilisers to non-intensive cropping systems, biological nitrogen fixation is important as a low-cost sustainable source of nitrogen to crops and pastures (Henzell, 1988; Bohlool *et al.*, 1992). The most important source of biological nitrogen fixation is the symbiotic association of *Bradyrhizobium* spp. (slow-growing nitrogen-fixing bacteria) and *Rhizobium* spp. (fast-growing) with legume crops, pastures, shrubs and trees. The amounts fixed by the four important tropical grain legumes treated in this book are shown in Table 3.7. When nitrogen is removed in plant residues as well as in the grain, the net nitrogen balance following a grain legume crop may be negative (George *et al.*, 1992), although Peoples & Craswell (1992) found that the balances were positive in 15 of 21 cases (see also Section 9.4). They further found consistent increases in yields of cereals following grain legumes. Other associations such as the actinorrhizal association between *Casuarina* spp. and *Frankia* spp. can contribute significant nitrogen (40–60 kg ha^{-1} yr^{-1}, Gauthier *et al.*, 1985) but they are exploited in few cropping systems. Other symbiotic sources of nitrogen include the cyanobacteria (blue-green algae); e.g., the *Azolla/Anabaena* symbiosis, which is of immense importance in wet rice production (Watanabe & Liu, 1992). There are also numerous non-symbiotic systems in which free-living diazotrophs fix nitrogen in association with roots of cereals (e.g., *Azospirillum* spp., *Klebsiella* spp. and *Enterobacter* spp. (Boonjawat *et al.*, 1991; Chalk, 1991). In Brazil, plant-associated *Acetobacter diazotrophicus* is believed responsible for 60–80% of up to 200 kg ha^{-1} nitrogen taken up in sugarcane crops (Boddey *et al.*, 1991).

Other nutrients and liming. In line with the extent of acid, leached soils in the tropics, other macronutrients are also low, two-thirds of the humid tropical soil area having generally low nutrient reserves (Table 3.3). As 74% of the humid tropical soil area is acid, with or without aluminium

Table 3.6. *The effect of inoculation with vesicular-arbuscular mycorrhizal (VAM) fungus and phosphorus application on dry weight (g per pot) of aerial parts of selected tropical crops grown in sterilised soil, a low phosphorus, acid (pH 4.3) Colombian Paleudult used for cassava production. The last column shows their relative mycorrhizal dependence calculated from the average yield at three phosphorus levels. (P_0, P_{100} and P_{500} indicate levels of 0, 100 and 500 kg P ha^{-1} applied as TSP)*

Plant species	Non-inoculated			Inoculated			Mycorrhizal dependency[a]
	P_0	P_{100}	P_{500}	P_0	P_{100}	P_{500}	
Cassava (*Manihot esculenta*)	0.34	0.72	0.54	4.33	14.21	16.36	95
Beans (*Phaseolus vulgaris*)	1.11	3.44	8.29	3.08	18.79	25.01	72
Cowpea (*Vigna unguiculata*)	0.96	0.64	13.65	2.60	20.68	36.32	74
Stylosanthes guianensis	0.08	0.08	2.74	1.25	9.33	12.20	87
Andropogon gayanus	0.15	0.39	34.24	1.26	16.67	32.18	30
Maize (*Zea mays*)	1.19	8.74	59.35	4.48	34.75	53.67	26
Rice (*Oryza sativa*)	3.79	26.63	30.60	3.83	22.36	31.23	-6

[a] Mycorrhizal dependency is the dry weight of mycorrhizal plants minus dry weight of non-mycorrhizal plants divided by dry weight of mycorrhizal plants multiplied by 100.
After Howeler, Sieverding & Saif (1987).

Table 3.7. *Estimates of the proportion and amount of plant nitrogen derived from symbiotic nitrogen fixation in four important tropical grain legumes*

Species	Location	Treatment variable	Total crop N ($kg\ N\ ha^{-1}$)	N_2 fixed Proportion	N_2 fixed Amount ($kg\ N\ ha^{-1}$ per crop)
Groundnut	Australia	Water supply	171–248	0.22–0.53	37–131
		Cultivar	254–319	0.55–0.65	139–206
		Rotation	181–247	0.47–0.53	85–131
	Brazil	Inoculation	147–163	0.47–0.78	68–116
	India	Cultivar	126–165	0.86–0.92	109–152
Common bean	Brazil	Cultivar	18–71	0.16–0.71	3–32
	Kenya	Phosphorus	128–183	0.16–0.32	17–57
Soybean	Brazil	Site/season	112–206	0.70–0.80	85–154
	Hawaii	Temperature	120–295	0.97–0.80	117–237
	Indonesia	Rotation	79–100	0.33	26–33
	Thailand	Cultivar	33–65	0.78–0.87	26–57
		Cultivar	121–643	0.14–0.70	17–450
		Water supply	157–251	0–0.45	0–113
Chickpea	Australia		109–194	0.17–0.85	37–97

Data collected from the literature by Peoples & Craswell (1992).

toxicity (Table 3.3), liming may be necessary to increase calcium or reduce toxic aluminium or manganese. A beneficial side effect of liming variable charge soils can be an increase in ECEC (Smyth & Sanchez, 1980), which improves the retention of fertiliser cations (e.g., potasssium; Goedert, Corey & Syers, 1975) and may greatly increase the rate of nitrification of soil organic matter and added urea fertiliser (Fig. 3.5). Rarely is lime applied in the tropics in quantities that raise pH to neutrality, since exchangeable aluminium and manganese are usually removed at pH 5.5–6.0. In Ultisols containing variable charge colloid, aluminium saturation (based on ECEC) values above 60% can indicate aluminium toxicity problems for maize (Kamprath, 1984). Silicates, which may be used as liming material in tropical soils (R.W. Pearson, 1975), also provide the silicon required particularly by grasses, such as sugarcane, on highly leached soils.

Micronutrient deficiencies occur in variable charge soils (e.g., zinc, boron, molybdenum and cobalt; Sanchez, 1977) as well as in other tropical soils (Cox, 1973*a,b*; Jones & Wild, 1975; Lombin, 1983*a,b*; Fagbami, Ajayi & Ali, 1985). The conditions which favour the occurrence of micronutrient deficiencies generally are presented in Table 3.8.

3.1.5 *Physical properties*

While the chemical fertility of variable charge soils is poor, they have many physical advantages. High iron oxide levels, as found in some Oxisols, Ultisols and Alfisols, promote the development of water-stable, sand-size aggregates in the 60–2000 μm range (Ahn, 1979; Bui, Mermut & Santos, 1989). Because of this, field texture may appear to be a loam or sandy loam while laboratory analyses reveal that clay content is high. A 'Kikuyu loam' in Kenya found to contain 60–80% or more clay is now called a 'Kikuyu friable clay' (Ahn, 1979). The agricultural benefits of aggregated, low bulk density, oxidic soils are several: they allow high rainfall acceptance rates, thereby reducing erosion; they allow rapid drainage so that the soil may be cultivated shortly after rains, and they provide good aeration. Highly weathered soils are usually deep so that root penetration is correspondingly deep and unhindered. As the aggregation into 'pseudo sand' (Ahn, 1979) is stable, surface crusting inhibitory to seedling emergence is also minimised.

The development of sand-size aggregates in oxidic soils also affects water content/water potential relationships: they behave like sands at high water potential owing to drainage of larger inter-aggregate pores, and like clays at lower water potential (i.e., more negative) where intra-aggregate water is released (Fig. 2.4). Vertisols, which may also have stable aggregates in the 20–50 μm range (N. Collis-George, 1982, personal communication) have much smaller inter-aggregate pores which do not drain at high water potential, and which decrease in size at lower water potential as the soil shrinks and inter-aggregate distances decrease. Simultaneously dry cracks form between conglomerations of aggregates. The water content (kg kg^{-1}) of Vertisol aggregates is higher than that of oxidic soils because of the higher surface charge density

Table 3.8. *Soils and environmental conditions that favour micronutrient deficiencies*

Element	Soils and conditions
Iron	Low soil Fe (alkaline and acid); neutral, alkaline and especially calcareous soils; high $CaCO_3$, HCO_3^-, P, Cu, heavy metals, and pH; cool weather; wet or flooded conditions; poor aeration; low organic matter; root damage; susceptible genotypes
Zinc	Low soil Zn (Haplaquepts, Haplaquolls, Fluvents, Udipsamments, Histosols); acid, sandy, neutral, alkaline, and calcareous soils; low organic matter in mineral soils; high rainfall areas; cool weather; high P, N, Ca ($CaCO_3$), Mg, Fe, Cu, heavy metals, and pH; restricted root zones (compacted soils or container-grown plants); wet and flooded conditions; subsoils exposed from land levelling or disturbances
Manganese	Low soil Mn, poorly drained soils (Histosols, Aquods (ortstein), Udents, Udipsamments, Haplaquepts, Limnic Medisaprists (marly)); slightly acid, sandy neutral, alkaline, and especially calcareous, peat, and muck soils; high $CaCO_3$, Fe, Cu, Zn, and pH; dry weather; low light intensity, low soil temperatures
Copper	Sandy, muck, and peat soils; mineral soils with < 5–6 μg Cu g^{-1} and organic soils with < 30 μg Cu g^{-1}; high P, N, and Zn
Boron	Low soil B (Fluvents, Spodosols, Histosols, Udipsamments, and Haplaquepts); soils formed from acid igneous rocks or fresh-water sedimentary rocks; leached acid, sandy, acid peat, or muck, some neutral, alkaline, and calcareous soils; low organic matter in some soils; moderate to heavy rainfall; dry weather; quantity (high) and quality of light
Molybdenum	Low soil Mo (Udipsamments, Spodosols, Histosols, Aquods (ortstein)); acid, sandy, highly podzolised, high in Fe and Al oxides, neutral, and calcareous soils; high pH and Fe soluble)

After Clark (1982).

(mol (e$^-$) m^{-2}) and higher specific surface (m^2 kg^{-1}) of smectite clays than of oxidic minerals. Andisols may also behave like oxidic soils because of their water-stable aggregates, but hold more water within the aggregates owing to the higher surface charge density and specific surface of allophane. Generally, in terms of the amount of available water, soils may be ranked in the order Andisols >Vertisols >oxidic soils > sandy soils. Values for the upper and lower bounds for available water have been questioned (e.g., Lal, 1979*a*), but available water may be considered as that between –30 J kg^{-1} (field capacity, pF 2.5) and -1500 J kg^{-1} (wilting point, pF 4.2) (Fig. 2.4). Andisols, in addition to favourable crop/soil water relations, often have large stable aggregates which contribute physical advantages similar to those of oxidic soils.

For deeply rooting crops such as perennial tree crops (e.g., coffee and rubber), a low amount of available water per unit depth is usually compensated for by the depth of soil explored, but annuals may suffer water stress. Toxic levels of exchangeable aluminium in the subsoil accentuate this problem in annuals such as maize by restricting rooting depth, but toxicity may be overcome by deep liming (Gonzalez-Erico *et al.*, 1979). Liming Oxisols on a 5% slope in the high-rainfall zone of Ghana has shown not only increased maize yields, but also marked reductions in rainfall runoff and soil loss, attributed to improved root development and soil physical properties (Bonsu, 1991).

3.1.6 *Soil organic matter*

Organic matter levels in virgin tropical soils are little different from those in temperate areas (Sanchez, 1976; Zinke *et al.*, 1984): the generally higher rates of organic matter decomposition in the tropics are balanced by higher rates of addition. In cultivated tropical soils, organic matter levels may be greatly reduced below that of the virgin state, owing to accelerated rates of decomposition and reduced inputs of organic matter from crop residues. Removal of crop residues for fuel or for feeding livestock hastens the loss even though roots are not removed; in soybean and corn they may constitute 40–50% of total residues (Buyanovsky & Wagner, 1986). In Puerto Rico, soils in moist climates exhibit greater variation in soil carbon content with changes in land use (in terms of both loss and recovery) than do soils in dry climates (Lugo & Sanchez, 1986).

The decline in soil organic matter is particularly evident in soils cleared of rainforest (Vitorello *et al.*, 1989). Twenty consecutive crops on an Ultisol in the upper Amazon resulted in a 27% decrease in topsoil organic matter (from 2.13%), most of the loss occurring in the first year (Bandy & Sanchez, 1986). Legume green manure crops may have a more important future role in maintaining soil organic matter levels and soil nitrogen availability (Bowen *et al.*, 1988; Lathwell, 1990). Soils under pasture may accumulate soil organic matter and in Puerto Rico often contain soil carbon in amounts similar to or greater than adjacent mature forest soils (60–150 t ha^{-1} in the top 25 or 50 cm) (Lugo & Sanchez, 1986).

Application of crop residues to acid Oxisols high in exchangeable aluminium has reduced the lime requirement (Adiningsih, Sudjadi & Setyorini, 1988). Thus soybeans grown in Ultisols treated with crop residues have exhibited reduced aluminium toxicity (Ahmad & Tan, 1986), probably owing to the effect of organic acids in complexing toxic aluminium (Tan & Binger, 1986; Kerven *et al.*, 1991). Increased organic matter inputs into weathered, infertile soils through the practice of minimum tillage or agroforestry may also increase phosphorus availability to crops and reduce the fixation of applied phosphorus (Datta & Scrivastva, 1963; Lee, Han & Jordan, 1990).

In perennial tree crop and agroforestry systems in the wet tropics, the release of nitrate-nitrogen and other nutrients from the decomposition of legume cover crops and trees is particularly important. The factors controlling the decay of fresh plant litter leading to the formation of soil organic matter are discussed by Melillo *et al.* (1989). Laboratory studies by Palm & Sanchez (1991) show that the net release of nitrogen by tropical legume leaves in acid soil was not correlated with leaf nitrogen or lignin concentration, but was negatively correlated with polyphenolic to nitrogen ratio (r=−0.75). Mineralisation in excess of the control soil was found only for materials with a polyphenolic to nitrogen ratio of <0.5. Leaf litter from leguminous trees may therefore not necessarily provide a readily available source of nitrogen. Estimates of nitrogen fixation by some tropical trees and shrubs are given in Table 9.2.

The release of nutrients through mineralisation of soil organic matter is also important in supplying nutrients to crops grown under shifting cultivation in both rainforest and savanna areas (Nye & Greenland, 1960). As the rate of mineralisation is controlled by soil moisture, temperature and depth in the soil profile (Cassman & Munns, 1980), rates in topsoils in the humid tropics can be rapid. Through mineralisation, ammonium-nitrogen, nitrate-nitrogen, sulphur, phosphorus, calcium, magnesium and micronutrients are made available. Nitrification, but not ammonification, in a tropical Ultisol appears to be influenced by soil pH and associated levels of exchangeable calcium and aluminium (Fig. 3.5). This indicates that the nitrifiers were very active at a pH above that of the untreated soil. In contrast, Bramley & White (1990) in temperate New Zealand soils found that the optimum pH for nitrification was close to that of the soil. Burning organic residues also releases nutrients for crop uptake (Iremiren, 1989).

In wet-and-dry climates the release of nutrients is markedly seasonal, there being a flush of nitrate-nitrogen at the onset of the wet season (Birch, 1964; Jones & Wild, 1975). In this environment mineralisation

Fig. 3.5. Effect of liming and soil pH on ammonification and nitrification of soil organic nitrogen in soil samples from the surface 15 cm of a Typic Paleudult at Onne, Nigeria, incubated at 30 °C for 65 days. Lime rates applied in the field 3 years before soil sampling are given in parentheses. (*a*) Soils incubated without addition of nitrogen. Regression analyses showed that the amount of nitrate-nitrogen accumulated at the end of the 65 day incubation period was significantly correleated with soil pH ($r=0.99$), exchangeable aluminium ($r=0.99$) and exchangeable calcium ($r=0.95$). (*b*) Soils treated with 100 ppm nitrogen as urea. Source: Arora, Mulongoy & Juo (1986).

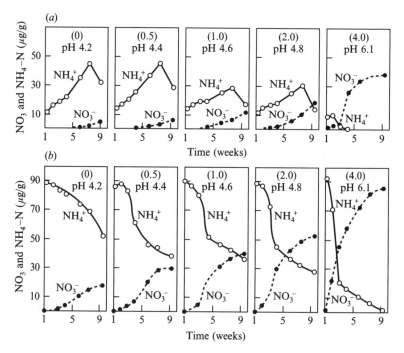

can continue below wilting point (-1500 J kg^{-1}) and nitrate-nitrogen may accumulate in the topsoil by capillarity from deeper layers during the dry season (Wetselaar, 1961*a,b*). Cropping systems in these areas need to be planned to take advantage of these phenomena and to reduce leaching of nitrate-nitrogen from the root zone of young seedlings when the wet season begins. During crop growth in the wet season mineralisation continues. The ratio of nitrogen mineralised annually to total nitrogen in the root zone (the mineralisation coefficient) may be as high as 12.5%, but is usually of the order of 2–5% (Wetselaar, 1967*a*).

Organic matter directly increases the ECEC of soils (Allison, 1973), which is particularly valuable in sandy soils, but the increase may be small or negligible in highly acid soils (Lopez & Cox, 1977). Organic matter also contributes to soil moisture retention and reduces soil erodibility by increasing aggregate stability (Lal, 1979*a,b*) and critical shear stress (Kandiah, 1979).

Variable charge soils, particularly those containing allophane, have a very high capacity to adsorb organic matter. Values in surface horizons of allophanic soils may typically reach 15–20%, contributing to low bulk density as well as to soil water retention and stable aggregates (Warkentin & Maeda, 1981). Some Oxisols may also have high organic matter values in the surface horizon; hence many of the important transformations affecting soil fertility occur in this zone. Conversely, the extremely low organic matter levels at depth in Oxisol profiles may explain the lack of reduction of iron during waterlogging (Couto, Sanzonowicz & Barcellos, 1985).

In spite of the frequent benefits of soil organic matter, as shown here, the relationship between soil organic matter level and soil fertility is not wholly clear, and in some soils high organic matter levels constrain productivity. For further information on soil organic matter in the tropics in relation to soil fertility and its dynamics the reader is referred to Sanchez & Miller (1986) and Parton *et al.* (1989) respectively.

3.2 Occurrence and properties of soils in the tropics

3.2.1 *Oxisols and Ultisols*

The soil orders of Oxisols and Ultisols can be considered together because they are commonly found adjacent to each other, though in different geomorphic positions in the landscape (Lepsch & Buol, 1974). In the tropics many of the Ultisols have oxidic families which can be considered as intergrades between Ultisols and Oxisols.

Oxisols and Ultisols are very extensive in the tropics, particularly in America and Africa (e.g., Amazon and Congo river basins; Fig. 3.2), where they account for 36% of the total tropical area (Table 3.2). In Indonesia about one quarter of the country is covered by Oxisols and Ultisols, where the annual rainfall is 2500–3000 mm (Adiningsih *et al.*, 1988). While most Oxisols and Ultisols are in presently wet climates, they are also found in wet-and-dry and dry climates of West Africa and Australia where they were developed under former wetter environments. They are usually found on very old stable landscapes where soil rejuvenation from solifluction and erosion was minimised. This allowed for long periods of weathering and leaching. Transported weathered sediments may also give rise to Oxisols on old fluvial terraces and pediments (Buol, Hale & McCracken, 1980), so that most Oxisols occur on surfaces with less than 5% slope.

Oxisols are highly weathered reddish, yellowish or greyish soils. Most are characterised by the presence of an oxic horizon at least 30 cm thick with a high content of low activity 1 : 1 clays and oxides. Profile differentiation into horizons is usually weak and there is only a gradual change in soil properties with depth below a well-defined humus-rich surface layer. The soils are normally very deep. As mentioned earlier, Oxisols usually have good physical properties, allowing high infiltration rates, but a low amount of available water, and are susceptible to compaction. In contrast to Oxisols, Ultisols have a differentiated profile: there is a clay B horizon of silicate clays. The profile is low in bases (<35% base saturation determined by sum of cations at pH 8.2). The surface horizon may be sandy, loamy or a clay loam, varying in colour between red, yellow and brown.

Owing to intensive weathering and leaching, these soils are acid and infertile with low ECEC values. The greater the degree of weathering, the more the chemical and physical properties affecting soil fertility will be determined by the behaviour of variable charge colloid, as described earlier, and the lower the availability of nutrients from weatherable primary minerals. Exchangeable aluminium levels may be high, particularly in the subsoil, and toxic to crops (Bouldin, 1979). Applied fertilisers may be leached rapidly (e.g., potassium, Ritchey, 1979) owing to the low ECEC. Applied phosphorus is liable to become unavailable (Sanchez & Uehara, 1980), although it may increase the availability of subsequently applied phosphorus by reducing the rate of its fixation (Hughes & Searle, 1964).

Oxisols and Ultisols constitute a major part of the block of acid, infertile tropical soils. This block, now considered marginal for food crop production, represents the largest area of potentially arable land in the world (Edwards *et al.*, 1991). Where these soils support rainforest which is cleared and cropped, care must be taken to minimise degradation of chemical properties (Sanchez, Villachica & Bandy, 1983; Gillman, 1984; Alegre, Cassel & Bandy, 1988) and favourable physical properties (Alegre, Cassel & Bandy, 1986*a*; Ghuman & Lal, 1989; Spaans *et al.*, 1990; Fig. 3.6), particularly through the inappropriate use of machinery (Dias & Nortcliff, 1985; Alegre, Cassel & Bandy, 1986*b*; Alegre, Cassel & Makarim, 1987; Onwualu & Anazodo, 1989) and on steep slopes (Siebert & Lassoie, 1991). Nevertheless, post-clearing tillage by chiselling and discing can be beneficial to crop production (Alegre, Cassel & Bandy, 1990). While it is commonly believed that cultivation degrades soils in the humid tropics, the results of Bandy & Sanchez (1986) on Ultisols in the upper Amazon show that soil properties improve with continuous cultivation systems that combine intensive cropping with appropriate fertilisation. After 20 consecutive crops topsoil pH increased from a very acid 4.0 before clearing to a favourable 5.7.

Where fertilisers are applied these soils can give high yields and their occurrence on gently sloping topography favours mechanisation; e.g., pineapples and sugarcane in Hawaii and soybeans in Brazil. However, disc tillage may adversely affect soil structure if carried out at high soil moisture, following rains (Stoner *et al.*, 1991). Nevertheless, over much of the area of these soils in the tropics, because of socio-economic limitations, low-input technology strategies will be the likely development pathway (Nicholaides *et al.*, 1985; Sanchez & Benites, 1987; Sanchez, Benites & Bandy, 1987). These strategies include the selection of crop species and varieties which can tolerate moderate levels of aluminium toxicity, low available phosphorus and calcium, and high manganese (Table 3.9); the use of the minimal quantities of lime required to reduce soil acidity; and maximum use of legumes as a nitrogen source. For further information on the management of acid soils in the humid tropics see Craswell & Pushparajah (1989).

3.2.2 *Entisols*

In the tropics, Entisols account for 12% of the soils (Table 3.2) and are most extensive in tropical Africa (Fig. 3.2). They show little or no development of pedogenic horizons.

The reasons for the absence of clear genetic horizons are diverse: the parent material may be inert, as in the case of quartz sand; the time of formation may have been too short, as in actively eroding landscapes or on flood plains where material is deposited faster than a soil profile can develop; and continuous saturation, as in mangrove swamps, may inhibit horizon formation. Man-made Entisols occur on rice terraces, such as the Banaue terraces of the Philippines, where the soil profile was created by bringing soil materials from elsewhere (Moormann & van

Fig. 3.6. Effect of land-clearing method on infiltration rate measured on three occasions. Open triangles, manual clearing; open circles, shear blade clearing; filled circles, forested control. Continuous curves were fitted using Philip's infiltration-rate equation. A is transmissivity in cm h⁻¹ and S is sorptivity in cm h⁻¹ᐟ². (a) Three months after clearing but before planting. (b) Two years after clearing and subsequent cropping. (c) Four years after clearing and subsequent cropping. Source: Ghuman, Lal & Shearer (1991).

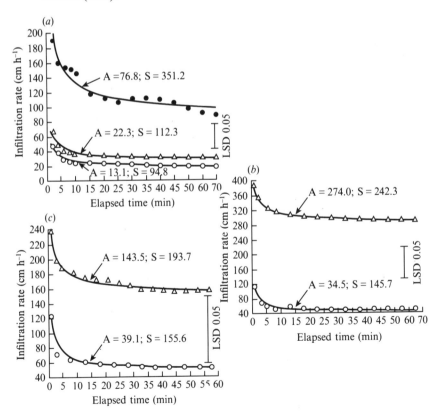

Breemen, 1978). Thus Entisols can be found from arid environments of the Namibian and Sahara deserts to littoral sands and flooded fertile alluvial soils in lacustrine situations in Asia. Aquents are a major suborder of Entisols for rice-growing, and it is important that the surface layer (to about 10 cm) dries out sufficiently each year to allow oxidation (Moormann & van Breemen, 1978). In Asia, Aquents suffer from salinity and potential acidity in coastal fringes, zinc deficiency or iron toxicity in inland areas, and recurrent deep flooding on river plains.

3.2.3 *Alfisols*

Alfisols are intensively used for agriculture in the tropics. Alfisols are most extensive in Africa (Fig. 3.2) and comprise 18% of the tropical land area (Table 3.2). About 62% of Alfisols in the semi-arid tropics are located in Africa and India (Sivakumar, Singh & Williams, 1987). In West Africa they are found where the rainfall is 1600 mm or less and in India they are associated with Vertisols (Krantz, Kampen & Russell, 1978). Where the mean annual soil temperature is between 15 and 22 °C or warmer, Alfisols tend to form a belt between the Aridisols of the arid regions and the Inceptisols, Ultisols and Oxisols of warm humid climates (Soil Survey Staff, 1975). In the Occidental Plateau of Sao Paulo State, Brazil, Lepsch, Buol & Daniels (1977) found a close association between Oxisols, Ultisols and Alfisols, but they were located in different geomorphic positions. Oxisols occupied high and low elevations where slopes were 0–5%; at intermediate elevations were Alfisols and Ultisols on slopes ranging from 5 to 20%.

Table 3.9. *Some important food crops considered to be generally tolerant of acid soil conditions in the tropics*

Generally tolerant species	Generally susceptible species with acid-tolerant cultivars
Rice (*Oryza sativa*)	Maize (*Zea mays*)
Groundnut (*Arachis hypogaea*)	Sorghum (*Sorghum bicolor*)
Cowpea (*Vigna unguiculata*)	Wheat (*Triticum aestivum*)
Pigeonpea (*Cajanus cajan*)	Soybean (*Glycine max*)
Cassava (*Manihot esculenta*)	Common bean (*Phaseolus vulgaris*)
Banana (*Musa* spp.)	
Potato (*Solanum tuberosum*)	

After Duke (1978) and Sanchez & Salinas (1981).

Like Ultisols, Alfisols also have an argillic B horizon, but whereas the Ultisols have a base saturation less than 35%, in Alfisols it is greater than 35%. While they are more fertile than Ultisols and are one of the more extensive soil resources of the tropics, the absolute quantities of cations per unit volume of soil are low, and yields are restricted by nutrient deficiencies and also by physical constraints such as water availability, crusting and erosion (van Wambeke, 1976). Crop yields on Alfisols are not only low but also unstable owing to climatic variability, but they are capable of improved productivity with appropriate soil water and crop management systems (Sivikumar, et al., 1987). Because of their poor water storage capacity they can be croppped only in the short rainy season in the semi-arid tropics, when over 80% of the annual rainfall occurs, and when they are liable to runoff and erosion (El-Swaify, Singh & Pathak, 1987; Stocking & Peake, 1986).

The texture of the surface varies from loamy sands to loams and clays. Overall, Alfisols tend to be less well-drained than Ultisols, which have a higher level of oxidic materials that favour drainage. Some Alfisols are oxidic and these have a higher phosphorus fixation capacity, but in the surface horizon of coarse-textured Alfisols it may be quite low (Juo & Maduakor, 1974).

In western Nigeria, Lal et al. (1975) found variability in soil physical properties and nutrient status over short distances within the same Alfisol series. They attributed this to human and biogenetic factors. The presence of termite mounds (sometimes as many as 60 per hectare) is also a source of variability because of the low organic matter and nutrient levels of the clay subsoil brought to the surface.

The annual rainfall of more than 1000 mm on Alfisols in western Nigeria can cause serious soil erosion on cleared land, resulting in changes in soil ·organic matter, clay content, soil structure, and in reduced water and nutrient uptake by annual crops (Lal, 1976a–c). High soil temperatures deleterious to the emergence and growth of crops have also been measured on these soils (Lal et al., 1975). Infiltration rates in some Alfisols in the forest region of West Africa have been found to be high and attributable to earthworm activity (Wilkinson & Aina, 1976). As for Oxisols and Ultisols, the method of clearing and tillage, whether manual or mechanical, has effects on Alfisol soil properties and crop yields (Opara-Nadi & Lal, 1987; Ojeniyi, 1990).

In India Alfisols do not store as much available water as adjacent Vertisols. Alfisols at Hyderabad were calculated to have a median growing season of only 17 weeks compared with 26 weeks for Vertisols at the same location (Krantz *et al.*, 1978; see Fig. 13.5). Because the surface 20 cm of soil may dry out between the intermittent rains at the start of the monsoon, sowing is usually delayed by at least 2 weeks compared with sowing in Vertisols. Alfisols in this area are susceptible to hard setting in the surface which is aggravated by tillage, particularly with a mouldboard plough (Awadhwal & Smith, 1990).

3.2.4 *Inceptisols*

Inceptisols are an important group of soils in the tropics, occupying 11% of the land area (Table 3.2). They are most extensive in tropical America (Fig. 3.2). Inceptisols are young soils that have available water for more than three consecutive months during the warm season. The kind of genetic horizon present varies with parent material and topographic situation. They vary greatly in their properties and fertility: two contrasting suborders are the Aquepts and Tropepts. Tropical Aquepts (Tropaquepts) are soils which have a reducing environment because they are saturated with water. Large areas are to be found in tropical river basins, such as that of the Amazon. Tropaquepts are also found on extensive alluvial sediments roughly between 17° N and 20° S, the largest areas occurring in south and southeast Asia (Moormann & van Breemen, 1978). They are extremely important for rice production. Most other Inceptisols in the tropics are Tropepts, which are most extensive in hilly areas. Only under conditions of high population pressure are they cropped, because of their general low fertility, shallow profile and liability to erosion.

Because of the lack of well-developed B horizons in Inceptisols formed on siliceous parent materials, water relations may be poor and base saturation and CEC low. Widespread deep siliceous sandy soils in the Cape York Peninsula area of north Queensland are quite infertile. In the wet-and-dry climate (mean rainfall 1140 mm per year) the soils have developed a weak spodic horizon. Attempts to improve fertility by application of phosphorus and sulphur fertiliser resulted in losses of 50% of the phosphorus from the top 30 cm and 70–90% of the sulphur in the top 100 cm of the profile over three wet seasons (Gillman, 1973).

3.2.5 *Vertisols*

In the tropics Vertisols are not extensive. They cover only 3% of the area (Table 3.2), mainly in India, Australia and Africa (Fig. 3.2), but with Alfisols are a very important soil resource in the semi-arid tropics.

Vertisols are clay soils that have deep wide cracks at some time of the year and high bulk density between the cracks. They are dark-coloured, with more than 30% clay and more than 50% 2:1 minerals in the clay fraction. The soils have a high CEC, high base status, mostly neutral-to-alkaline reaction and high moisture-holding capacity. The surface is usually granular, resulting from its self-mulching characteristics: large surface clods break down naturally on drying into small aggregates.

The most common feature of Vertisols is that they have a seasonally dry profile (Buol *et al.*, 1980); wet-and-dry climates typify the environment favourable to their development. The land surface is usually flat to gently undulating but the surface micro-relief consists of a network of depressions and mounds called 'gilgais' or 'crabholes'. These are promoted by the formation of cracks during the dry period into which the surface soil falls: in the wet season soil swelling leads to surface 'buckling'.

Vertisols are usually quite fertile, but are sometimes low in nitrogen, phosphorus and zinc. Their physical properties have a profound influence on crop growth and management. Krantz *et al.* (1978) at Hyderabad showed that the initial infiltration rate of monsoon rains in Vertisols and Alfisols was very high – around 75 mm hr^{-1} (Table 2.1). Within 2 h large cracks in the Vertisols began to close, reducing the infiltration rate nineteen-fold compared with only five-fold in adjacent Alfisols. However, the Alfisols developed a surface crust early in the rainy season so that they lost much more water by surface runoff. Vertisols, which were also deeper, did not show comparable runoff rates until the profile was saturated in the second half of the rainy season. The result was that Vertisols stored much more water than Alfisols (see Fig. 13.5). Tillage and crop residue management can substantially affect soil organic matter, microbial activity in the surface layers, and water relations to at least 1.2 m depth (Dalal, 1989).

During the dry season Vertisols are very hard to cultivate; when the rains come, the surface soon becomes sticky and again difficult to work. The moisture range over which they can be cultivated is small compared with many other soils.

3.2.6 *Aridisols*

Aridisols are very extensive in the tropics (16%, Table. 3.2), particularly in North Africa (Fig. 3.2), but they make only a small contribution to total crop production.

Aridisols are mineral soils of the deserts and semi-deserts with a period of adequate soil water availability of less than 90 days while the soil is warm enough for plant growth, but which have horizon differentiation.

The climatic regime in the tropics in which Aridisols are found can be characterised as one in which potential evaporation greatly exceeds precipitation during most of the year (Buol *et al.*, 1980). Vegetation is usually sparse and comprises ephemeral grasses and broad-leaved plants and scattered xerophytic perennials. The soils cannot be cropped without irrigation. Their low level of soil organic matter is associated with the lack of vegetation. Because of the low rainfall, chemical reactions, leaching and illuviation are very slow. Hence these soils are typically high in bases and potentially quite productive if irrigated, unless they happen to be too saline.

3.2.7 *Mollisols*

Mollisols are not widespread in the tropics, comprising less than 2% of the area (Table 3.2). Most of the area is in tropical America, particularly Paraguay and Mexico (Fig. 3.2).

Mollisols are dark-coloured, well-structured soils with high base saturation throughout and a deep surface horizon. The bases are predominantly divalent. The surface does not harden when dry as in many Vertisols and Alfisols. They are very fertile soils which are usually found at high altitudes in the tropics.

3.2.8 *Andisols*

Andisols are not extensive in the tropics (Table 3.2, Fig. 3.2) and are to be found where there are surface deposits of volcanic ash. However, because of their agricultural potential they are important. They are most extensive in tropical America – in the Caribbean, in Colombia, Ecuador, Peru and Bolivia (Sanchez & Isbell, 1979). In tropical Asia and the Pacific they are to be found in Indonesia (Tan, 1965), Papua New Guinea (Bleeker, 1983), including New Britain (Hartley, Aland & Searle, 1967) and Bougainville (Scott *et al.*, 1967), and Hawaii.

The predominant mineral of the clay fraction of these soils is allophane (Wada, 1981). The soils are usually free-draining, low in bulk density, and high in pH-dependent charge (Warkentin & Maeda, 1981). They may be cindery or contain vitric (glassy) volcanic ash. They are generally fertile and already under cultivation. Where clearing is necessary, erosion is the most serious problem, especially in mountainous areas (Daage, 1987). Fertility depends to a large degree on the extent of weathering; desilication often proceeds readily, resulting in higher concentrations of oxidic materials and high phosphorus fixation rates (Nunozawa & Tanaka, 1984; Alvarado & Buol, 1985; Table 3.5). Highly weathered Andisols may be stripped of most nutrients and become infertile residues of iron and aluminium oxides and hydroxides. Organic matter levels may become quite high owing to adsorption by allophane, providing a sink for both nitrogen and sulphur (Fox, 1981). When phosphorus fertiliser is applied crops may respond also to nitrogen mineralised as a result of improved phosphorus availability to the soil microorganisms (Munevar & Wollum, 1977). Physically these soils have few problems because of their excellent structure, stability on slopes and high infiltration rates (e.g., Martini & Luzuriaga, 1989). In these respects they may behave like Oxisols; however, they have a much greater available water range (Fig. 2.4).

3.2.9 *Histosols*

Histosols or peat soils are of only minor significance in the tropics although in some countries they are very important. Their estimated area in the tropics ranges from 32 m ha (Driessen, 1978) to 36 m ha (Sanchez & Salinas, 1981; Table 3.2). Estimates vary according to the criteria used to define the soil. In the US soil taxonomy, organic soil material must be more than 40 cm thick where it overlies an Alfisol or Ultisol before it is classified as a Histosol. The criterion of Driessen (1978) is that it be deeper than 50 cm. Most tropical peats are in southeast Asia, where there are 18 m ha of coastal swamp covered with deep oligotrophic forest peat. Histosols are of major importance in Indonesia, where swampy lands high in peat cover 27% of the total area of the country, excluding Irian Jaya (Collier, 1979). Peat soils are also locally important agriculturally in highland areas of Papua New Guinea (Bleeker, 1983).

In the tropics most Histosols have developed under anaerobic conditions caused by the presence of water in low-lying maritime and lacustrine situations and in inland swamps and marshes (Andriesse, 1974). Virgin tropical peats vary in properties depending on the conditions under which they are formed, but most have a water table close to the soil surface. Bulk densities are often low, but increase on drainage. Nutrient levels, including nitrogen, calcium, magnesium, sulphur, copper, boron, zinc, manganese, molybdenum and iron, may be low (Driessen, 1978; Ambak, Bakar & Tadano, 1991). Copper deficiency may be a particular problem for wet rice (Ambak & Tadano, 1991) and the deep oligotrophic peats of southeast Asia are invariably associated with spikelet sterility (Bouman & Driessen, 1985). There is a danger that on draining shallow peats overlying pyrites the profile will acidify and there will be a loss of peat depth through mineralisation. Nevertheless, with good management of 60 cm deep forest peat over pyritic sediments in Kalimantan, rice farmers are able to obtain yields of up to 3 t ha^{-1} (Bouman & Driessen, 1985). Acid deep woody peats in Malaysia (pH 3.8–4.0) have given maize and tomato responses up to 8 t ha^{-1} to lime and micronutrients which increased pH to 5.0–5.3 (Ambak *et al.*, 1991). Soil shrinkage may lay bare the roots of perennial crops such as rubber, so they are prone to fall, but the looseness of the organic material aids in the harvest of root crops such as cassava.

3.2.10 *Spodosols*

Spodosols are another minor soil order in the tropics (Table 3.2). They are found in Indonesia, on the east coast of the Malay Peninsula, in Sri Lanka (Williams & Joseph, 1970) and Surinam (Janssen & van der Weert, 1977).

Spodosols are highly leached, acid, coarse-textured and infertile soils. Typically they have a bleached A2 horizon above a spodic horizon, which is an amorphous mixture of organic matter and aluminium with or without iron. In the tropics these are found characteristically on coastal dunes where groundwater, often brackish, may be found. These soils have all the problems associated with inherently low nutrient levels, low CEC and low water-holding capacity.

3.3 Problem soils

There are millions of hectares of land in the tropics which are topographically and climatically suited to crop cultivation but where soil problems limit production. They include salt-affected soils, acid sulphate soils and peats. Acid sulphate soils and peats account for some 46% of the 39 m ha of tidal swampland in three of the main islands of Indonesia (Noorsyamsi & Sarwani, 1989). Peats have been described in Section 3.2.9; salt-affected and acid sulphate soils are dealt with below.

3.3.1 *Salt-affected soils*

Soils affected by salt may be classified as follows:

	Electrical conductivity (dS m^{-1})	Exchangeable sodium percentage (ESP, %)
Saline non-sodic	> 4	< 15
Saline sodic	> 4	> 15
Non-saline sodic	< 4	> 15

Saline soils have high levels of neutral soluble salts such as chlorides and sulphates of sodium, magnesium and calcium. They occur in coastal areas where salinity is caused by the intrusion of seawater and in arid and semi-arid areas where it is caused by evaporation of groundwater or surface water. If the electrical conductivity of the saturated soil extract is 4 dS m^{-1} or higher at 25 °C, the soil is said to be saline. At high soil salinity levels sodium chloride is the dominant salt, but most saline soils have sufficient calcium for crop nutrition. Small quantities of boron and other toxic elements may also be present in the soil solution. The saturated paste pH of saline soils is nearly always less than 8.2; many have a pH less than 7.0 and some are extremely acid (e.g., acid sulphate soils). Soil physical properties are good, they are well aggregated, but the high osmotic pressure due to the salinity reduces soil water availability to plants. In coastal areas of India high soil salinity delays rice planting until rains have leached accumulated salts from the topsoil (Abrol, Bhumbla & Meelu, 1985).

Soils with an ESP of more than 15 are usually classified as sodic or alkaline. They generally lack neutral soluble salts but contain substantial amounts of sodium carbonate. Their electrical conductivity varies but is often less than 4 dS m^{-1}. Saturated soil paste pH is 8.2 or higher and may be as high as 10.0. Physically the soils are dispersed and poorly permeable to water and air, the extent depending on factors which include soil mineralogy and level of soil organic matter. In wet and dry climates soil chemical and physical properties, including pH and clay dispersion, can vary with the season (Topark-Ngarm *et al.,* 1990).

In south and southeast Asia saline soils (sodic and non-sodic) occupy at least 62 m ha (Ponnamperuma & Bandyopadhya, 1980). In India alone about one quarter of the estimated 12 million hectares of saline/sodic soils, most on the otherwise fertile Indo-Gangetic alluvial plain, are unproductive, mainly owing to high sodicity (Yadav & Gupta, 1984). In such soils high Na/Ca ratios may suppress calcium uptake (Muhammed, Akbar & Neue, 1987). Because rice is relatively tolerant of high sodicity (Table 3.10), wet rice is used to reclaim saline and sodic soils, the standing water reducing salinity through leaching and dilution (Abrol *et al.*, 1985).

3.3.2 Acid sulphate soils

Acid sulphate soils are extensive in the tropics: 5.4 m ha in south and southeast Asia, 3.7 m ha in Africa and 2 m ha in Venezuela (van Breemen, 1980). They are formed where large quantities of pyrite (FeS_2) accumulate in intertidal sediments. The soils have the potential to acidify if they are aerated by draining for agricultural production; FeS_2 is oxidised, releasing

Table 3.10. *For crop plants grown in salt-affected, medium-textured (sandy loam) soils of increasing exchangeable sodium percentage (ESP): the threshold ESP (ESP_T) at which yield decline was first observed, the slope of the line relating decline in yield to ESP, and the ESP value at which yield was reduced to 50% of maximum (ESP_{50})*

	ESP_T	Slope	ESP_{50}	Note
Cereals				
Rice (*Oryza sativa*),				
transplanted	24.4	0.9	80.0	
	32.7	1.3	71.2	(P)
Pearl millet (*Pennisetum glaucum*)	13.6	2.6	32.8	
Wheat (*Triticum aestivum*)	16.4	2.1	40.2	
	20.4	2.6	39.6	(P)
Legumes				
Groundnut (*Arachis hypogaea*)	8.0	2.2	29.7	(1)
Soybean (*Glycine max*)	8.0	3.5	22.3	(1)
Chickpea (*Cicer arietinum*)	7.7	5.0	17.7	
Guar (*Cyamopsis psoraloides*)	11.9	3.2	27.5	
Lentil (*Lens esculentum*)	4.9	5.5	14.0	
Cowpea (*Vigna unguiculata*)	13.5	9.1	19.0	
	9.0	3.0	25.7	(1)

From Gupta & Sharma (1990).
The ESP values were averages over the 0–15 cm layer in the field.
(P) The studies were conducted in pots.
(1) The ESP_T values are the lowest ESP at which trials were conducted.

Table 3.11. *Major criteria used in assessing the suitability of soils for crops in Malaysia*

Crop	Slope (deg)	Effective soil depth (cm)	Soil texture and structure	Drainage	Salinity (dS m⁻¹ at 25 °C)	pH	Depth to acid sulphate zone (cm)	Thickness of peat (drained)
Wet rice	0–2	>25	Sandy clay loams or finer	Drainage control necessary	<4 in the top 25 cm	>4.0	>25	No peat
Maize	0–6	>50	Exclude sands and clays	Good to imperfect	<2 in the top 50 cm	>5.0	>125	No restriction
Sorghum	0–6	>50	Exclude sands	Good to imperfect	<4 in the top 50 cm	>5.0	>125	No restriction
Groundnut	0–6	>25	Exclude sands and clays	Good to moderately good	<4 in the top 50 cm	5.5–7.0	>50	No peat
Soybean	0–6	>25	Exclude clays and soils with poor structure	Good to imperfect	<4 in the top 50 cm	5.5–6.5	>50	<25 cm
Cassava	0–6	>50	Exclude clays and soils with poor structure	Exclude poorly drained	<2 in the top 100 cm	4.3–7.3	>50	No restriction
Sweet potato	0–6	>50	Exclude clays and soils with poor structure	Exclude poorly drained	<2 in the top 100 cm	4.3–6.0	>50	No restriction
Banana	0–12	>125	Exclude loamy sands or coarser	Good to imperfect	<2 in the top 100 cm	5.0–7.0	>125	<25 cm

After Protz (1981).

H_2SO_4 and Fe^{2+} (Moore, Attanandana & Patrick, 1990). Thus the dominant adverse factor in these soils is iron toxicity (Moore & Patrick, 1989*a*), sometimes phosphorus deficiency and aluminium toxicity, and probably calcium and magnesium deficiency (Moore & Patrick, 1989*b*). pH values may be between 3 and 4 (Charoenchamratcheep *et al.*, 1987) and their unfavourable properties greatly restrict crop productivity, including the efficiency of nitrogen fixation by blue-green algae in wet rice. Wet rice is one of the crops most suited to acid sulphate soils as the reducing conditions associated with flooding suppress acidity.

3.4 Cropping tropical soils

In the foregoing sections, soil chemical and physical properties that affect crop growth have been identified, and comments have been made on their relative importance in the eleven soil orders. For the cropping potential of tropical soils to be fully exploited, an inventory is required of both soil properties and crop edaphic requirements. Some soil constraints can of course be ameliorated by management and fertiliser, but optimal land use is attained only when the edaphic criteria of a climatically-adapted crop match the properties of the soil. As an example, Table 3.11 collates, for Malaysia, edaphic criteria for eight of the crops dealt with in this book.

Further reading

International Board for Soil Research and Management (IBSRAM) (1987). *Soil Management under Humid Conditions in Asia* (ASIALAND). Proceedings of the First Regional Seminar on Soil Management under Humid Conditions in Asia and the Pacific, Khon Kaen, Phitsanulok, Thailand, 13–20 October 1986, 466 pp.

International Board for Soil Research and Management (IBSRAM) (1987). *Land Development and Management of Acid Soils in Africa*. Proceedings of the Second Regional Workshop on Land Development and Management of Acid Soils in Africa, Lusaka and Kasama, Zambia, 9–16 April 1987, 339 pp.

International Crops Research Institute for the Semi-Arid Tropics (ICRISAT) (1989). *Soil, Crop, and Water Management Systems for Rainfed Agriculture in the Sudano-Sahelian Zone*. Proceedings of an International Workshop, 11–16 January 1987, ICRISAT Sahelian Centre, Niamey, Niger. Patancheru, India: ICRISAT, 385 pp.

IRRI (1985). *Wetland Soils: Characterization, Classification and Utilization*. Proceedings of an International Workshop, 26 March – 5 April, 1984, IRRI, Los Baños, Philippines: International Rice Research Institute, 558 pp.

Mulongoy, K., Gueye, M. & Spencer, D.S.C. (eds.) (1992). *Biological Nitrogen Fixation and Sustainability of Tropical Agriculture*. Chichester: Wiley.

Theng, B.K.G. (1991). Soil science in the tropics – the next 75 years. *Soil Science,* **151,** 76–90.

II

CEREALS

4

Cereals in tropical agriculture

4.1 Introduction

The four most important cereals grown for human food in the tropics are rice, maize, sorghum and pearl millet. There are other minor cereals of warm temperate or tropical origin which are of local importance in the tropics. These include finger millet (*Eleusine coracana*), a staple food in parts of East and Central Africa; barnyard millet (*Echinochloa frumentacea*), cultivated in India and southeast Asia; foxtail millet (*Setaria italica*), grown in parts of India; and teff (*Eragrostis tef*), which is confined to the Ethiopian highlands.

The temperate cereals wheat and barley are also grown to a limited extent in the tropics, largely at high altitudes: for example, wheat in Kenya and barley in Ethiopia. Wheat and barley are important crops in the Indian subcontinent, and are grown there in the cool season at low altitudes, but the bulk of the crop area lies outside the tropics.

4.2 Regional production

Table 4.1 gives the area, yield and production of cereals in the tropics, and is approximate in two aspects. First, in the FAO data from which it is derived pearl millet is conflated with finger millet, barnyard millet, foxtail millet and proso millet (*Panicum miliaceum*) under the general title of 'millets', and in some countries millets are reported together with sorghum. Second, the definition of the tropics is not exact, in that the data are available only on a national basis and some important cereal-growing nations straddle the tropics. In Tables 4.1, 9.1 and 14.1, a nation has been designated tropical where the major proportion of its land area lies within the tropics. This approximation accounts, for example, for the large area and production figures for 'other cereals' in tropical Asia: they are mainly for wheat and barley in subtropical India.

Table 4.1 shows that the tropics contribute 30% to the world's cereal production. Rice is the most important tropical crop: it is grown on one-third of all land cropped to cereals and it currently contributes 52% to all tropical cereal production because of its relatively high productivity per hectare. The relative importance of tropical cereals continues to increase: the contribution of 30% to the world's cereal production is a substantial increase over the 22% of a decade ago. This is because the percentage of both world area and world productivity per hectare has increased for all cereal crops during the past 10 years.

Tropical Asia produces nearly twice the tonnage of cereals produced by tropical Africa and America combined. This superiority is due mainly to the large Asian area of rice with a yield level substantially higher than the staple maize of tropical America or the maize, sorghum

Table 4.1. *Area, yield and production of cereals in the tropics, in 1991*

	Tropical Africa	Tropical America	Tropical Asia	Total Tropics	World	Tropics as % of world
Area (m ha)						
Rice	6.15	6.15	94.7	107.04	148.37	72
Maize	16.80	23.37	16.04	56.21	129.15	44
Sorghum	18.68	2.59	16.05	37.32	44.70	83
Millets	15.41	0	15.93	31.34	37.12	84
Other	17.63	4.79	56.80	79.22	344.40	23
Total	68.52	36.90	199.56	304.88	703.73	43
Yield (t ha^{-1})						
Rice	1.61	2.72	2.76	2.69	3.50	77
Maize	1.16	1.86	1.64	1.59	3.71	43
Sorghum	0.76	2.59	0.72	0.87	1.29	67
Millets	0.70	–	0.60	0.65	0.78	83
Other	0.95	1.96	1.79	1.61	2.15	75
Total	1.04	2.07	2.06	1.83	2.68	68
Production (m t)						
Rice	9.89	16.77	261.43	288.09	519.87	55
Maize	19.57	43.43	26.25	89.25	478.78	19
Sorghum	14.22	6.72	11.53	32.47	57.76	56
Millets	10.79	0	9.49	20.28	28.97	70
Other	16.69	9.37	101.81	127.87	740.75	17
Total	71.16	76.29	410.51	557.96	1883.89	30

After FAO (1992).

and millets of tropical Africa. The low yield of sorghum and millets in Africa and Asia is because they are grown largely as rainfed crops in semi-arid regions. Sorghum is not of major importance in tropical America, and millets are scarcely grown there at all.

Even though wheat and barley are important crops in non-tropical India and (for reasons given above) appear in 'other cereals' for tropical Asia, total 'tropical' production (as defined) of cereals other than rice, maize, sorghum and millets is less than 7% of world production. The latter, of course, includes the large temperate cereal production of the USA, Russia, Ukraine, Kazakhstan, Canada, China and others.

4.3 Role in tropical human nutrition

4.3.1 *Cereals as an energy source*

Cereals are the prime source of energy for tropical peoples. Although root and tuber crops or bananas are the main carbohydrate staple in certain regions where ecological conditions are less favourable to the growth of cereals, the total area in the tropics sown to such crops is less than one-tenth of the area sown to cereals (Table 4.1, cf. Table 14.1).

de Vries, Ferwerda & Flack (1967) compared tropical cereals and non-cereal carbohydrate crops in terms of their yield of 'edible energy'; that is, the consumable portion of harvested yield. Table 4.2 is a modification, using more accurate yield data, of their table. All cereals have approximately the same energy value (about 15 MJ kg^{-1}) and except for rice, where yield is normally expressed as 'paddy', harvested yield approaches edible yield. The table shows that the edible energy per hectare of an average crop of tropical rice or maize is more than twice that of tropical sorghum or millet.

On a per crop basis, edible energy yields of cereals compare unfavourably with those of non-cereals. However, the time required to accumulate the high yields of non-cereal crops is correspondingly long, and in terms of edible energy per hectare per day of crop growth, rice and maize are superior to non-cereals. (The approximations of Table 4.2 do not permit anything but the broadest comparisons.)

Thus the advantage of cereals over non-cereal energy crops is not their yield of edible energy per crop, but the fact that a moderate energy yield may be obtained in a short cropping period. Cereals are hence of particular importance in sustaining human populations in the semi-arid tropics. Where rainfall or irrigation permit more than one cereal crop to be grown each year, the annual output of edible energy per hectare from cereals may readily exceed that from non-cereal energy crops.

Most studies on the energetic efficiency, or energy output/input ratio, of crop production have been concerned with mechanised agriculture and have emphasised the high fossil fuel inputs of the cropping systems of advanced nations. A very limited amount of research has been carried out on non-mechanised tropical food production systems, where inputs are wholly in the form of human energy, supplemented in some regions by draft animal energy. Black (1971), using man-hour data from Africa, southeast Asia and Mexico and assuming an average value of 0.63 MJ h^{-1} for net energy expenditure of tropical cultivators engaged in farm work (see Phillips, 1954), calculated the energetic efficiency of rainfed hoe cultivation of a range of cereal crops. The average food energy output was 10.5 × 10^3 MJ ha^{-1} and the human energy input 654 MJ ha^{-1}, with an unweighted average efficiency ratio of 17.5 and a range of 9–34. Norman (1978), assuming an average net energy expenditure rate of 0.75 MJ h^{-1}, examined man-hour data from Clark & Haswell (1970) on African rainfed cereal production and arrived at an average efficiency ratio of 18.5, with a range of 7–39. Out of the 21 sets of data

Table 4.2. *Comparative energy yield of cereals and non-cereal energy crops*

Crop	Average tropical yield (t ha^{-1})	Edible energy value (MJ kg^{-1})	Proportion of edible energy (%)	Edible energy per ha (MJ × 10^3)	Average crop growth period (days)	Edible energy per ha per day (MJ)
Cereals[a]						
Rice	2.69[b]	14.8	70	27.9	140	199
Maize	1.59	15.2	100	24.2	130	186
Sorghum	0.87	14.9	90	11.7	110	106
Millet	0.65	15.0	100	9.8	100	98
Non-cereals[a]						
Cassava	9.63	6.3	83	50.3	330	152
Sweet potato	5.86	4.8	88	24.8	140	177
Yams	7.00	4.4	85	26.2	280	94
Banana	13.00	5.4	59	41.4	365	113

After de Vries *et al.* (1967).
[a] Cereals, air-dry; non-cereals, fresh.
[b] Paddy.

examined by Black and Norman, 8 gave an efficiency ratio of between 15 and 20.

Superficially, these figures suggest that non-mechanised subsistence cereal production in the tropics is a comfortably secure mode of life: a farmer can expect to produce from a cereal crop 15–20 times more food energy than he expends in growing it. However, in assessing the energy balance of the subsistence farmer a more appropriate denominator for the calculation of an efficiency ratio is the total energy requirement of the farm family throughout the year, which may be up to 5 times that of the net energy actually expended in crop production (Norman, 1978). Furthermore, crop energy output, particularly under rainfed conditions, can vary greatly from season to season owing to climate and attack by pest or disease, and crop storage losses may be high. Norman suggests that an average ratio of crop energy output to net human energy expended in crop production of less than 10 represents a vulnerable situation for the subsistence farmer.

4.3.2 *Cereals as a protein source*

The proportion of total protein intake by tropical peoples that is derived from plants varies from region to region according to the importance of domestic animals in the farm economy and the availability of fish, but the general range is 70–90%. On average, about two-thirds of plant protein intake in the tropics is derived from cereals. The digestibility of these proteins by humans varies substantially: FAO (1991) lists the digestibility of polished rice, maize and millet as 88%, 85%, and 79% and suggests that the digestibility of diets in some developing countries is only 54–75%.

Litzenberger (1973) compared per hectare protein production from a range of tropical crops. Table 4.3 is a modification of his table, substituting average tropical yields for average world yields and restricting the comparison to cereals and legumes. The general range of protein content in threshed cereal grain ready for human consumption is 9–12%. The table shows that an average tropical maize crop has a protein yield of the same order as that of an average crop of beans or chickpeas, but that the average protein yield of groundnuts and soybeans exceeds that of the cereals. (As with Table 4.2, the approximations do not allow for any more detailed comparison.) However, it must be remembered that though the substitution of cereals for pulses in a cropping system may not result in an immediate reduction in protein output, the main source of nitrogen for protein formation in the legume crops is the atmosphere and is inexhaustible, rather than the soil, where it is not.

The most common deficiencies in the amino acid spectrum of cereal protein are lysine, methionine and tryptophan. The lysine content of maize is particularly low, and a major effort has been made recently to breed maize cultivars of higher lysine content. Rice milled by modern methods is deficient in thiamine, which is largely present in the endosperm; lack of thiamine leads to beri-beri. A high reliance on maize as a source of energy may induce pellagra, caused by niacin deficiency; maize has adequate niacin, but it is chemically bound in a non-available form (Brouk, 1977).

4.4 Place in tropical cropping systems

The place of cereals in tropical cropping systems has been reviewed by Rao (1986). Regional accounts of tropical cropping systems in general – Okigbo & Greenland (1976; Africa), Pinchinat *et al.* (1976; Latin America), Harwood & Price (1976; Asia), ICAR (1972; India) and ICRISAT (1974, the semi-arid tropics) – include much relevant information on cereal-based systems. Cropping systems based on wet rice are different in character from upland cereal-based systems; they are considered separately in Chapter 5.

Throughout the tropics, rainfed cereals (including upland rice) are more likely to be found growing as a component of mixed or intercrop-

Table 4.3. *Comparative protein yield of cereals and legume crops*

Crop	Average tropical yield (t ha^{-1})	Protein content (%)	Average protein yield (kg ha^{-1})
Cereals			
Rice	2.69[a]	7.5[a]	202
Maize	1.59	9.5	146
Sorghum	0.87	10.5	91
Millet	0.65	10.5	68
Legumes			
Soybean	1.34	38.0	509
Groundnut	0.89	25.5	227
Beans	0.60	22.0	132
Chickpea	0.66	20.0	132

After Litzenberger (1973).
[a] Paddy.

ping patterns than as a sole crop. (This is not true of irrigated cereals.) Thus in Nigeria only 10% of the area of millet is sole crop; for maize the figure is 14%. In Uganda, 54% of the sorghum but only 16% of the maize is grown as a sole crop (Okigbo & Greenland, 1976). In north Nigeria, where the dominant cereals are sorghum and pearl millet, the most frequent crop mixtures are millet/sorghum (25% of all mixed crops), millet/sorghum/cowpea, sorghum/groundnut and millet/sorghum/ groundnut/cowpea (Norman, 1974). In tropical America the staple cereal, maize, is commonly grown with beans or with cassava (Pinchinat, Soria & Bazan, 1976). Rainfed cereals in semi-arid India are frequently grown in intercrop patterns with legumes: e.g., sorghum/pigeonpea, pearl millet/groundnut. Upland rice in southeast Asia is often inter-cropped with maize and cassava; in Sierra Leone 91% of the upland rice is sown with maize, cassava, cotton, pigeonpea, peppers or okra (Spencer, 1973, quoted by Okigbo & Greenland, 1976).

There are numerous advantages to mixed cropping. Agronomic research is steadily and cumulatively confirming the traditional belief of the tropical farmer that in many instances the combined crop yield is greater than the sum of the individual crops grown in monoculture on the same total land area (Willey, 1979). That is, the land equivalent ratio (LER), the total land area required in monoculture to produce the same yield as a given area of mixed crop, expressed as a ratio, is greater than unity. Table 4.4 provides some examples of LERs of intercrops which are above unity. This is the result of more efficient utilisation of environmental resources, particularly radiation and soil water, and in some cases reduced damage from pests and diseases. In addition, the normally greater degree of ground cover may give greater protection from soil erosion (Aina, Lal & Taylor, 1979; see Fig. 2.2) and weed invasion.

Intercropping also gives greater yield stability between years and between sites. Figure 4.1, which gives data from 94 experiments on sorghum and pigeonpea sole crops and intercrops in India, converted to monetary terms, illustrates this clearly.

While mixed cropping is one means of risk reduction (Fig. 4.1), another is to leave soil fallow during the first part of the rainy season, to accumulate soil moisture and ensure success of a cereal planted to utilise soil moisture at the end of the wet season. On deep Vertisols in India, about 12 m ha are left fallow during the first half of the wet season, thereby under-utilising the potential for crop growth, especially where average annual rainfall exceeds 750 mm (Singh & Reddy, 1988).

Table 4.4. *Land equivalent ratios (LERs) and income equivalent ratios (IERs; LERs converted to economic returns) from intercrops at several locations*

Type	Crops A/B	Location	Single stand yields (t ha⁻¹)		Intercropped yields (t ha⁻¹)		LER	IER
			Crop A	Crop B	Crop A	Crop B		
Relay	Maize/beans	Mexico	2.05	0.84	1.53	0.48	1.33	1.34
Mixed	Maize/peanuts	Tanzania	1.73	0.83	1.68	0.42	1.48	–
Row	Maize/mungbeans	Philippines	2.43	1.17	1.85	0.90	1.53	1.45
Row	Sugarcane/cowpeas	Brazil	1.41	1.36	1.41	1.35	2.00	–
Row	Maize/cotton	Kenya	3.05	1.51	2.65	0.56	1.24	1.06
Row	Maize/pigeonpea	Trinidad	3.13	1.87	2.61	1.85	1.82	–

After Sanchez (1976).

The effective growing season, determined by rainfall and soil type, permits variations in cropping systems from single crops (in India, usually pearl millet or sorghum) to intercropping (a cereal with pigeonpea or a herbaceous legume such as chickpea), or to two relay crops (at above 750 mm) (Table 4.5).

Being the staple food of the great majority of tropical peoples – the basis of subsistence – cereals usually occupy a favoured position in crop sequences. Thus in shifting cultivation or short-fallow semi-intensive systems, cereals are usually grown as the first crop after the fallow period, when available soil nutrients are highest. There are exceptions: in the Guinea savanna zone of West Africa, where the initial decomposition of savanna fallow vegetation temporarily immobilises soil nitrogen, yams and legumes are grown in the first crop year and cereals in the second (Vine, 1968).

Similarly, the staple cereal crop usually occupies the central position in relay cropping, with additional crops, often also cereals, being included in the rotation if they are supported by the length of growing season (e.g., Rao, 1986). Table 4.5 illustrates a broad positive relation between complexity of cropping pattern and length of growing season. Systems with maize as the main cereal component tend to be more com-

Fig. 4.1. Income stability from sole crop and intercropped sorghum and pigeonpea in south India. Source: ICRISAT (1980).

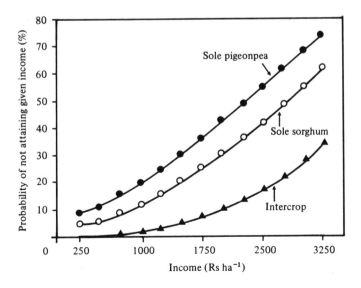

Table 4.5. *Potential cropping systems in relation to rainfall and soil type, for arid and semi-arid zones in India*

Rainfall (mm)	Soil type	Effective growing season (weeks)	Suggested cropping system
350–600	Alfisols and shallow Vertisols	20	Single rainy-season crop
350–600	Deep Aridisols and Entisols	20	Single cropping with either a rainy-season or post-rainy-season crop
350–600	Deep Vertisols	20	Single post-rainy-season crop
600–750	Alfisols, Vertisols and Entisols	20–30	Intercropping
750–900	Entisols, deep Vertisols, and Entisols	20–30	Intercropping
750–900	Entisols, deep Vertisols, deep Alfisols and Inceptisols	30	Double cropping with monitoring
900	Entisols, deep Vertisols, deep Alfisols and Inceptisols	30	Double cropping

plex than those which feature sorghum and millet (which require less water than maize). This is well exemplified by the cropping pattern diagrams for various rainfall zones in west Africa (Figure 4.2). .

The depleting effects of continued cereal cropping on soil fertility may be mitigated by growing the cereals in combination with crop legumes, as seen above, but are often accentuated by the removal and utilisation of stubbles for forage or fuel. The rate of decline in cereal yield with continuous cropping without manure or fertiliser varies greatly with soil type, as indicated in Fig. 1.2. A less dramatic fall in yield was demonstrated in a classic experiment on an Alfisol in semi-arid north Nigeria (Dennison, 1961), in which the yield of pearl millet cropped continuously for 25 years fell from about 900 to 300 kg ha^{-1} and that of sorghum from 500–600 kg ha^{-1} to less than 100 kg ha^{-1}. Research in a comparable environment in north Australia has shown the superiority of pearl millet over sorghum in exploiting deep soil layers for available nitrogen (Wetselaar & Norman, 1960).

Emphasis has been placed on upland cereal cropping without animal manure or mineral fertiliser because in the tropics this is the normal situation. Animal manure tends to be reserved for the more specialised

Fig. 4.2. Cereal-based rainfed cropping systems in Nigeria. Source: Okigbo & Greenland (1976).

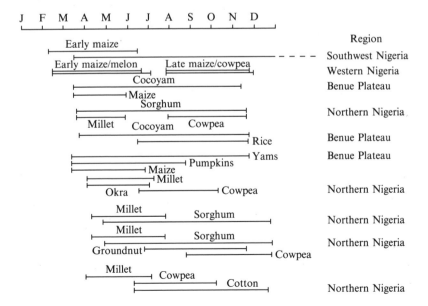

Fig. 4.3. Influence of soil type on cereal-based rainfed cropping systems in semi-arid India. Source: Rao (1986).

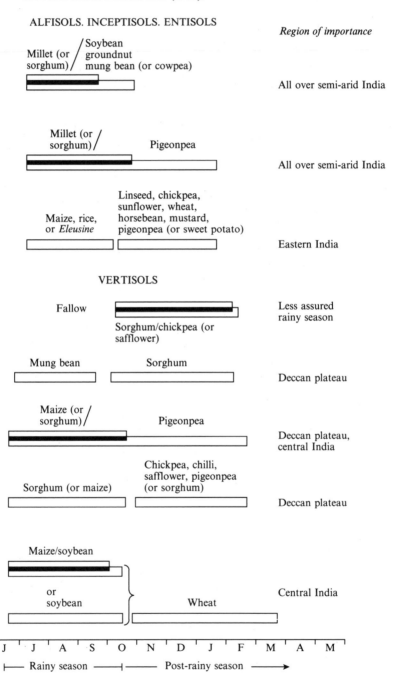

food crops grown around the domicile and for cash crops, and the amount of mineral fertiliser applied to rainfed upland cereals in the tropics is very small. The main focus of fertiliser application to tropical cereals is wet rice and, in continental locations with cool dry seasons just outside the tropics, irrigated wheat.

The wide array of cereal-based upland cropping patterns, governed by climate, soil and food preferences, makes further useful generalisation difficult. Figure 4.2 shows the shift from maize-based systems in the high-rainfall regions of west and southwest Nigeria to sorghum and millet in drier northern Nigeria, and Fig. 4.3 demonstrates the influence of soil type in semi-arid India, where Vertisols support a longer cropping period than Alfisols. These examples illustrate the complexity of the topic.

Further reading

Fussell, L.K. (1992). Semi-arid cereal and grazing systems of West Africa. In *Field Crop Ecosystems*, ed. C.J. Pearson, pp. 485–578. Amsterdam: Elsevier.

Rao, M.R. (1986). Cereals in multiple cropping. In *Multiple Cropping Systems*, ed. C.A. Francis, pp. 96–132. New York: Macmillan.

5

Rice (Oryza sativa *and* O. glaberrima)

5.1 Taxonomy

The genus *Oryza* belongs to subfamily Bambusoideae, tribe Oryzae (Clayton & Renvoize, 1986). There are two cultivated species: *O. sativa*, of Asian origin, and *O. glaberrima*, of African origin. The former is far more important than the latter.

Chang (1976) and Nayar (1973) give general accounts of the systematics of the genus. Chang lists 20 species. The 'sativa complex', all of which are diploid (2n = 24), includes *O. sativa* and its wild relatives (*O. rufipogon* and *O. nivara*) and *O. glaberrima* and its wild relatives (*O. barthii* and *O. longistaminata*). Nine of the other wild species are tetraploid or exist in both diploid and tetraploid forms.

O. sativa is differentiated into three races or subspecies: *indica*, *japonica* and *javanica*. Their morphological and agronomic characteristics are summarised by Purseglove (1972). Intraspecific hybrids are unimportant in nature because of sterility barriers but synthetic interspecific hybrids with high yield potential (15 t ha^{-1}) are now grown in Asia.

5.2 Origin, evolution and dispersal

There appears to be general agreement that *O. sativa* and *O. glaberrima* represent the end-points of independent and parallel domestication. It seems likely that both species developed from annual progenitors: in the case of *O. sativa* a species resembling *O. nivara*, and in the case of *O. glaberrima* one resembling *O. barthii*.

The wild relatives of *O. sativa* have long awns, high cross-pollination frequency, shattering spikelets and high seed dormancy, and are strongly photoperiodic. Domestication to *O. sativa* involved a reduction in awns, frequency of cross-pollination, shattering, seed dormancy and photoperiodic response, together with agronomic improvements such as greater panicle size and number of leaves, a bunch habit and lengthening of the period of grain-fill (Chang, 1976).

The question of whether, in *O. sativa*, upland rice evolved from flooded rice or vice versa is still a source of controversy. Chang (1976) and Harlan (1977) support the concept of domestication in shallow swamps (*O. nivara* is found in ditches and on the edges of ponds), but Gorman (1977) considers that rice was first cultivated as an upland crop in piedmont regions. It appears that irrigation of flooded rice was a comparatively late development.

Most authorities believe *Oryza sativa* was domesticated independently at several locations in south and southeast Asia. Archaeological evidence from India (Vishnu-Mittre, 1977) dates back to about 5000 BP, from China (Ho, 1977) to about 5500 BP, and from Thailand, though this is less certain, also to about 5500 BP (Gorman, 1977). Ho (1977) concluded that 'while existing comparative data are in favour of China, rice was probably independently domesticated in the southern half of China, the southeastern Asian mainland, and the Indian subcontinent'. From this region, cultivated rice spread to the Yellow River valley, where the *japonica* race evolved, and from thence to Korea and Japan. The *indica* race spread to the Yangtse valley about 2000 BP; the *javanica* race developed in Indonesia at a later date (IRRI, 1978*b*).

The modern genetic base of *O. sativa* has its origins in Taiwan in the 1950s, with the breeding of photo-insensitive short-strawed *indica* types, profusely tillering and highly responsive to nitrogen fertiliser. One of these was crossed with an *indica* type from Indonesia by the International Rice Research Institute to produce IR8, the first of the 'green revolution' rice cultivars (IRRI, 1972). The subsequent spread of IR8 and later IRRI releases throughout the developing world is documented by Dalrymple (1978) and Anden-Lacsina & Barker (1978).

O. glaberrima is grown only in West Africa in a triangle from Senegal to Lake Chad to the Nigerian coast (Portères, 1976). Its closest wild relative, *O. barthii*, is actually a semi-arid zone species found across the Sahel and savanna zones from Sudan to the west coast. Domestication probably began in Nigeria about 3500 BP.

Table 5.1 gives area, yield and production data for countries producing more than 5 m t of rice annually. All but four – China, Japan, Korea and the USA – are wholly or largely tropical nations. Both China and India produce rice within the tropics although the greatest area of rice in both countries is associated with their river systems in the subtropics. Average yields of the tropical countries are substantially lower than those of temperate countries: they range from 2.0 to 4.3 t ha^{-1}, but productivity of rice in south China averages 5.7 t ha^{-1}, compared with 5.6–6.7 t ha^{-1} in temperate countries. Average productivity of rice in the top producing tropical countries is now 2.8 t ha^{-1} and has increased by 36% during the past decade.

5.3 Crop development pattern

Rice shows a wide range of variation in development pattern through its adaptation to the climatic factors of temperature, depth of flooding and daylength. Temperature adaptation is illustrated by the importance of the crop in the wet, wet-and-dry and cool tropics (Huke, 1976). Adaptation to depth of flooding and degree of water control has given rise to three cultural classes of rice: upland rice; wet or padi rice, grown in water less than 1 m deep; and deep-water rice, grown in water 1–6 m deep. With respect to daylength, traditional wet rice cultivars grown in the wet season of monsoon climates are usually responsive to short days. High-yielding modern cultivars and many quick-maturing, dry-season or upland genotypes are only weakly sensitive to daylength (Vergara, Puranabhavung & Lilis, 1965; Vergara, 1976; Chang & Oka,

Table 5.1. *Major rice-producing countries: area, yield and production in 1991 of countries producing more than 5 m t per annum*

Country	Area (m ha)	Yield (t ha^{-1})	Production (m t)
China	33.10	5.66	187.45
India	42.20	2.63	110.94
Indonesia	10.19	4.35	44.32
Bangladesh	10.94	2.61	28.57
Thailand	10.00	2.00	20.04
Vietnam	6.29	3.09	19.43
Myanmar	4.83	2.73	13.20
Korea[a]	1.88	6.69	12.58
Japan	2.05	5.86	12.00
Philippines	3.42	2.83	9.67
Brazil	4.14	2.29	9.50
USA	1.11	6.30	7.00

After FAO (1992).
[a]Republic and DPRP combined.

1976). Deep-water rices may be highly sensitive or relatively insensitive to daylength (HilleRisLambers, 1977).

These adaptations have led to an almost infinite range of development patterns, particularly as temperature–daylength interactions occur within each cultural class (Owen, 1971). Common development features have been reviewed by de Datta (1981) and Murata & Matsushima (1975). In the tropics, seeds are usually pregerminated by 24–48 h imbibition. The first leaf appears 3 days after sowing pregerminated seed. Leaves are then produced at 0.25–0.3 d^{-1} (de Datta, 1981) and roots are produced three nodes below that of the emerging leaf (Fujii, 1961, cited by Murata & Matsushima, 1975). Tillering begins at the four to five leaf stage and tiller production may be synchronous with leaf production on the main stem. Attainment of maximum tiller number may coincide with floral initiation (Dingkuhn *et al.*, 1992a) or be reached before, or after, the beginning of panicle growth (Fig. 5.1a,b). Floral initiation is discernible after the initiation of tillers in wet rice; initiation may be synchronous with tiller elongation in short-duration genotypes (Fig. 5.1a) or lag behind elongation in long-duration, daylength-sensitive types (Fig. 5.1b). Floral initiation can occur as early as 6–9 days after seedling emergence in daylength-sensitive deep-water rice (HilleRisLambers, 1977).

Stansel (1975) divided the development of wet rice into three periods; *vegetative*, ending at floral initiation; *reproductive*, ending at anthesis; and *grain-filling*, ending at maturation. The ends of the vegetative and reproductive periods were at development index (DI) values of 0.4 and 0.6 respectively for short-duration genotypes in Texas, latitude 32° N. de Datta (1981) gives DI values for floral initiation of 0.4 for short-duration types and 0.55 for a long-duration, daylength-sensitive type growing at Los Baños (14° N) (Fig. 5.1). These values correspond well with development events for wet rice in Taiwan, latitude 23° N. Likewise, floral initiation and anthesis are relatively predictable in those upland and deep-water rices that are only weakly sensitive to daylength, amounting to approximately two-thirds of the genotypes tested by de Datta & Vergara (1975) and Ikehashi (1977), but the relative duration of vegetative growth is variable in daylength-sensitive types in these classes. In all cases, the period from anthesis to maturity seems relatively constant (e.g., Fig. 5.1), and represents from one-quarter to one-sixth of the life cycle of various genotypes under contrasting environments.

5.4 Crop/climate relations

Sowing of rice takes place in a variety of ways: e.g., dibbling seed into a dryland seedbed, broadcasting pregerminated seed in mud, and sowing into a nursery bed with highly controlled water availability. Germination has a very broad temperature optimum from 18 to 40 °C (Table 5.2). Under aerobic conditions, the coleoptile emerges before the coleorhiza and early growth of shoot and root are typical of gramina-

Fig. 5.1. Development (*a*) short-duration (105–120 days) daylength-insensitive cultivars and (*b*) long-duration (150 days) cultivars. *Note*: in long-duration cultivars there is a so-called lag vegetative period during which maximum tillering, stem elongation and panicle initiation occur in succession. Source: de Datta (1981).

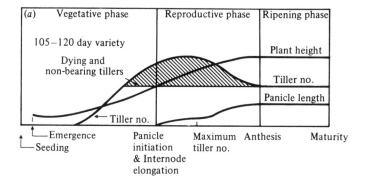

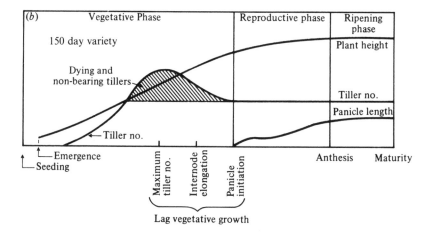

ceous seedlings; at low oxygen concentrations, in either saturated soil or under water, leaf and root emergence are retarded and growth is reduced (e.g., Kordan, 1972; Opik, 1973). Lack of oxygen should not affect seedlings, provided the initial concentration in water is at least 5–6 ppm (Chapman & Petersen, 1962). Increasing depth of water (to 20 cm) reduces the establishment of direct-seeded rice; surviving seedlings produce fewer tillers and are more susceptible to lodging than those established in shallow water (de Datta *et al.*, 1973).

As mentioned in Section 5.3, leaf production is relatively constant (about 0.3 d^{-1}) for wet rice in the lowland tropics. Rate of production is, however, sensitive to temperature: each leaf requires about 100 day-degrees to develop before floral initiation and about 170 day-degrees during reproductive growth (Yoshida, 1977). Thus rates of leaf production may be as low as 0.07–.0.1 d^{-1} above 1500 m altitude. Leaf elongation rate has a relatively marked optimum at about 30 °C (Table 5.2). It also provides a sensitive indication of plant water status, elongation ceasing at pre-dawn water potentials of –0.5 to –1.2 MPa (Steponkus, Cutler & O'Toole, 1980). Net photosynthetic rates of leaves are typical of C3 species: 0.8–1.6 mg CO_2 m^{-2} leaf s^{-1} (Yoshida, 1968; Dingkuhn *et al.*, 1992*b*) and 1.1–2.2 mg m^{-2} field s^{-1} (Tanaka, 1976; Schnier *et al.*, 1990*a,b*). Various workers, mostly using detached leaves, have found that net photosynthesis is relatively unaffected by temperature in the range 20–35 °C but that it diminishes rapidly beyond 40 °C (Tanaka, 1976); the dependence of photosynthesis on leaf nitrogen is illustrated in Fig. 2.8.

Table 5.2. *Effect of temperature on various processes of rice*

Growth stage	Critical temperature (°C)		
	Low	High	Optimum
Germination	16–19	45	18–40
Seedling emergence and establishment	12–13	35	25–30
Rooting	16	35	25–28
Leaf elongation	7–12	45	31
Tillering	9–16	33	25–31
Initiation of panicle primordia	15	—	—
Panicle differentiation	15–20	30	—
Anthesis	22	35–36	30–33
Ripening	12–18	> 30	20–29

After various authors, collated by Yoshida (1977).

Moderate rates of net photosynthesis emphasise the importance of leaf area index (LAI) and canopy architecture in determining dry matter productivity of rice; there is evidence that LAI limits canopy photosynthesis during vegetative growth (Schnier *et al.*, 1990*a,b*). Most management practices affect LAI, whereas canopy extinction coefficient, leaf thickness, internode and tiller length depend primarily on genotype–environment interactions. Above-ground growth is greatest when LAI exceeds 4. However, there are clear genotype and environment differences in the relationship between maximum crop LAI (usually attained at about development index (DI) = 0.6) and both growth rate and final grain yield. As Fig. 5.2 shows, the optimum LAI is lower

Fig. 5.2. Relationship between grain yield and leaf area index (LAI) (upper) and between above-ground crop growth rate and LAI (lower) for a traditional tall cultivar Peta (left) and a short-stature 'improved' cultivar IR8 (right) in the Philippines. *Note*: observe the marked difference between dry-season (high radiation) and wet-season (low radiation) crops. Source: IRRI (1969).

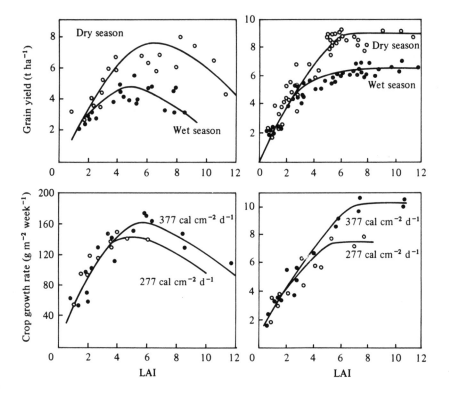

for cv. Peta, typical of tall traditional types, than for dwarf improved types, and lower in the wet (low radiation) season than in the dry season. Likewise, yield responses to nitrogen fertiliser are less in the wet than the dry season (Fig. 5.3). Further, direct-seeded rice, which is replacing nursery-bedding-and-transplanting throughout the tropics, has relatively high tiller numbers and LAI since growth is not checked by transplanting (e.g., Schnier *et al.*, 1990*a*,*b*).

High leaf areas do not, however, necessarily translate into higher grain yield: within-plant shading and depletion of nitrogen concentrations during grain-filling have led some (Dingkuhn *et al.*, 1992*b*) to advocate that varieties for direct seeding should have lower LAI and

Fig. 5.3. Effects of nitrogen and radiation on grain yield of a traditional tall cultivar Peta and 'improved' cultivars IR8 and IR20 in the Philippines. Solar radiation in dry season: IR20, 93.2 kJ cm^{-2}; IR8, 91.6 kJ cm^{-2}; Peta 90.7 kJ cm^{-2}. Solar radiation in wet season: IR20, 67.6 kJ cm^{-2}; IR8, 68.0 kJ cm^{-2}; Peta, 68.5 kJ cm^{-2}. Source: de Datta & Malabuyoc (1976).

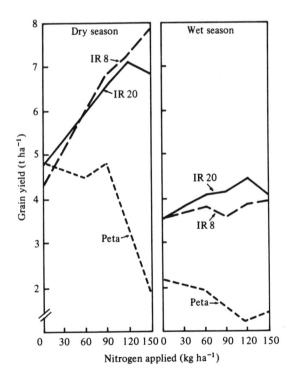

increased concentrations of nitrogen in upper leaves during reproductive growth.

At the same LAI, genotypes with small, erect leaves may have an advantage in terms of potential dry matter production. The disadvantages of tall tillers are high respiration and liability to lodging. In wet rice this leads to the highest-yielding cultivars being erect semi-dwarfs, but in situations where flash flooding or slow inundation is likely (as in 20–30% of the world's rice-growing area), and in deep-water regions, the rate of stem elongation is an important criterion for survival. In deep-water rice, normal rates of elongation are 2–10 cm d^{-1}, but rates of up to 25 cm d^{-1} and tillers 6 m long have been recorded (Vergara, Jackson & de Datta, 1976).

The micrometeorological characteristics of wet rice fields are, naturally, most noteworthy for the high proportion of radiation used for evaporation and the high total water use relative to upland cropfields. The daily Bowen ratio for a rice padi (0.18) lies between those of shallow water (0.08) and maize (0.55) (Uchijima, 1976). Tomar & O'Toole (1980) reviewed seasonal crop water use and showed that although canopy conductance may be quite high ($>10^{-1}$ m s^{-1}) crop transpiration approaches total evaporation only near maturity. The 'crop' water use factor (E/E_p) is 1 at transplanting, 1.15 at the stage of maximum tiller number (about DI = 0.7) and about 1.3 at flowering and during grain development; the average seasonal E/E_p ratio is 1.2.

Rice roots are relatively shallow: in wet rice 80% of root weight may be in the top 20 cm of soil, and in upland rice, 40–60% (IRRI, 1979a). Even in upland rice the effective rooting depth, estimated from neutron moisture measurements, is less than that of other crops (IRRI, 1979a).

Flowering and grain formation begin at about DI 0.75–0.85. Spikelet sterility results from temperatures at about anthesis below 20 °C (Satake, 1969), from high air temperature or vapour deficit (IRRI, 1978a), and from low soil moisture (IRRI, 1979c). During stress, when midday water potential was about −2 MPa, spikelet fertility was reduced to 20% that of unstressed plants (IRRI, 1979a). After removal of stress, grain yield was reduced from 4 to 1 t ha^{-1}. The temperature optima for spikelet number and fertility are in the range 22–25 °C (Yoshida, 1973, 1977).

Once grain number is determined, final yield is the product of the length of the effective grain-filling period (EGFP) and the rate of individual grain growth. Duration of filling appears relatively short and constant in the wet tropics and there are few studies of EGFP in other tropical environments. In Japan, EGFP is inversely related to tempera-

ture (Yamakawa, 1962; Sato & Takahashi, 1971). There are also genotype–temperature interactions, as shown for cv. IR20 and Fujisaka 5 at
temperatures from 20/12 to 32/24 °C (Yoshida, 1977). Rates of grain-
filling may reach 1.7 mg per grain d^{-1} (Egli, 1981) and rates increase
with increasing temperature.

Final yield thus depends on radiation, temperature and water availability, particularly, but not exclusively, during the period of anthesis
and grain filling. Hanyu, Uchijima & Sugawara (1966) proposed:

$$Y = 4.14D - 0.13 \ (21.4 - T)^2,$$

where D is the sum of sunshine hours and T mean air temperature during the grain ripening period. The importance of radiation throughout
the life of the rice crop is summarised in Evans & de Datta (1979).
Radiation in the 20 days after panicle initiation shows the highest correlation with number of panicles per hectare, and radiation during grain-
filling shows the highest correlation with final grain number. In their
survey data, radiation had little influence on individual grain weight,
presumably because of physiological compensation. Low radiation may,
through associated low temperature, reduce the rate of grain-filling but
increase the length of EGFP. From the equation above and other analyses (Yoshida & Parao, 1976), predicted yields are highest at about
22 °C. This moderate temperature optimum indicates that maximum
yields per crop will be found in the subtropics or cool tropics. However,
because of shorter crop duration and the potential for year-round cropping, yields per year are highest for wet rice in the tropics: e.g., 24 t ha^{-1}
from three or four crops in the Philippines, in contrast to 6–8 t ha^{-1}
from one crop in Australia (latitude 35° S) or Hokkaido, Japan (43° N)
(Yoshida, 1977).

Rice is sensitive to water stress and also to anaerobiosis associated
with total submergence. Water deficit causes leaf water potentials and
rates of elongation to decline more rapidly in rice than in maize or
sorghum, so that dry matter accumulation and nutrient uptake decline
or cease (Tanguilig *et al.,* 1987). Water deficit during the vegetative
stage may have relatively little effect on grain yield (O'Toole & Moya,
1981) perhaps owing to compensatory growth or changed partitioning
of dry matter after the stress is relieved. However, grain yields are quite
sensitive to stress at anthesis (IRRI, 1979*a*) and during grain-filling
(O'Toole & Moya, 1981). Such stresses occur particularly with upland
rice. For three varieties in the Philippines, yield reduction related to
water deficit after anthesis was due equally to reduced panicle numbers

and increased sterility, and in small part (minus 7%) to reduced individual grain weight (O'Toole & Moya, 1981).

Submergence also reduces or stops leaf and stem elongation (Pearson & Jacobs, 1986; Tanguilig *et al.*, 1987). Khan *et al.*, (1987) have shown elongation capacity (when unstressed) to be correlated with ethylene production during submergence, while Pearson & Jacobs found leaves were relatively protected from submergence through differential loss of weight of roots and stems. When rice plants are only partially submerged, however, there may be no deleterious effects of water to 1 m depth because hollow tissue (aerenchyma) facilitates the transport of atmospheric carbon dioxide and oxygen to submerged plant parts. In floating rice in deeper water, concentrations of oxygen in internodes decrease from 16–20% at the water surface to 5% at depth while carbon dioxide increases from 1–3% to 5–10% at 1.8 m depth, causing Setter *et al.*, (1987) to suggest that low internal oxygen may restrict respiration and plant development and, near the soil, nutrient uptake.

5.5 Crop/soil relations

5.5.1 *Landforms and soil types*

Most wet rice is grown in coastal plains and fluvial lowlands. The relationship between these landforms and soil types is shown in Fig. 5.4. On rapidly aggrading coastal plains close to the shoreline, the soils are mainly Aquents, many of which lack distinct pedogenetic development; further inland the main soils are Aquepts, with more conspicuous horizon differentiation but with variable groundwater depth depending on surface topography. On slowly aggrading or stationary coastal plains the soils are similar but contain concentrations of pyrites in sediments near the shoreline so that Sulfaquents and Sulfaquepts are present. These require careful cultivation and water control to overcome their agricultural constraints. Halaquepts may also be present (Driessen & Moormann, 1985). On the 2.5–3.0 m ha of coastal soils on which rice is grown in India, salinity is a serious yield constraint (Abrol *et al.*, 1985).

In inland valleys of hilly and mountainous areas the soils at the bottom are likely to be variable in parent material and pedogenetic development. Where sediments are added faster than horizon differentiation, Aquents and Fluvents are found, but most valley-bottom soils include Aquepts and Aquic subgroups of Tropepts and Ochrepts. Aqualfs,

Ustalfs and Vertisols occur where there is a pronounced, extended dry season. Valley slopes, sometimes terraced for wet rice, are mainly Inceptisols; Driessen & Moormann (1985) provide a more detailed description.

River fans differ widely in their sediment and mineral composition and age. For this reason their soils may range from very young Entisols to well-developed Alfisols and Ultisols. Small, levelled, and bunded seasonal rice fields dominate upper fan areas; larger, wetter fields are common near the base.

River floodplains comprise three main components: riverbed, levee and basin. Riverbeds are used in seasonally dry climates and are of limited extent. Levees are not used for wet rice except adjacent to river basins, where they are levelled and bunded. Upstream levee deposits are coarser-textured than those downstream, and become finer with increasing distance from the riverbed. Most often levee soils are Fluvaquents, Tropaquents/Haplaquents, and Tropepts/Ochrepts with aquic properties; where levees grade into adjacent basin areas, Aqualfs or Aquults

Fig. 5.4. Summary of occurrence of the commonest soil orders in various landforms: na, not applicable; –, absent or rare; +, common; ++, abundant. Source: Driessen & Moormann (1985).

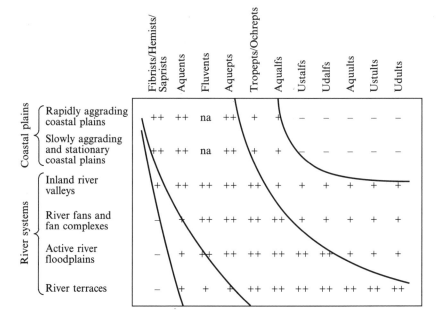

sometimes occur. While basin soils may vary enormously, they are typically fine-textured and hydromorphic.

River terrace soils vary in age, mineral composition, height above erosion base level, and topography. Rice-growing soils on younger terraces include Tropepts/Ochrepts and, in more poorly drained areas, Tropaqualfs/Ochraqualfs. Soils on older Pleistocene terraces, which accompany many rivers, particularly in southeast Asia, are often Tropaquults and Tropudults.

A wide range of well-drained soils in orders such as Inceptisols, Entisols, Alfisols, Ultisols and Vertisols can be used for wet rice if irrigation water is available. On the Deccan Plateau of India for instance, Usterts are so used (Moormann & van Breeman, 1978). Upland rice cropping is carried out in southeast Asia mostly on Ultisols and Alfisols, in Latin America mostly on Oxisols (de Datta & Feuer, 1975), and in Africa mostly on Alfisols, Ultisols and Oxisols (de Datta, 1981). Salt-affected soils, acid sulphate soils and peat soils, dealt with in Chapter 3, occur most extensively in rice-growing areas of south and southeast Asia and greatly constrain productivity.

5.5.2 *Soil physical properties*

While rice is grown on a wide range of soil texture classes, the studies of Kawaguchi & Kyuma (1974) on wet rice lands of tropical Asia indicate that most have a medium (fine loam) to fine (fine clay) particle size distribution. Coarse soils, unless underlain by impervious horizons, lead to inefficiency in use of water and nutrients owing to high drainage rates.

Certain particle size distributions predispose the soil to the development of 'traffic pans' when puddled – compacted 5–10 cm thick subsurface horizons at 10–40 cm depth. These develop through pressure of human and animal feet rather than through ploughing: animal hooves alone may reduce total porosity in the 10–30 cm layers while increasing it in the 0–10 cm layer (IRRI, 1988). Traffic pans are more likely to occur on fine loamy soils than on more sandy or more clayey soils (Sanchez, 1973a). In modal wet rice-growing soils the traffic pan is underlain by iron and manganese pans resulting from redox effects. Incipient pan formation is destroyed during drying in self-mulching soils such as Vertisols, in those with a stable structure (Andisols and Oxisols), and in soils of high organic matter content (Moormann & van Breemen, 1978).

Soil structure is of little importance to wet rice and is deliberately broken down in the process of puddling. This results in a reduction in macropores and an increase in micropores and compaction. The consequences of this are a reduction in hydraulic conductivity (Table 5.3), which improves water use efficiency and reduces nutrient loss, and reduced soil strength, which improves rice root development. The usual outcome is higher yield (e.g., Sharma & de Datta, 1985*a*). Post-flooding puddling increased the depth to which soil strength was below 0.5 MPa (the strength at which rice root extension was much impeded) from 4 cm for nil treatment to 12 cm for ploughed and harrowed soil (IRRI, 1988). Reduced hydraulic conductivity also promotes anaerobic conditions that influence changes in redox and pH usually favourable to nutrient availability, and it helps to control weeds. Nevertheless puddling does not always improve the nutrient uptake of rice (Sanchez, 1973*b*; Sharma & de Datta, 1985*b*), perhaps because phytotoxins are less likely to be leached from puddled soil (Sharma, de Datta & Redulla, 1989).

Tilth after puddled, flooded rice is usually unfavourable for following crops. On draining and drying the soil may break into hard, medium-to-large clods. If the soil does not dry, it may become a tough paste after tillage (Prihar *et al.*, 1985). Upland crops grown in rotation with wet rice may suffer from impaired aeration, poor water relations (IRRI, 1988), and low nitrogen and phosphorus availability. As re-oxidation of flooded soils yields ferrihydrite, which has a large capacity to absorb phosphorus, responses to phosphorus fertilisers in upland crops following wet rice are generally poorer than in crops grown in the same soil previously unflooded (Willett, 1986).

Table 5.3. *Effect of puddling on drainage rates of six Philippine rice soils*

Soil	Mineralogy	Clay (%)	Drainage (cm d^{-1})	
			Aggregated soil	Puddled soil
Psamment	Siliceous	9	267	0.45
Fluvent	Mixed	24	215	0.17
Aquept	Montmorillonitic	30	183	0.05
Aqualf	Montmorillonitic	40	268	0.05
Ustox	Kaolinitic	64	155	0.05
Andept	Allophanic	46	214	0.31
Mean			217	0.18

After Lal (1985).

5.5.3 *Soil chemical properties*

When soil is flooded for rice growing there is a fairly rapid depletion of oxygen through the activity of obligate and facultative aerobes in the soil. The greater the concentration of incorporated decomposable organic substrate the more rapid the depletion. In most soils oxygen is reduced to relatively low concentrations within a day or even within hours of flooding (Sanchez, 1976; Reddy, Rao & Patrick, 1980). Thereafter aerobic respiration ceases in the soil except in the surface 10 mm or so, where dissolved oxygen can still penetrate. The ratio of living, photosynthetic, oxygen-producing organisms to decaying organisms and plant material in the water, and the rate of water percolation, will influence the depth of the oxidised zone. Below this, in the reduced zone, decomposition of organic matter will continue through the activity of anaerobes.

Within the reduced zone the following chemical and electrochemical changes take place (Ponnamperuma, 1972, 1976):

1. Decrease in redox potential.
2. Increase in pH of acid soils and decrease in pH of calcareous and sodic soils.
3. Increase in specific conductance.
4. Reduction of Fe^{3+} to Fe^{2+} and Mn^{4+} to Mn^{2+}.
5. Reduction of NO_3^- and NO_2^- to N_2 and N_2O.
6. Reduction of SO_4^{2-} to S^{2-}.
7. Increase in supply and availability of nitrogen.
8. Increase in availability of phosphorus, silicon and molybdenum.
9. Decrease in concentrations of water-soluble zinc and copper.
10. Generation of CO_2, CH_4 and toxic reduction products such as organic acids and H_2S

The time course of these changes and the extent to which they occur, as determined by soil and management factors, have been discussed by de Datta (1981) and Patrick & Reddy (1978).

Applied fertilisers should be available within 20 cm of the surface for maximum growth rates, as 80–95% of roots are found within this depth (Beyrouty *et al.*, 1988). Oxygen diffuses through aerenchyma in the shoots and lysigenous channels in the root (e.g., Armstrong, 1971), allowing roots to respire aerobically for nutrient uptake and modifying the normal redox effects of flooded soils. The rate of diffusion diminishes with depth in relation to the decline in partial pressure of oxygen (IRRI, 1977c). This explains why upland rice is generally more deeply rooting than wet rice.

Improved high-yielding varieties take up large quantities of nutrients, as shown in Table 5.4. Table 5.5 contrasts uptake over 5 years in traditional and high-yielding varieties and illustrates the need to fertilise high-yielding varieties if they are to achieve their yield potential.

Nitrogen is the main limiting nutrient in the production of modern rice varieties in Asia. Its availability to wet rice is higher than to upland rice on the same soil, not because it is mineralised more rapidly, but because less nitrogen is immobilised (de Datta, 1981). The availability of nitrogen in flooded soils increases with the nitrogen content of the soil, soil pH, temperature, and duration of previous desiccation of the soil (Ponnamperuma, 1965). An interesting feature is that in spite of the lack of inputs of nitrogen from legumes or fertiliser, sustained (though low) yields of wet rice have been obtained from the same land over many centuries. In the Philippines, for instance, after 24 wet rice crops over 12 years, the nitrogen budget showed that, exclusive of nitrogen from rainfall and irrigation, there had been an input of 103 kg of nitrogen per year, with no decrease in soil total nitrogen level (App *et al.*, 1984). This input has been attributed to a number of nitrogen sources, the most important of which is nitrogen fixation.

Nitrogen fixation in wet rice systems is attributable to four main groups of organisms. The first group includes a wide range of blue-green algae (e.g., *Tolypothrix*, *Nostoc*, *Schizothrix* and *Calothrix*), which live in the water above the soil surface and the population densities of which are usually correlated with soil pH and available phosphorus (IRRI, 1988). The annual amount of nitrogen fixed can exceed 70 kg ha^{-1} (Stewart *et al.*, 1979). The second group contains *Azolla* spp., of which *Azolla pinnata* is the most common in tropical Asia. This small floating fern with a blue-green algal symbiont (*Anabaena azollae*) has

Table 5.4. *Nutrient removal of a rice crop (cultivar IR8), Philippines, 1979 dry season (kg of nutrient removed per tonne of straw or grain)*[a]

	Nutrient element												
	N	P	K	Ca	Mg	S	Fe	Mn	Zn	Cu	B	Si	Cl
Straw	5.3	0.8	13.6	3.9	2.6	0.7	0.2	0.6	0.03	0.00298	0.0089	74.0	1.8
Grain	10.9	2.0	3.1	0.5	1.1	1.0	0.04	0.05	0.01	0.00506	0.0038	16.8	1.6

After de Datta (1981).
[a] Yield of crop was 7.9 t ha^{-1} grain and 7.0 t ha^{-1} straw.

Table 5.5. Change in nutrient balance as a result of intensive cropping over a five-year period, averages of three locations and three varieties

Cropping system and fertiliser application	Total grain yield in 5 years (kg ha⁻¹)	Nutrients added (kg ha⁻¹)			Nutrients removed (kg ha⁻¹)			Balance		
		N	P	K	N	P	K	N	P	K
Traditional culture, one crop, no fertiliser	7 500	—	—	—	126	281	164	− 126	− 28	− 164
HYV^a culture, double cropping.										
No fertiliser	32 400	—	—	—	545	122	710	− 545	− 122	− 710
105 kg N ha⁻¹ per crop	50 100	1050	—	—	840	190	1100	+ 210	− 190	− 1100
105 kg N + 26 kg P ha⁻¹ per crop	54 700	1050	262	—	920	208	1199	+ 130	+ 55	− 1200
105 kg N + 26 kg P + 50 kg K ha⁻¹ per crop	59 300	1050	262	498	995	225	1299	+ 55	+ 37	− 801

After von Uexküll (1978).
^{a}High-yielding variety.

fixed 70–110 kg nitrogen ha^{-1} in the Philippines (IRRI, 1979). In the Niger Basin *Azolla pinnata* increased grain yield by 27%, equivalent to an application of 40 kg N ha^{-1} as urea (Kondo, Kobayashi & Takahashi, 1989). Fertilisation of paddy water with phosphorus can stimulate the growth and nitrogen yield of both groups (IRRI, 1988). The third group includes rhizosphere organisms such as free-living *Azospirillum* spp., *Klebsiella* spp. and *Enterobacter* spp. (Boonjawat *et al.*, 1991), and symbiotic *Azorhizobium* spp (IRRI, 1988). The fourth group are non-rhizosphere heterotrophs, which may fix 6–25 kg N ha^{-1} yr^{-1}, depending on the amount of rice straw returned (Wetselaar, 1981). Significant differences occur between rice cultivars in the amount of nitrogen fixed by these groups, even when crop growth duration, biomass at heading and nitrogen uptake are similar (IRRI, 1989). The contribution of fixation to the nitrogen balance of wet rice is discussed by Roger & Ladha (1992).

Other sources of nitrogen for wet rice are in rainfall (1–38 kg ha^{-1} yr^{-1}; Wetselaar & Ganry, 1981), irrigation (6–16 kg ha^{-1} per crop; Wetselaar, 1981), groundwater (0.1 kg ha^{-1} yr^{-1}; Brown, 1981), atmospheric ammonia (Hanawalt, 1969), soil organic matter and crop residues. Crop residues are potentially valuable sources of nitrogen for rice but require careful management to minimise nitrogen losses. Crop residues include those from green manure legumes such as *Sesbania rostrata*, which is tolerant of waterlogging and which has been recorded as fixing 198 kg N ha^{-1} in 60 days, increasing rice grain yield by 2 t ha^{-1} (Morris *et al.*, 1989). IRRI (1989) has shown that the availability of nitrogen from green manure and inorganic fertiliser applied at transplanting is the same. However Buresh *et al.*, (1989) found virtually all of the 77 kg nitrate-nitrogen (NO_3^-N) ha^{-1} in the surface 0.6 m of a Vertic Tropaquept following mungbean (*Vigna radiata*) was lost by denitrification or leaching when the soil flooded for rice production. The apparent recovery of fertiliser nitrogen in rice is low: ^{15}N recovery rarely exceeds 30–40% in wet rice (de Datta, Fillery & Craswell, 1983) and no more than 60% in upland rice (Mitsui, 1954). It is influenced by solar radiation and cultivar (Fig. 5.3).

The transformations and losses of nitrogen in the oxidised and reduced zones of flooded soils, which affect the efficiency of nitrogen fertiliser use, are shown in Fig. 5.5. The extent to which ammonia volatilisation, denitrification, runoff and leaching contribute to nitrogen loss may vary greatly with the particular situation and have been discussed by de Datta (1987), Mikkelsen (1987), Buresh & de Datta (1991) and Jayaweera & Mikkelsen (1991). de Datta (1987) summarised estimates for ammonia volatilisation from applied fertiliser of 5–47% and for denitrification of 28–33%. Mikkelsen (1987) considered that while average fertiliser denitrification losses were also of this order (25–35%), they could vary from 0 to 70%. Rice straw burning, a common practice in Asia, results in losses of around 70% of the nitrogen in the straw (IRRI, 1988).

Low soil phosphorus is widespread in wet rice growing areas, particularly on highly weathered soils such as Oxisols and Ultisols (Haynes, 1984). However, as flooding increases phosphorus availability, wet rice may show no response to phosphorus fertiliser while upland rice on the same site does. An important reason for the increase in phosphorus availability is the reduction of ferric phosphate ($FePO_4.2H_2O$) to more soluble ferrous phosphate ($Fe_3(PO_4)_2.8H_2O$). Other reasons for increased phosphorus availability are given by Sanchez (1976) and Willett (1986). There are significant differences between cultivars in the efficiency with which they utilise low levels of soil phosphorus (Fagaria, Wright & Baligar, 1988).

Fig. 5.5. Fate of fertiliser nitrogen in wet rice soil: 1, urea hydrolysis; 2, ammonia volatilisation; 3, nitrogen immobilisation; 4, runoff; 5, ammonium fixation; 6, dentrification; 7, leaching; 8, plant uptake. Source: Craswell & Vlek (1979).

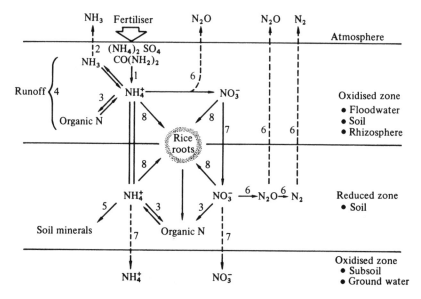

Where nitrogen and phosphorus fertiliser applications have been used to increase yields of improved varieties, there has also been a move to forms with lower sulphur content – from ammonium sulphate (21% N, 24% S) to urea (46% N, 0% S) and from single superphosphate (9% P, 13% S) to triple superphosphate (20% P, 1% S). This has resulted in an increase in the incidence of sulphur deficiency in tropical wet rice (Ismunadji *et al.*, 1983).

Next to nitrogen and phosphorus, zinc deficiency is perhaps the most important nutritional factor limiting the grain yield of wet rice (de Datta, 1978). As much as 50% of the zinc in rice straw is lost by burning (IRRI, 1988). Zinc deficiency is found in Histosols, sodic, calcareous and sandy soils, and in soils wet for prolonged times (Ponnamperuma, 1977; Table 3.8).

The requirement of rice for potassium is second only to its requirement for silicon (Table 5.4). However, the response to applied potassium is variable, dependent to a large extent on soil type. When soil is flooded there is an increase in soluble potassium, owing to its displacement from exchange sites by soluble ferrous and manganous ions. While this increases potassium available for uptake, it also increases its liability to leaching (Chang, 1971).

Silicon is an important nutrient for the growth of rice, crop removal being as much as 90 kg per tonne of grain plus straw (Table 5.3). Though the silicon level in rice straw is high, the short-term availability of it to a wet rice crop is likely to be low (Ma & Takahashi, 1991). Silicon deficiency may occur more frequently in upland rice on weathered soils, such as Ultisols, than in wet rice on the same soil receiving an inflow of water containing the element (Yamauchi & Winslow, 1989). Deficiency may also occur on Histosols (Snyder, Jones & Gascho, 1986). Leaves of silicon-deficient plants become soft and droopy so that the erect habit, important to high photosynthetic rates, is lost.

Although rice is not tolerant of high salinity (Abrol *et al.*, 1985), it is highly tolerant of high sodicity (Table 3.10). While 6–7 dS m^{-1} salinity of the saturated soil paste extract can halve rice yield (Mass & Hoffman, 1977), maintenance of surface water in wet rice culture, which leaches and dilutes salts, can give satisfactory yields where the measurement on the topsoil is 20–25 dS m^{-1} (Abrol *et al.*, 1985). Irrigation with groundwater with a salinity level of 4.0 dS m^{-1} has depressed grain yield by 25% (Beecher, 1991).

5.6 Place in cropping systems

In discussing the place of rice in farming systems, a clear distinction may be drawn between upland rice and flooded or 'wet' rice. This is because upland rice is grown like any other upland seed crop and takes its place in normal rainfed upland crop patterns, whereas the local environment of the wet rice crop – a flat floodable field or bay – dictates the character of the cropping system and distinguishes it from all other types.

5.6.1 *Upland rice cropping systems*

Upland or rainfed rice occupies about 13% of the total rice area of the world (17 m ha out of 143 m ha). Locally, upland rice is particularly important in West African and tropical American rice cropping systems. In West Africa, about 60% of the total rice area is upland, and in Latin America 75%. For Laos, Bangladesh, Indonesia and the Philippines the proportions are 29%, 22%, 21% and 20% respectively; the average for tropical Asia as a whole is 10% (Hanfei, 1992).

Upland rice is a component of many shifting and semi-intensive cropping systems in southeast Asia, tropical America and West Africa. In central Sumatran rainforest, shifting cultivators plant upland rice and follow it with rubber, which is tapped, somewhat irregularly, when the trees have developed. A group of shifting cultivation systems based on upland rice practised by tribal peoples in the hills of northern Thailand is described in Kunstadter *et al.* (1978): the Hmong grow rice, maize and opium, the rice normally in a separate area at a lower altitude than the other crops (Keen, 1978); the Karen combine upland rice grown on hillslopes under shifting cultivation with permanent wet rice in the narrow valleys below (Hinton, 1978). Examples of shifting and semi-intensive systems in West Africa including upland rice are given by Okigbo & Greenland (1976).

In southeast Asia some clearly defined cropping patterns involving continuous annual cultivation of upland rice on fertile land have developed. In regions of Indonesia with a rainfall regime that permits cropping for most of the year, upland rice is sown with quick-maturing maize early in the wet season. This mixture may be planted about a month later with cassava, or be followed sequentially by a second crop of maize (McIntosh, Effendi & Syarifuddin, 1977). However, the most concentrated region of continuous upland rice growing is in the savannas of central-west Brazil, where cultivation of the crop is partially mechanised and 70% of the total Brazilian crop is produced (de Datta, 1975).

5.6.2 Wet rice cropping systems

The techniques of growing rice in a near-flat field with impounding banks evolved, as we have seen, in south and southeast Asia, and this region remains by far the most important rice-growing area of the tropics (Table 5.1) and grows the highest proportion of wet rice. In the following brief account, attention is confined to tropical Asian cropping systems, if only for the reason that published information on wet rice cropping is overwhelmingly Asian in origin. African wet rice (and upland rice) systems are discussed by Chabrolin (1977).

In Chapter 1, a broad classification of wet rice systems was proposed: (1) shallow-water rice, (2) shallow-water rice with upland crops, (3) long-standing floodwater rice. The degree of water control in groups 1 and 2, which include both irrigated and rainfed systems, varies widely, but in all cases an attempt is made to maintain water depth at less than 1 m, and normally substantially less than that. About 60–65% of the shallow-water rice grown in south and southeast Asia is rainfed, not irrigated. Deep-water rice, which accounts for about 25% of the world's rice-growing area, is characterised by the fact that at some period in the life of the crop it is flooded to a depth of 1 m or more (Vergara, 1992).

Local topography has a profound influence on the character of rainfed wet rice cropping systems. The drainage of water from one rice field to its neighbour, or the run-on from catchment areas to rice fields below them, is an important hydrological element governing the cropping pattern. The effect of topographical position on the date of feasible transplanting and the duration of floodwater in a region in the Philippines is shown in Table 5.6.

Table 5.6. *Relation between topographical position and period of flooding of rice fields, Philippines*

Topographical position	No. of weeks from Jan. 1 to crop establishment	No. of weeks from Jan. 1 to last date of standing water	No. of weeks flooded
Bottomland	25	46	21
Plain	27	45	18
Plateau	27	41	14
Sideslope	28	41	13

After IRRI (1979*b*).

The definition of cropping season in relation to rainfall and to topographic and soil characteristics has been taken further by Morris & Zandstra (1979) through the use of the concept of a flooded status day (FSD): that is, a day when there is standing water in the field. In any given rainfall regime, the soil-related factors governing the seasonal pattern of FSD are water table depth, soil texture, water source and relative elevation.

Shallow-water rice-only systems. In wet-and-dry climates with less than about 6 months of rain, with limited late run-on and no irrigation, only one crop may be grown each year. In a more extended wet season, or with irrigation water available for a limited period after the wet season, two crops may be grown with a short fallow period in late dry season. In wet tropical climates or where irrigation water is available year-round, three crops are often feasible, though peak labour stresses and periodic excessive rain mean that, in practice, local average cropping frequencies rarely exceed 2.5. General accounts of wet rice cropping patterns in Asia in relation to rainfall regime are given by IRRI (1975, 1979c), Oldeman & Suardi (1977), de Datta (1981) and Rao (1986).

The above postulates a simple positive relationship between duration of available water and cropping frequency, but it must be emphasised that this only holds good if water from rainfall and run-on is not excessive and surface drainage is good. Inundation of the cropping area for extended periods does not permit a succession of rice crops, and under such conditions the cropping system may revert to a single crop of deep-water rice, as is common in the deltas of the great rivers of southeast Asia.

Intensive irrigated cropping, with two or three crops a year, is usually associated with advanced agronomic technology, and fertiliser, particularly nitrogen, is applied to sustain yields. However, even without nitrogen fertiliser, if other nutrients are not limiting, intensification from one to two to three crops a year may not result in a major fall in yield per crop, since at a yield level of 2000–3000 kg ha^{-1} the individual rice crop may be virtually independent for nitrogen as a result of fixation (see Section 5.5.3 and Walcott *et al.*, 1977). However, intensive cropping increases the possibility of pest and disease outbreaks.

Systems of shallow-water rice with upland crops. In the majority of cropping patterns in which wet rice is the main crop and there is opportunity to grow a sequential crop, that second crop is also rice. However, there are four sets of circumstances in which other second crops are grown:

1. Limited duration or quantity of water.
2. Excessive percolation.
3. Low temperature.
4. Greater demand for other crops.

These circumstances are enlarged upon below.

1. With little or no late-season rain or irrigation water or a poor topographical position, the water available after the main rice crop may be inadequate, in duration or volume or both, for a second rice crop. Kung (1971), quoted by Oldeman & Suardi (1977), gives the following ranges for total water consumption by rice and alternative upland crops:

Rice: 380–880 mm, or 2.9–6.3 mm d^{-1}
Soybeans: 300–350 mm, or 2.3–3.5 mm d^{-1}
Maize: 350–400 mm, or 2.9–3.5 mm d^{-1}
Groundnuts: 400–500 mm, or 2.7–3.5 mm d^{-1}

2. Preferentially, wet rice is grown on soils with low percolation rates, either natural or influenced by puddling. A rate of not more than 1 mm d^{-1} is ideal but in some rice areas it may be up to 10 mm d^{-1} (Oldeman & Suardi, 1977). Cropping with wet rice on freely percolating soils may be feasible in the period of the year when rainfall or irrigation is adequate to maintain water levels, but at other times of the year it may only be possible to grow upland crops.

3. In most lowland areas of the tropics temperature during the low-sun period remains high enough for rice, but in some continental regions (e.g., north India) winter temperature is too low and temperate crops such as wheat, barley, mustard, etc., are grown in sequence with wet-season rice.

4. If the expected return from upland crops is higher than that from rice, then (even if ecological conditions are suitable for rice) other second crops may be grown, and the inputs to create favourable drainage conditions for such crops may be quite high. For example, in northern Thailand drainage ditches are dug after the wet-season rice harvest to form beds about 1.5 m wide on which dry-season tobacco is grown. Around the major cities of southeast Asia wet-season rice and dry-season vegetables, on beds or ridges, is a common crop sequence. In subtropical Taiwan, more intensive sequences involving two rice crops followed by an upland crop are followed (Ruthenberg, 1980), e.g.:

February–June: First rice crop
July–October: Second rice crop

Table 5.7. *Characteristics of rice growing in long-standing floodwater or deep water*

Country	Area (10^5 ha)	Start of flooding	Rate of flooding increase (cm d^{-1})	Maximum depth (cm) at given date	Broadcasting/ transplanting	Flowering date
Mali	1.32–1.56	June–Oct	3–10	300 Oct–Nov		Nov–Dec
Niger	0.50	June–Oct		300 Oct–Nov		
Myanmar	4.70	June	5–30	300 Oct–Nov	Apr–May	Sept–Nov
Vietnam	9.00	Aug	1	300 Nov		Dec–Jan
Thailand	5.00–18.00	Sept–Oct	6–8	400 Nov–Dec	Apr	Nov–Dec
Sri Lanka				120	Aug	Jan
Bangladesh	20.00	May–June	5–8	100–1000 Aug–Sept	Mar–Apr	Oct–Nov
Indonesia	1.00				Oct–Nov	Feb
India						
Uttar Pradesh	14.00	June–July		600	Mar–June	Oct–Nov
Bihar	12.00–20.00	June–July		25–400 Dec–Jan	Feb–Apr June–July	Oct–Nov
West Bengal	13.50			165–300 Dec	Apr–May	Oct–Dec
Orissa	10.00			300		
Assam	1.00–11.70	May–June		600	Apr–May	
Andhra Pradesh	3.50			150–180		
Tamil Nadu	1.00			150–180		
Madhya Pradesh	1.25					

Source: Vergera (1992)

October–February: Sweet potato, tobacco, wheat, maize, vegetables, etc.

The upland crops may be relay-planted into the maturing second crop of rice.

Long-standing floodwater or deep-water rice-growing systems. In areas subject to deep seasonal flooding in wet-and-dry climates, particularly the deltas of the major rivers of south and southeast Asia – the Ganges, Brahmaputra, Irrawaddy, Chao Phraya and Mekong rivers – rice is grown in water which attains a depth of 1 m or more, the maximum being about 6 m. The characteristics of rice-growing in long-standing floodwater or deep-water rice are given in Table 5.7.

After land preparation, seed is broadcast either dry or pregerminated, though some deep-water rice is transplanted. Cultivars have evolved in specific locations, adapted to latitude and average date of floor recession, with a daylength response that brings them to maturity as the water level subsides. Normally only a single crop of rice is grown each year, though in Bangladesh some dry-season crops – legumes, potatoes, chillies, wheat – are grown on residual soil water.

Further reading

Datta, S.K. de (1981). *Principles and Practices of Rice Production*. Los Baños, Philippines: International Rice Research Institute.

Hanfei, D. (1992). Upland rice systems. In *Field Crop Ecosystems*, ed. C.J. Pearson, pp. 183–204. Amsterdam: Elsevier.

IRRI (1985). *Soil Physics and Rice*. Los Baños, Philippines: International Rice Research Institute, 430 pp.

Lal, R. (1985). Tillage in lowland rice-based cropping systems. In *Soil Physics and Rice*, pp. 284–307. Los Baños, Philippines: International Rice Research Institute.

Vergara, B.S. (1992). Tropical wet rice systems. In *Field Crop Ecosystems*, ed. C.J. Pearson, pp.167–82. Amsterdam: Elsevier.

6

Maize (Zea mays*)*

6.1 Taxonomy

Maize is classified as belonging to the subfamily Panicoideae, tribe Andropogoneae or, by others, tribe Maydeae. The tribe includes two genera of New World origin: *Tripsacum*, with numerous wild species, and *Zea*. The taxonomy of the genus *Zea* which is given here was proposed by Iltis & Doebley (1980) and supported by isozyme and chloroplast DNA studies (Doebley, 1990). It divides *Zea* into four species and *Zea mays* into four subspecies, of which one is cultivated maize:

Section Luxuriantes	Section Zea	
Z. diploperennis	*Z. mays*	subsp. *mexicana*
Z. perennis		subsp. *parviglumis*
Z.. luxurians		subsp. *huehuetenangensis*
		subsp. *mays* (maize)

The species and subspecies other than maize are called teosinte. Section Zea is wholly annual and diploid (2n=20) and the wild teosintes in it intercross readily with maize and their progeny are fertile.

Races within maize were first distinguished solely on endosperm characteristics, in the late nineteenth century, though for the past 40 years attempts have been made to develop a more natural classification. Brown & Goodman (1977) have reviewed this complex topic. The endosperm characters, which still form part of the primary description of races, are given below:

1. Pop: small smooth kernels with hard endosperm.
2. Flint: large smooth kernels, mainly hard endosperm but often with a small floury centre.
3. Floury: large smooth kernels with floury endosperm.
4. Dent: large kernels with a central core of floury endosperm which,

on drying, shrinks more than the surrounding hard tissue, denting the kernel.

5. Sweet: large kernels with carbohydrates stored largely as sugars; kernels wrinkled and translucent when dry.

6.2 Origin, evolution and dispersal

The origin of maize has been the source of much controversy. Until recently, there were two major schools of thought: one that teosinte was the ancestor of maize, the other that the cultigen was derived from a hypothetical wild maize. However, over the past decade or so the former theory has become generally accepted (Galinat, 1983; Iltis, 1983; Doebley, 1990). Although the teosinte subspecies *mexicana* is a weed of maize fields – and hence a likely ancestor crop – while subspecies *parviglumis* is a predominantly wild plant, molecular systematic studies by Doebley, Goodman & Stuber (1987) suggest that *parviglumis* was the ancestral taxon.

The earliest archaeological evidence is from the Tehuacan Valley, Mexico, dated about 7000 BP (Mangelsdorf, MacNeish & Galinat, 1967). The cobs are thought to be from an early domesticate (Galinat, 1977). Although geographically separate regions of maize domestication have been postulated, it now seems likely that domestication occurred only in Mexico (Heiser, 1979; Doebley, 1990).

It is possible that maize was introduced by the Portuguese to the West African coast in the early sixteenth century (Miracle, 1965), but linguistic evidence suggests that many areas of tropical Africa received maize across the Sahara from introductions to the Mediterranean region. It was established as a primary crop in the interior of West Africa by the late eighteenth century, and in the Congo Basin after 1930. Maize was unknown in Uganda until after 1861, and not until well into the twentieth century did it become a staple in tropical East Africa.

There is linguistic evidence for maize in the Indian subcontinent in AD 1590, but it appears that it was not widely cultivated at the time (Sarkar *et al.*, 1974). The general consensus is that maize first came to south Asia in the sixteenth century via the Portuguese trade route from Brazil to Goa in south India. However, Johannessen & Parker (1989) have drawn attention to features of twelfth and thirteenth century south Indian sculpture that resemble maize ears, though the fact that no remains of pre-Columbian maize have been found in the subcontinent casts doubt on their interpretation. The crop was introduced to south-east Asia in the sixteenth century by the Spanish via the Philippines.

Maize is now grown largely in warm temperate regions between latitudes 30° and 47° (Shaw, 1977) and only 4 of the 15 countries producing more than 5 m t a year are mainly tropical: Brazil, Mexico, India and Indonesia (Table 6.1). These countries have average yields of 1.4–2.1 t ha^{-1}. The area of tropical maize has changed little in the past decade: increases in production of 35–50% by tropical countries have been through increases in productivity per hectare.

6.3 Crop development pattern

Early development stages of maize were delineated by Hanway (1963) in terms of the appearance of leaf pairs. This developmental index has subsequently been refined and recently has been related to above-ground growth rates (Swan, Brown & Coligado, 1981). Leaf production on a tiller ceases usually after 8–10 leaves, when the apical meristem initiates primordia that form a staminate flower (tassel). Before this conversion to a reproductive apex, and coinciding with internode elongation, initia-

Table 6.1. *Major maize-producing countries: area, yield and production in 1991 of countries producing more than 5 m t per annum*

Country	Area (m ha)	Yield (t ha^{-1})	Production (m t)
USA	27.85	6.81	189.86
China	20.49	4.56	93.35
Brazil	11.89	1.90	22.60
Mexico	7.05	1.92	13.52
France	1.77	7.24	12.80
Romania	2.57	4.08	10.49
Former Yugoslavia	2.30	3.83	8.80
India	5.70	1.44	8.20
South Africa	3.03	2.71	8.20
Argentina	1.96	3.96	7.76
Hungary	1.13	6.67	7.50
Canada	1.08	6.75	7.31
Indonesia	3.01	2.13	6.40
Italy	0.86	7.24	6.20
Egypt	0.91	5.78	5.27

After FAO (1992).

tion of ear primordia occurs by development of buds in the leaf axils. Classic descriptions of the development of the tassel and ear are given by Bonnett (1940, 1966). The axillary buds are capable of forming either an ear shoot or a tiller; thus the primary tiller is determinate, in that it terminates in a tassel, but the maize plant is essentially indeterminate. The uppermost axillary buds commonly develop most rapidly and form ears. In highly developed cultivars dominance of these upper one or two ears is such that lower ears and tillers regress, but many tropical land races show near-synchronous development of three to five ears and appreciable axillary tillering.

Fertilisation requires pollen shedding from the tassel and development of receptive silks through elongation of carpels of female spikelets, starting from the base of the ear. In the lowland tropics, cultivars commonly reach anthesis at about 50–60 days (DI = 0.6) and come to maturity, i.e., black layer formation at the hilar region of the seed (Daynard, Tanner & Duncan, 1971), in 80–120 days. However, since development, particularly before anthesis, is highly dependent on temperature, and since maize is grown from the lowland wet tropics to over 3500 m altitude, it is not possible to generalise about development patterns and time to maturity. Events such as floral initiation and anthesis probably occur at similar DI values in 100-day lowland genotypes and their 300-day counterparts in the cool tropics (Y. Aitken, personal communication). Table 6.2 shows that when maize is grown at various altitudes but

Table 6.2. *Development of maize, cv. H6302, at three sites in Kenya; the sites were within 1°4' of the equator*

	1270	1890	2250
Altitude (m)	1270	1890	2250
Mean soil temperature throughout growth (°C)	23.3	19.2	16.7
Planting to emergence (d)	5	7	11
Emergence to 12th leaf (d)	19	33	41
12th to 18th leaf (d)	17	25	27
Emergence to 50% tassel (d)	62	96	109
50% tassel to maturity (d)	69	83	96
Number of grains per cob[a]	422	461	464
Filling of potential grains (%)[a]	56	66	68
Individual grain weight (g)[a]	0.33	0.35	0.41
Number of grains per plant	401	481	594
Grain yield (g per plant)	131	166	239

After Cooper (1979).
[a]Refers to primary cob.

at about the same latitude (at the equator in Kenya), the duration to maturity increases at a mean rate of 7.6 days per 100 m, from 131 days at 1270 m to 205 days at 2250 m altitude. Such altitudinal effects are now taken into account in classifying maize genotypes.

6.4 Crop/climate relations

The large seed of maize (usually 0.2–0.4 g) leads to rapid radicle and epicotyl growth after imbibition. The radicle appears from the seed before the epicotyl (i.e., at a lower seed water content) and both radicle and shoot elongate linearly with time, sharing a temperature optimum of about 30 °C and showing negligible elongation at less than 9 °C or above 40 °C (Blacklow, 1972). The advantages of seedling vigour in the tropics may be partly offset by the relatively large water requirement at imbibition (making maize reputedly more sensitive to low soil water at sowing than, say, pearl millet) and by the absence of dormancy or germination inhibitors (so that seeds may germinate on reaching maturity even when attached to the parent inflorescence).

Experiments by Hesketh and others (e.g., Hesketh & Musgrave, 1962; Hesketh, 1963; Hofstra & Hesketh, 1969) established the carbon dioxide exchange characteristics of maize leaves, which led to recognition of the ecology and biochemistry of the C4 photosynthetic pathway. Photosynthetic rates are increased by topping (removal of upper leaves, usually to feed livestock) (Pearson *et al.*, 1984). Rates peak at 30–40 °C and are negligible at 44–50 °C (Cooper, 1975). Rates of leaf emergence and lamina expansion also peak at about 30 °C (Swan *et al.*, 1981). Thus, as Duncan (1975) postulated, 'one would expect greatest maize growth in environments conducive to leaf temperatures of 30–33 °C during the day but with cool nights'. Within the tropics one would therefore expect higher dry matter yields in the wet-and-dry and the cool tropics than in the wet tropics, which 'usually have less diurnal variation and might be expected to produce less total growth'.

Short-term above-ground crop growth rates may reach 520–530 kg ha^{-1} d^{-1} (Kowal & Kassam, 1973). These are achieved at modest leaf area indices of approximately 4. The efficiency of conversion of photosynthetically active radiation (PAR) into dry matter averages 5.1–7.2% in the tropics (Fischer & Palmer, 1980). High growth rates are usually achieved at populations of about 50 000 plants ha^{-1} (e.g., Kowal & Kassam, 1973). Higher populations will, at least during early growth, increase rates of dry matter accumulation, but under rainfed conditions long-term growth, water use and particularly grain yield may or may not respond to population, depending on soil water availability, as shown in Nigeria (Babalola & Oputa, 1981; Lucas, 1986)

Total above-ground dry matter yield at maturity is commonly between 12 and 20 t ha^{-1} for experimental crops in the tropics, but grain yields range from national averages of 1 to 1.5 t ha^{-1} (Table 6.1) to 5–8 t ha^{-1} with good management and 10–12 t ha^{-1} in experiments at 1500–2000 m altitude (CIMMYT, 1980; Fischer & Palmer, 1980). Low grain yield in the tropics is attributed to dry matter distribution within the crop, the formation of grain primordia and asynchronous flowering, and to the sensitivity of both net photosynthesis and partitioning to environmental stress, particularly water deficit. Effects of water deficit on vegetative development and growth have been widely reviewed from the viewpoint of physiology (e.g., Boyer, 1976, p. 182) and water requirement (Downey, 1971). Figure 6.1 shows schematically the paths to grain primordia production and to final yield, and the range of values for each attribute at various levels of nitrogen.

The physiological mechanisms of dry matter distribution, i.e., short- and long-distance translocation, are relatively well documented for maize (e.g., Eastin, 1969; Hofstra & Nelson, 1969). It is surprising then that so little is known of quantitative partitioning to roots in field environments, given their function in mineral uptake and the provision of support, particularly in the case of the aerial or 'brace' roots that usually emerge from lower above-ground nodes near the end of vegetative growth. In northern Nigeria root weight increased fastest during rapid ear growth (development index (DI) = 0.6–0.75); there were distinct time differences in the attainment of maximum leaf, root and whole plant weight, at DI values of 0.5, 0.75 and 0.9 respectively (Kowal & Kassam, 1973).

Flowering in tropical maize is accelerated by short days – critical daylengths are 14.5–15 h – whereas maize of temperate origin is less daylength-sensitive (Francis, 1972). Time to flowering is accelerated by increasing temperature. After flowering there is a 'lag phase', which is of 3–8 days duration in open-pollinated tropical cultivars but up to 18 days in maize in temperate climates (Johnson & Tanner, 1972). The actual grain-filling period is commonly 20–30 days but may be as much as 51 days in temperate climates. Maximum grain weight is attained and growth terminated at black layer formation (Daynard & Duncan, 1969).

The number of grains that fill depends on temperature (Fig. 6.2), both directly, through fertilisation and photosynthate production, and indirectly, through an increase in axillary tillering at low temperature. It appears that, in the tropics, number of grains set per primary cob varies by only 10% according to temperature, but tillering may change the number of grains per plant by 50% (Table 6.2). Filled grain number is linearly related to radiation received after floral initiation (Fig. 6.2). Likewise, it may be related, presumably through radiation, to above-ground growth rate, e.g. in Kenya:

$$G = 727R + 1,$$

where G is grain number per plant at harvest and R is above-ground growth per unit thermal time (g per plant per day-degree; calculated above a base temperature of 9 °C). This relationship ($r^2 = 0.92$) was fitted from 8 site-years of data (Hawkins & Cooper, 1981).

Fig. 6.1. The sequence of developmental events which culminate in the formation of grain. Numbers in brackets are mean values or ranges found under varying levels of nitrogen; percentages are the reductions caused by nitrogen stress. Source: Jacobs and Pearson (1991).

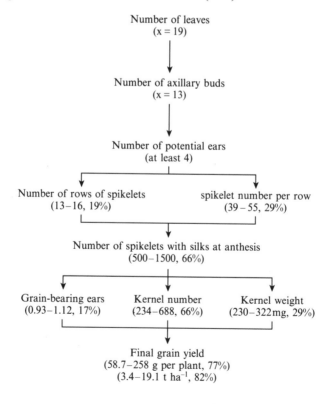

We might expect, by analogy with pearl millet, that individual grain growth at relatively high temperatures in the tropics would be rapid but of short duration; field data in Zimbabwe support this (Wilson, Clowes & Allison, 1973). Egli (1981), summarising data from various sources, found that rate of grain filling ranged from 3.6 to 9.7 mg per seed d^{-1} for inbreds and 7–10 mg d^{-1} for hybrid maize. There are few direct measures of individual grain growth within the tropics; at Malang, Indonesia (altitude 550 m, latitude 6° S) a local cultivar, Kretek, usually has rates of filling of 8 mg d^{-1}.

Grain growth is not directly related to the current photosynthetic rate of the plant (Duncan, Shaver & Williams, 1973); it may be limited either by 'sink effects' or by total photosynthate production after flowering (Tollenaar, 1977). Fischer & Palmer (1980) distinguish three strategies of dry matter partitioning to grain and relate these to climatic zones where maize is grown. The tall leafy tropical landraces are characterised by short growth duration, a low partitioning index (ratio of grain yield to net above-ground dry matter gain after flowering) of approximately 0.8, and a harvest index of 0.3–0.4.

It has long been recognised that maize is particularly sensitive to water deficit at flowering (Salter & Goode, 1967). Drought reduces leaf area, leaf photosynthetic rate (during the stress period, although leaves may recover completely), delays silking and reduces grain yield components, particularly grain number (Hall *et al.*, 1980; Hall, Lemcoff & Trapani, 1981). Reduction in grain number is due to increased asynchrony in

Fig. 6.2. Effect of mean minimum and maximum temperatures and total (accumulated) radiation from floral initiation to anthesis on final grain number in 12 populations of tropical maize grown in 15 environments in Mexico. Source: Fischer & Palmer (1980).

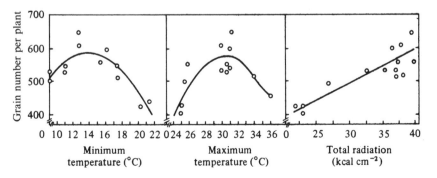

flowering: water deficit reduces the rate of pollen production during the period silks are receptive and reduces the period when silks are exposed to pollen (Fig. 6.3), but does not affect pollen viability (Herrero & Johnson, 1981). Grain survival is also reduced by water stress, as shown in potted plants when sensitivity appeared to be greatest during 2–7 to 22 days after silking (Grant *et al.*, 1989).

To conclude, it seems that maize in the tropics has low yield relative to temperate maize, and that this is related to supra-optimal temperature (Fig. 6.2; Chang, 1981), water deficits (Fig. 6.3; Grant *et al.*, 1989 and others) and to inefficient redistribution of dry matter to grain. In tropical maize, stem sugars increase from flowering to maturity (e.g., from 16 to 22% of stem dry weight; CIMMYT, 1975), whereas in temperate climates stem carbohydrates are depleted during the second half of grain filling (Daynard, Tanner & Hume, 1969). The other limitation to yield, sensitivity to water stress, influences both vegetative and reproductive growth. Where water stress cannot be avoided, maize is replaced by sorghum or pearl millet.

Fig. 6.3. Effect of water deficit on synchrony of flowering and pollen production in maize. Percentage of populations with extruded anthers (triangles) and silks (circles) and temporal distribution of pollen production (continuous line) in control (left) and stressed (right) plants of cv. Dekalb 2F10. Horizontal bar shows period of intense stress, when mean nightfall leaf water potential was –12 MPa. Source: Hall *et al.* (1982).

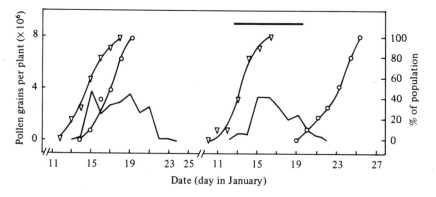

6.5 Crop/soil relations

6.5.1 *Soil physical properties*

While maize is adapted to a wide variety of soils in the tropics, ranging from sands to heavy clays, most maize is grown on well-structured soils of intermediate texture (sandy loams to clay loams), which provide adequate soil water, aeration and penetrability. In the tropics as a whole Oxisols, Ultisols, Alfisols and Inceptisols have the greatest potential for maize production. Vertisols and Mollisols are excellent cereal soils but are of limited extent in the tropics.

Seedbed properties affect seedling development (Alexander & Miller, 1991) and nutrient uptake. Total root length of maize seedlings grown in 6.4–12.8 mm sized aggregates were less than 60% of those grown in 1.6–3.2 mm aggregates (Donald, Kay & Miller, 1987). In a silt loam optimum volumetric moisture content for elongation of maize seedling roots was 0.24–0.27 m^3 m^{-3} and bulk density <1.22 Mg m^{-3} (Logsdon, Reneau & Parker, 1987).

Poor soil structure restricts root development and depresses yields of maize in western Nigeria, where over half the land surface has a gravel horizon or stone line at a shallow depth (Smyth & Montgomery, 1962). The total porosity and water-holding capacity of the ferruginous Alfisols and Ultisols of this area are consequently low, and root elongation rates and root volume are low (Babalola & Lal, 1977*a,b*), rendering the plants more susceptible to drought. In such soils at Kurmin Biri in Nigeria, maize grain yield was increased 6% by ridging, and 24% by subsoiling, which increased moisture storage by 15–30% in the 0–45 cm depth (Adeoye & Mohamed-Saleem, 1990). Prior cropping to a deep-rooted legume, pigeonpea (*Cajanus cajan*), in similar soils at Ibadan, Nigeria, has shown higher maize yield, associated with improved root development to a greater depth (Hulugalle & Lal, 1986). Nicou & Chopart (1979) have reported the effect of soil properties on maize yield in the extensive sandy and sandy clay loams of Senegal: maize root density increased by over 300% over the total porosity range 43–48%, resulting in a virtual doubling of grain yield to around 4000 kg ha^{-1}. A comparison of maize root production with those of other crops in these soils is shown in Table 6.3.

High soil bulk density that affects the growth of maize may occur naturally, as in the case of three soils in Tanzania, one a Vertisol, reported by Northwood & Macartney (1971), but sometimes it is induced by human activity. The use of machinery to clear vegetation resulted in an increase in bulk density in sandy soils (Entisols) in Surinam and reduced maize yields (Janssen & van der Weert, 1977).

Soil erosion of Nigerian Paleustalfs was shown by Lal (1979c) to lead to deterioration in infiltration rate and soil structure. Maize yields on eroded slopes declined with slope angle and quantity of soil eroded. For a cumulative soil loss of 50 t ha⁻¹, yields were reduced by 56%, 14%, 8% and 8% on 1%, 5%, 10% and 15% slopes respectively. Where excessive erosion occurred, yield reductions could not be corrected by the addition of fertiliser. On acid Oxisols, liming has been shown to promote maize root development and thereby reduce soil degradation (Bonsu, 1991).

In Nigeria, clay subsoil material brought to the surface by mound-building termites (*Macrotermes bellicosus* and *M. subhylinus*) can have deleterious effects on maize yields. While this material is usually less fertile (Lal *et al.*, 1975), the main adverse effects appear to be physical: crust formation, which reduces emergence, and increased bulk density, which affects root growth (Kang, 1978).

Maize is intolerant of waterlogging (Chaudhary, Bhatnagar & Prihar, 1975); more so than sorghum (IRRI, 1976). Flooding to a depth of 5–15 cm for 48 h at different growth stages reduced yield by 31–63% (Palvadi & Lal, 1976). Germination and seedling growth is more tolerant to prolonged waterlogging than to alternating periods of stress and non-stress (VanToai, Fausey & McDonald, 1988).

Table 6.3. *Estimates of root weight (0–30 cm depth) for four cereal crops growing in sandy and sandy clay soils of Senegal under a variety of field environments and management conditions*

	Root weight (kg ha⁻¹)		
	Mean	Minimum	Maximum
Upland rice	1200	300	1950
Maize	1500	450	2700
Sorghum	1550	350	3700
Millet	800	350	1700

After Nicou & Chopart (1979).

6.5.2 Soil chemical properties

Nutrient deficiencies commonly limit maize yields on Oxisols and Ultisols. When nutrient deficiencies are corrected and improved cultivars are used, high yields are attainable (see Section 6.4). Grove (1979) observed that 95% of maximum maize yield was obtained at 80–120 kg ha^{-1} applied nitrogen on Oxisols and Ultisols in Brazil, Puerto Rico and Ghana. Responses and recovery of applied nitrogen in tropical and temperate areas were similar and independent of soil properties (Grove, 1979; Grove, Ritchey & Naderman, 1980; Fig. 6.4). Nevertheless, maize does not always respond to nitrogen on Oxisols. Semb & Garburg (1969) recorded no response to 90 kg N ha^{-1} applied to an East African Kikuyu friable clay; as much as 184 kg N ha^{-1} can be mineralised in such soils during a growing season (Semb & Robinson, 1969). In shifting agriculture, nutrient availability to maize is usually strongly determined by the period of accumulation of nutrients before clearing (Table 6.4). Mineralisation of organic nitrogen accumulated in this way would suit the nitrogen uptake pattern of maize; 60–75% of total uptake occurs after the onset of tasselling (Friedrich, Schrader & Nordheim, 1979; Mills & McElhannon, 1982).

Maize yields have been increased by inoculation by *Azospirillum* spp., but the relative effects of nitrogen fixation, nitrate reductase, improved nitrogen uptake *per se* and auxin production by the organisms have yet to be fully clarified (Boddey & Dobereiner, 1988).

Fig. 6.4. Effect of fertiliser nitrogen (*a*) grain yield and (*b*) apparent nitrogen recovery of maize at temperate and tropical locations. *N*, number of experiments. Source: Grove (1979).

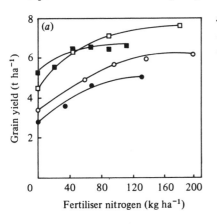
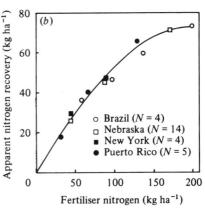

Maize is one of the more demanding crops, particularly during early growth, for soil solution phosphorus as evaluated by the soil test of Fox & Kamprath (1970) (Table 6.5). Yield, as determined by kernel number, is particularly dependent on phosphorus nutrition up to the six-leaf stage; at this stage shoot phosphorus concentration should be at least 0.5% (Barry & Miller, 1989).

Oxisols and Ultisols are commonly deficient in phosphorus: e.g., in

Table 6.4. *Maize grain yield in the first and second year after clearing three vegetation types on an oxic Paleustalf in Nigeria*

	Grain yield (kg ha^{-1})		
	Year 1	Year 2	
Previous vegetation	Nil NPK	Nil NPK	+ NPK[a]
9 year secondary forest	2460	3640	4220
4 year regrowth thicket	1695	2160	4094
1 year cassava	1482	1340	3334

After Kang & Moormann (1977).
[a] N = 120, P = 40, K = 50 kg ha^{-1}.

Table 6.5. *Summary of reported external phosphorus requirements (phosphorus in soil solution required to produce 95% maximum yield) of different crops*

Crop	Growth stage	Field or pot	Soils	External P requirement (μmol l^{-1})
Rice (flooded)	Early	Pots	Vertisols	3.2
Rice (upland)	Maturity	Pots	Oxisol	3.2
Maize	Early	Field	Ustox, Andisol (Hydrandept)	6.5
Maize	Maturity	Field	Ustox, Andisol (Hydrandept)	1.9
Sorghum	Maturity	Field	Andisol (Hydrandept)	1.6
Millet	Early	Pots	Ultisols	6.5
Millet	Early	Pots	Andisols (Hydrandept)	0.6–1.9
Millet	Early	Pots	Oxisols, Andisol (Dystrandept)	1.9

After Sanchez & Uehara (1980).

the Amazon (Smyth & Cravo, 1990). Lathwell (1979*b*) summarised results of extensive work with maize and other crops in Puerto Rico, Ghana and Brazil. For satisfactory growth, up to 440 kg phosphorus (P) ha^{-1} (on fine-textured soils) or as little as 18 kg P ha^{-1} (on coarse-textured soils) was required. While Ultisols and Oxisols are liable to fix applied phosphorus most rapidly (Sanchez & Uehara, 1980), residual phosphorus may sometimes be quite satisfactory for maize on these soils (Lathwell, 1979*b*; Yost *et al.*, 1981).

Liming can improve uptake of phosphorus by maize where soil phosphorus is adequate but where root growth may be restricted by high exchangeable aluminium, low calcium or low magnesium (Mikkelsen, Freitas & McClung, 1963; Jones & Fox, 1978). However, Lathwell (1979*a*) reported that in Ghana and Peru responses by maize to applied phosphorus on Ultisols and Oxisols were reduced where lime was applied. This was attributed to an increased rate of mineralisation of soil organic phosphorus. In the rainforest region of Nigeria, crop trash burning has shown both an increase in soil pH, due to the cation content of the ash, and an increase in yield attributed to improved availability of nutrients, particularly of phosphorus (Iremiren, 1989). Even in the absence of liming or burning, substantial quantities of phosphorus may be mineralised. Adepetu & Corey (1977) found that 25% of the organic phosphorus in the surface 0–15 cm disappeared during two maize crops grown through the rainy season, following clearing of a secondary forest in Nigeria. The quantity of phosphorus released was about three times as much as that taken up by the crops.

Maize roots have phosphatase activity (Juma & Tabatabai, 1988) which could increase the availability of phosphorus from organic sources, particularly under phosphate stress (Sachay, Wallace & Johns, 1991). Mycorrhizal fungi increase the uptake of phosphorus by maize in low-phosphorus soils (Lambert, Baker & Cole, 1979). Khan (1972) demonstrated a twelve-fold increase in grain yield and grains per ear in field-grown corn when inoculated with a mycorrhizal fungus. Spore production by several species of mycorrhizal fungi was greater in sorghum than in maize but was reduced in most species by water stress in the host (Simpson & Daft, 1990). Flooding maize fields for rice production depletes the number of mycorrhizal propagules (Ilag *et al.*, 1987).

Studies of potassium on Oxisols and Ultisols in Brazil, Puerto Rico, Ghana and Peru led Ritchey (1979) to conclude that for maize and other crops, the critical level of soil potassium was 0.13–0.20 cmol K kg^{-1} when measured by the North Carolina double acid extractant. Jones & Wild

(1975) give values for West Africa of 100–200 kg K ha⁻¹ total uptake to produce 6–8 t ha⁻¹ total dry matter, the grain containing only 25–30 kg K ha⁻¹. Thus considerably larger quantities of potassium may be lost if maize stalks and leaves are removed from the field in addition to the grain. Other work in West Africa has shown that even where potassium fertilisation of forest and savanna Alfisols does not result in higher yield, lodging may still be significantly reduced (Esechie, 1985; Kayode, 1986).

Aluminium toxicity due to high levels of exchangeable aluminium is a problem of considerable magnitude for maize on acid tropical soils. Aluminium toxicity can of course be overcome by liming. R.W. Pearson (1975) concluded, in a review of soil acidity and liming in the humid tropics, that a response to liming by maize could be expected whenever soil pH fell below 5.0 and exchangeable aluminium exceeded about 15% saturation. Deep lime incorporation in the subsoil of some Oxisols has overcome aluminium toxicity, thereby improving rooting depth in maize and tolerance to dry periods (Lathwell, 1979*a*).

6.6 Place in cropping systems

Broadly speaking, maize is an important component of cropping systems in the middle rainfall zones of the tropics, say from 750 to 1750 mm annual rainfall. In more arid regions the average period of available water is too short for genotypes of high yield potential and the incidence of within-season dry spells too severe: in Africa and south Asia at least it is replaced by sorghum and pearl millet (Table 6.6). In high-rainfall zones maize tends to be replaced as a staple by tuber crops or bananas, which can take advantage of a long growing season for the production of food energy, though maize may still form a subsidiary component of the cropping system. The adverse features of the more humid zones for maize are low radiation, a high incidence of pests and diseases, and high night temperatures (Kassam et al., 1975).

Maize is commonly grown in mixed cropping situations. Seventy-five per cent of the area of maize in Nigeria and 84% in Uganda is in association with other crops (Okigbo & Greenland, 1976). In the Northern Guinea savanna of Nigeria 70% of the maize is in two- to four-crop mixtures and 3% in five to six-crop mixtures (Norman, 1972). Francis, Flor & Temple (1976) estimate that at least 50% of the maize area in tropical America is in mixed crop patterns. In southeast Asia, except in some Thai cropping systems producing maize for export, the crop is normally grown as a component of a mixture.

6.6.1 *Maize in cropping systems of tropical Africa*

Miracle (1967), Ruthenberg (1980) and Okigbo & Greenland (1976) are useful sources of information on African cropping systems based on maize. Okigbo & Greenland (1976), summarising data from a number of sources, give the following associated crops in examples of mixed cropping patterns that include maize:

Nigeria: Cassava, yams, cocoyam

Liberia: Upland rice, banana, cassava

Sierra Leone: Upland rice, cassava, okra, peppers

Central African Republic: Sweet potato, cassava, squash, tobacco

Tanzania: Beans, groundnuts, sweet potato, sorghum, legumes, upland rice

In most of the African crop mixtures which include maize and tuber crops, maize is one of the earliest crops to be planted and is the first to mature, though in some regions of extended rainfall, e.g., Nigeria (Okigbo & Greenland, 1976), Tanzania and the Ivory Coast (Ruthenberg, 1980), the complete season's cropping may include a second sequential crop of maize. Since maize makes relatively high demands on soil nutrients, it is more generally grown in the early phases of the cropping break in shifting cultivation or semi-intensive systems than in the late phase (Braun, 1974).

Maize is adapted to those extended rainfall regimes capable of

Table 6.6. *Principal cereal crops in four rainfall zones of West Africa*

Region	Mean annual rainfall (mm)	Length of growing season (d)	Main crops[a]	Secondary crops[a]
A	350	60	SS millet	—
B	550	100	SS, MS sorghum MS millet	SS sorghum SS millet
C	800	120	MS, LS sorghum MS, LS millet	SS maize SS sorghum SS millet
D	1000	145	LS maize LS sorghum LS millet	SS maize SS sorghum SS millet

After Cochemé & Franquin (1967).
[a]SS, MS, LS: short-, mid- and long-season cultivars.

supporting perennial tree crops, and in such regions it is commonly grown either in permanent association with perennials or during their establishment phase (Ruthenberg, 1980); for example, with oilpalm and coconut in West Africa, with coffee in Tanzania, and in the establishment of coconuts in Tanzania and coffee on the Ivory Coast.

6.6.2 *Maize in cropping systems of tropical Asia*

Maize has a subsidiary place in some Asian cropping systems based on wet rice – as a second crop in association with legumes or vegetables following a wet-season rice crop (McIntosh, *et al.*, 1977) – but it is more frequently grown under rainfed upland conditions.

Under shifting cultivation in high-rainfall zones it is commonly associated with upland rice. Ruthenberg (1980) quotes examples of forest fallow systems in Assam (2500 mm rainfall) based on maize and upland rice grown for 2 years, followed by a 10-year fallow, and in the Philippines (also 2500 mm rainfall) based on maize, upland rice and sweet potatoes for 2–4 years followed by 8–10 years of fallow. One well-known, if not to say notorious, shifting cultivation system in Asia is the maize/opium/upland rice cropping pattern of the mountain regions of Thailand, Burma and Laos (Keen, 1978). Maize is planted in early wet season and relay-planted with opium in mid-wet season; the upland rice is grown in separate fields.

More intensive upland cropping systems that include maize are characteristic of Indonesia. McIntosh *et al.* (1977) give two main patterns: maize/upland rice/cassava, and maize/upland rice followed sequentially by maize. The first of these is a good example of a multiple cropping pattern adapted to an extended single-peak rainfall regime. Quick-maturing maize is planted first in early wet season (September) and is interplanted shortly afterwards with upland rice; cassava is interplanted 4–6 weeks later. In December–February, the period of maximum rainfall, all three crops are making a heavy demand on soil water. Maize is harvested in January, rice in February, and cassava continues to utilise end-of-season rainfall and water stored in the root zone until June.

6.6.3 *Maize in cropping systems of tropical America*

By far the most widespread upland cropping pattern involving maize in tropical America is maize in association with common beans (*Phaseolus vulgaris*). According to Francis *et al.* (1976), about 80% of the bean area and more than 50% of the maize area is intercropped. In a summary of shifting cultivation practice in Latin America, Grigg (1974) noted that maize, maize/bean and maize/bean/squash cropping patterns are characteristic of elevated regions and/or wet-and-dry climates of medium rainfall duration: in the wet tropical lowlands the main energy crops are cassava and, to a lesser extent, upland rice and sweet potatoes, though maize may be a subsidiary crop.

Maize/bean intercrops have been studied extensively at CIAT, Colombia (Francis *et al.*, 1976; Francis, Flor & Prager, 1978a; Francis & Sanders 1978; Altieri *et al.*, 1978) and at Turrialba in Costa Rica (Pinchinat *et al.*, 1976). The mixtures include maize with bush beans (i.e., free-standing plants of short stature) and climbing beans (which climb up the maize stems). Francis *et al.* (1978), in an extensive series of experiments, found that when maize was planted at the relatively low densities that are common in farming practice, the presence of associated beans (both bush and climbing) had little or no detrimental effect on the grain yield of maize (Table 6.7). Land equivalent ratios (LERs) ranged from 1.21 to 1.68. Association with beans also reduced the incidence of insect attack on maize (Francis *et al.*, 1978a; Altieri *et al.*, 1978).

Table 6.7. *Productivity of maize/bean intercrop mixtures in Colombia*

| Experiment number | Crop yield (kg/ha^{-1}) | | | | Land Equivalent Ratio |
	Maize sole	Bean sole	Maize assoc. with bean	Bean assoc. with maize	
7501[a]	6535	2148	7318	429	1.32
7509	5674	2815	7175	1180	1.68
7510	5500	3486	6794	517	1.39
7511	5445	2165	6518	1443	1.23
7513	5096	2574	5923	1030	1.56
7515	5600	2688	4177	1275	1.21
7516[a]	3729	2014	3193	870	1.29
7517	4435	3623	4089	1765	1.41
7518	4739	4307	4934	2075	1.52
7525[a]	5165	3023	4702	1150	1.29

After Francis *et al.* (1978).
[a] In experiments 7501, 7516 and 7525 different combinations of cultivars were tested. To reduce the size of the table, one result has been chosen at random from each of these experiments.

Maize and beans are grown together in a wide range of environments. In the experiments quoted above, carried out at CIAT in the Cauca valley (1000 m altitude), beans matured in 100 days and maize in 145 days. In the same region but at altitudes of 2500 m, maize and climbing beans are grown in a relay pattern over a period of 10–11 months. Maize is sown first; after 5–6 months beans are planted at the base of the maize stalks, and the two crops mature together about 5 months later.

Further reading

Sprague, G.F. & Dudley, J.W. (eds.) (1988). *Corn and Corn Improvement*, 3rd edn. Madison, Wisconsin: American Society of Agronomy.

Wedderburn, R.N. & DeLeon, C. (eds.) (1987). *Proceedings of the Second Asian Regional Maize Workshop*. El Batan, Mexico: CIMMYT (International Maize and Wheat Improvement Centre).

7

*Sorghum (*Sorghum bicolor*)*

7.1 Taxonomy

Sorghum is a member of the subfamily Panicoideae, tribe Andropogoneae (Clayton & Renvoize, 1986). Harlan & de Wet (1972) recognise two species only of *Sorghum*, the diploid *S. bicolor* (2n = 20), including all cultivated and wild sorghums, and the tetraploid *S. halepense* (2n = 40), a perennial weed of arable land. Doggett (1988) also recognises a wild diploid species, *S. propinquum*, and pasture agronomists usually accord specific status to the cultivated forage types sudan grass (*S. sudanense*) and columbus grass (*S. almum*).

Doggett (1988) recognises three subspecies of the annual, *S. bicolor*: *bicolor*, which is cultivated, *drummondii*, which is a weed of sorghum in Africa; and *verticilliflorum*, which is naturalised widely throughout savannas. Harlan & de Wet (1972) divide the cultivated *S. bicolor* subspecies *bicolor* into five races:

1. *bicolor*: the least specialised, with the widest distribution. Now sporadically cultivated across the African savanna and in south and southeast Asia.
2. *guinea*: the dominant race of the West African savanna, India and southeast Asia.
3. *caudatum*: important in central tropical Africa.
4. *kafir*: the dominant race in Africa south of the equator.
5. *durra*: the dominant race in the Near East and parts of India.

Harlan & de Wet also postulate ten 'intermediate' cultivated races, being all combinations in pairs of the five basic races. They classify all the wild races as subspecies *arundinaceum*.

7.2 Origin, evolution and dispersal

Although all are agreed that the domestication of sorghum took place in Africa, there is controversy concerning its focus. Harlan (1986) believes the crop is 'non-centric'; in his classic paper on centres and non-centres, Harlan (1971) considered that the crop was first domesticated across an east–west belt of broad-leaved savanna from Lake Chad to the Sudan, whereas Doggett (1976, 1988) regards the Ethiopian highlands as the primary centre of domestication. Archaeological evidence is scanty, and dates are vague: Doggett estimates 5000–6000 BP.

According to Harlan & Stemler (1976), the earliest cultivated sorghums most closely resembled the current race *bicolor*. Early types of *bicolor* moved westward to West Africa before 3000 BP, where the *guinea* race evolved (Doggett, 1976; Harlan & Stemler, 1976). The development of the *durra* race is disputed: Doggett maintains that it evolved in the Ethiopian–Sudan region and then spread to the Near East and to India, whereas Harlan & Stemler believe that early types of *bicolor* reached the Sind–Punjab area of India before 3000 BP, evolved there to the *durra* race, and then returned to East Africa. However, they also mention the possibility of *durra* evolving in both Ethiopia and India through interchange of germplasm by sea.

Both Doggett and Harlan & Stemler support the movement of early sorghum domesticates to southern tropical Africa and their evolution to the *kafir* race. The *guinea* race either evolved in West Africa (Doggett) or differentiated first in East Africa and then moved west (Harlan & Stemler). It also spread to southeast tropical Africa and was taken by sea to India. According to Doggett, the earliest archaeological record in India is from Rajasthan, about 3500–4000 BP.

Early *bicolor* types spread to Burma and Indonesia and thence north to China where they evolved to the *kaoliang* type, at about 2000 BP according to Doggett. However, Harlan & Stemler suggest the possibility of germplasm moving overland through Central Asia to China before 3500 BP.

Sorghum was taken to the American continent from West Africa with the slave trade, the first introductions being of the *guinea* race. Both *durra*, from northern tropical Africa, and *kafir* races, from southern tropical Africa, reached America later, in the 1870s. The basis of modern sorghum improvement, initiated in the USA, has been the milo (race *durra*) × *kafir* cross.

Area, yield and production of sorghum for nations producing more than 500 000 t a year are given in Table 7.1. Total production from the major tropical sorghum producer, India, is double that of the next tropical country, Nigeria. Indian production has declined marginally during the past decade as small increases in productivity per hectare have been more than offset by reduction in the area sown. In the next-ranked tropical countries, Nigeria and Mexico, total production has increased (Nigeria), or remained static (Mexico) owing to increases in productivity being offset by reductions in area. Productivity of sorghum in tropical countries is highly variable, reflecting irrigation and levels of technological inputs.

7.3 Crop development pattern

Sorghum is an erect grass usually 1–4 m tall. Most but not all cultivars are freely tillering. Tillers are terminated by an inflorescence, which is a panicle with a central rachis carrying primary, secondary and sometimes tertiary branches which bear the racemes of spikelets. The first and second tillers are produced from the base of the plant and later axillary tillers arise from further up the elongated primary tiller. Unlike maize and pearl millet, sorghum may regrow or ratoon following tiller decapitation.

Table 7.1. *Major sorghum-producing countries: area, yield and production in 1991 of countries producing more than 500 000 t per annum*

Country	Area (m ha)	Yield (t ha^{-1})	Production (m t)
USA	3.97	3.70	14.72
India	15.00	0.72	10.80
China	1.60	3.51	5.61
Nigeria	4.60	1.04	4.80
Mexico	1.38	3.16	4.37
Sudan	4.69	0.63	2.94
Argentina	0.68	3.33	2.25
Burkina Faso	1.30	0.86	1.11
Australia	0.40	2.22	0.89
Ethiopia	0.80	1.01	0.81
Colombia	0.26	2.87	0.74
Mali	0.75	0.97	0.73
Egypt	0.14	4.71	0.66
Venezuela	0.28	2.24	0.62

After FAO (1992).

Although each succeeding ratoon produces fewer tillers, grain yields may be sufficient to make ratooning economic in the wet tropics (Escalada & Plucknett, 1975).

The seedling has a juvenile phase (of less than 15 days at 32/21 °C day/night; Caddel & Weikel, 1972) during which it produces five leaves and will not become reproductive. Thereafter most sorghums, particularly the tropical races, are sensitive to short days. Daylength sensitivity is not absolute: it is modified by daylength trend (whether increasing or decreasing) and stage of vegetative development (the larger the number of leaves, the earlier the initiation) (Kassam & Andrews, 1975). Nonetheless, it is a most important feature which helps ensure that locally adapted wet-season cultivars reach anthesis at about the end of the rains in the wet-and-dry tropics. Thus, for local West African types, a delay of 1 week in sowing (after the start of the wet season, in May) may delay head emergence by only 1 day (Curtis, 1968), whereas for improved daylength-sensitive cultivars the delay at heading may be 2 days (Kassam & Andrews, 1975). In India, where 40% of sorghum is sown as an early dry-season crop, adapted short-duration genotypes show genotype × sowing date interactions which suggest that they also have some daylength sensitivity (ICRISAT, 1978).

Given such sensitivity, it is clearly difficult to generalise about developmental events in tropical sorghums. Temperate sorghums, mostly derived from cv. Milo, are less daylength sensitive and reach anthesis at about development index (DI) $= 0.5$ in the tropics and at $DI = 0.65$ under temperate conditions (Anderson, 1979). Floral initiation in short-duration crops occurs between $DI = 0.2$ (Eck, Wilson & Martinez, 1975) and $DI = 0.3$ (Vanderlip, 1972). In India the average time to floral initiation for 49 genotypes was 33 days, at $DI = 0.27$, and anthesis took place after a further 42 days at $DI = 0.6$ (ICRISAT, 1976). In the West African savanna, where a crop duration of 110–120 days is considered ideal for yield, grain-filling of adapted cultivars occupies 33–40% of the crop's life (Kowal & Kassam, 1978). Grain maturity is apparent when a black closing layer forms on the grain (Quinby, 1971; Eastin, Hultquist & Sullivan, 1973).

7.4 Crop/climate relations

The ecology of sorghum shows many attributes intermediate between those of maize and pearl millet. Seeds are intermediate in size, weighing between 15 and 40 mg. Seed dormancy, imbibition and crop establishment are reviewed by Peacock & Wilson (1984). Seedlings at germination and emergence are more sensitive to low soil temperature than maize but less sensitive than millet (Launders, 1971). Field temperatures of 18 °C are considered suitable for germination (Anderson, 1979) and time to emergence decreases by 1 day for each 0.9 °C increase in mean soil temperature (at 7.6 cm depth) from 13 to 21 °C (Adams, 1967); soil temperatures of 40–48 °C are lethal (Knapp, 1966; Kailasanathan, Rao & Sinha, 1976).

The optimum temperature for seedling growth, about 33 °C (Vinall & Reed, 1918), reflects the optimum for net carbon dioxide exchange rather than for vegetative development: leaf emergence increases with temperature but tillering has a relatively low temperature optimum of about 20/15 °C (Downes, 1968). The optimum soil moisture for germination is 20–50% of field capacity (Fawusi & Agboola, 1980).

Net photosynthetic rates are higher in sorghum (a C4 species) than in, say, wheat, and they increase with temperature, probably reaching an optimum at 35–40 °C (Downes, 1970). The C4 pathway and associated absence of net respiration in the light are supplemented by 'efficient' leaf anatomy – small individual cell size and a high ratio of internal cell surface to volume compared with other species – to give high net photosynthetic rates and high water use efficiency (e.g., El-Sharkawy & Hesketh, 1965). This is illustrated by field data in Fig. 7.1, where diurnal carbon dioxide fixation per unit leaf area was 2–3 times greater in sorghum than in soybean and concurrent transpiration losses were smaller owing to high residual or mesophyll conductance in sorghum (Rawson, Turner & Begg, 1978). Transpiration accounts for 50% of evaporation (E) at leaf area index (LAI) 2 and 95% of E at LAI 4; the ratio of transpiration to E reaches a plateau at about DI = 0.4, whereas LAI peaks at DI = 0.7 and then declines (Brun, Kanemasu & Powers, 1972). Rates of attainment of maximum LAI and complete radiation interception vary with genotype and plant population, but complete interception appears to be reached at least 4 weeks before head emergence, even at low (24 000 plants ha^{-1}) populations (Goldsworthy, 1970a).

In north Australia (latitude 14° S) root mass reached a maximum halfway between floral initiation and anthesis, at about DI = 0.4 (Myers, 1980). It was found that 76–79% of root mass and 60–63% of root length were in the top 0–20 cm of soil. Maximum root dry weight was 1 t ha^{-1} (see also Table 6.3). Root/top ratios declined throughout the life of the crop, from 0.75 at floral initiation to 0.6 at DI = 0.4 and to 0.1–0.2 at maturity.

Sorghum displays a tolerance to water deficit regarded as intermediate between maize and pearl millet. The crop adapts to water deficit by a range of processes. Photosynthetic rates are maintained, or decline more gradually under slow stress, owing to stomatal conductance remaining high at lower leaf water potentials than, say, maize (Turner, 1974). Photosynthesis and growth are also maintained by leaf cell osmotic adjustment: under water stress, differences among drought-sensitive and drought-tolerant sorghums are associated with leaf osmotic and water potentials (Fig. 7.2) while the importance of hydraulic resistances is contentious (Kirkham, 1988; Bawazir & Idle, 1989). In sorghum, osmotic adjustment disappears within 10 days of relief of stress (Turner & Jones, 1980).

Fig. 7.1. Daily carbon dioxide fixation and energy fluxes of the penultimate leaf of (*a*) sorghum, cv. TX610, and (*b*) soybean, cv. Bragg, in northern New South Wales, Australia (latitude 30° S). F, net photosynthesis; ^{14}F, $^{14}CO_2$ uptake; Q, transpiration; I, quantum flux density; $Q.VPD^{-1}$, transpiration per mbar vapour pressure deficit; r_1, gas phase resistance; r_m, residual resistance. Measurements were made during flowering and early grain-filling. Note for sorghum the close relation between net photosynthesis and gross photosynthesis as estimated by $^{14}CO_2$ uptake, in contrast to soybean. Source: Rawson *et al.* (1978).

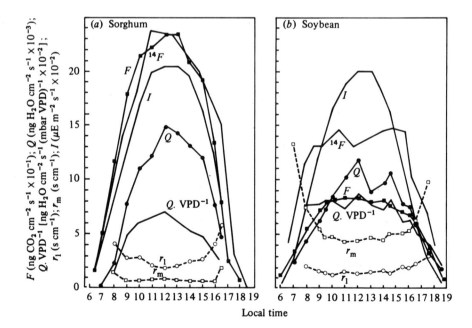

Water deficit, even of a continuous mild nature (–0.5 MPa), reduces plant size and root length: for example, from 4.8 km per plant under daily watering to 4 km per plant after 80 days under –0.5 MPa (Merrill & Rawlins, 1979). Deficits have a marked effect on root distribution, root length and density shifting to greater depth under stress (Fig. 7.3; also Blum & Ritchie, 1984). Moreover, sorghum has the ability to continue root growth for a relatively long period. Roots and above ground vegetative organs may, under favourable conditions, continue to grow after anthesis: 'non-senescent' genotypes and dwarf hybrids may thereby exploit end-of-season water (McClure & Harvey, 1962; Duncan, Bockholt & Miller, 1981).

In addition to its tolerance of drought, sorghum is also more tolerant of waterlogging than is maize (see Table 7.2). At IRRI in the Philippines relative crop tolerance in seedlings to 1–5 days of soil submergence was sorghum > groundnut > maize (IRRI, 1976). Waterlogging suppresses the development of seminal roots more than

Fig. 7.2. Water potential and osmotic potential of a drought-resistant (squares) and drought-sensitive (circles) sorghum grown in soil that was not watered. The matric potential of the soil also is given. Vertical bars represent the standard deviation. Source: Kirkham (1988).

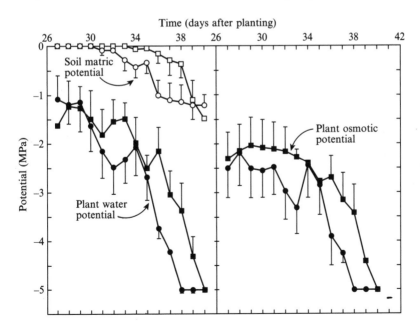

nodal roots, but after cessation of waterlogging nodal roots develop relatively more quickly than seminal roots (Pardales, Kono & Yamauchi, 1991, Table 7.2). Early waterlogging reduces normal tillering and yield (through a reduction in seed number) but stimulates late tillering (Orchard & Jessop, 1984).

High potential photosynthetic rates, efficient water use and drought tolerance lead to sorghum being concentrated in areas of rainfall of 600–1000 mm AAR (Kowal & Kassam, 1978) in the wet-and-dry tropics. Seasonal growth rates in excess of 300 kg ha^{-1} d^{-1} have been recorded in the subtropics (Fischer & Wilson, 1975a) and rates ranging from 60 to 250 kg ha^{-1} d^{-1} in the wet season in India (ICRISAT, 1976). However, cultural traditions restrict productivity to well below these levels. There appear to be three main restrictions upon vegetative growth:

Season. Since sorghum is grown in the wet season but on the dry margins of maize, or as an early dry-season crop utilising stored soil water, growth rates may be constrained by low radiation and waterlogging in the middle of the wet season (Goldsworthy, 1970b), by water deficits and, in early dry season, by low temperatures (ICRISAT, 1978). Wet-season crops may use in transpiration only about the same proportion of total available water as is lost through drainage (35% transpired compared with 29% lost

Fig. 7.3. Root length density of sorghum cv. Pioneer 887 as affected by severe water deficit: (*a*) irrigated when matric potential at 25cm soil depth reached –1 MPa; (*b*) mild deficit (–0.05MPa); or (*c*) no deficit (daily watering). Successive samples are labelled (1) to (5). Source: Merrill & Rawlins (1979).

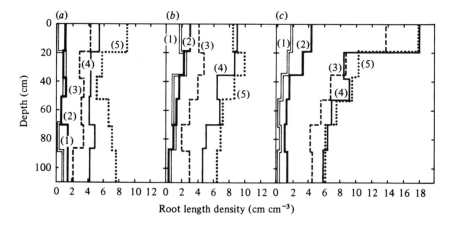

Table 7.2. *The effect of waterlogging sorghum on plant parameters*

	No waterlogging; drained 0–33 days	Drained days 0–24 waterlogged days 24–33	Drained days 0–12 waterlogged days 12–33	Drained days 0–12 waterlogged days 12–24; drained days 24–33
Nodal root				
Number	22	26	14	24
Length (cm)	686	469	187	450
No. of first-order laterals	604	626	458	1008
Length of first-order laterals (cm)	3257	1683	1859	4509
No. of second-order laterals	2762	1002	1145	3714
Seminal root				
Length (cm)	47	143	34	39
No. of L type first-order laterals	52	28	20	31
Length of L type first-order laterals (cm)	381	283	140	292
No. of second-order laterals	488	353	124	304

After Pardales, Kono & Yamauchi (1991).
Pots were subject to waterlogging or draining at the times shown (in days) after sowing. All plants were harvested at day 33.
First order laterals were grouped into long (L) and short (S); L type laterals normally branch to give second order laterals while S type laterals do not.

by drainage on an Alfisol at Hyderabad, India), whereas in post-monsoon crops, which use a higher proportion of available water (53%), the ratio of actual to potential transpiration may fall to less than 0.5 during flowering and grain-filling (Singh & Russell, 1979).

Population. Plant populations are commonly low (less than 100 000 ha⁻¹), yet yield is responsive to increasing density (e.g., Goldsworthy, 1970*a*, *b*; Fischer & Wilson, 1975*a*; Burnside, 1977). Since planting time and genotype are chosen to ensure grain maturation during the post-rainy period, to avoid diseases such as covered smut (*Spacelotheca sorghi*), loose smut (*S. cruenta*) and head smut (*S. reiliana*), the period of growth after canopy closure can best be increased through increasing population. Such increases may, but do not necessarily, lead to increased stress after anthesis.

Weed control. High populations are associated with narrow row spacings: wide rows (200 cm) are appropriate for subsistence farming in the tropics where grain yields are below 1 t ha⁻¹, but a row spacing of 25 cm is optimal if yields exceed 3.7 t ha⁻¹ (Myers & Foale, 1981). Narrow rows require herbicides for effective weed control, but as yet there are few herbicides which are selective in sorghum against closely related weeds such as *S. halepense.* Weed control appears important only in the first 4 weeks of crop life (e.g., Burnside, 1977).

Floret number is, of course, determined before head emergence. Numbers are higher under wet conditions. In India, grain number in dry-season, irrigated dry-season and wet-season crops was 9800, 16 400 and 21 400 m⁻² respectively (ICRISAT, 1978). Grain number is also sensitive to temperature extremes: low night temperatures (10 °C) cause male sterility (Brooking, 1976), and high temperature (33/28 °C) after initiation will terminate floret development prematurely since immature florets at the base of the panicle abort (Downes, 1972). Grain number per head is low in traditional tropical genotypes (Goldsworthy, 1970*b*). Late planting to ensure grain maturation in dry weather reduces grain number (Fig. 7.4). Stern (1968) made daily sowings in one season and twice-weekly sowings throughout the next season in north Australia (latitude 14° S) and found that delayed planting reduced grain number and yield. There was a negative correlation between saturation deficit 12 days after anthesis (when florets might be committed to develop or abort) and grain yield.

The grain-filling period of tropical sorghums is 30–40 days (ICRISAT,

1978; Kowal & Kassam, 1978) and the rate of grain-filling ranges from 0.4 to 2.3 mg per seed d^{-1} (ICRISAT, 1976). Pre-anthesis assimilates contribute only a small proportion (*c.* 12%) to final grain weight (Goldsworthy, 1970*c*; Fischer & Wilson, 1971), so that rate of grain-filling and final grain size are closely tied to concurrent photosynthesis. Fischer & Wilson (1975*b*), growing sorghum at various populations in the subtropics, showed that grain size increased almost linearly with

Fig. 7.4. Effect of sowing date on yield components of sorghum at Samaru, north Nigeria (latitude 11° N). Bars show LSD at $P = 0.05$ for weight of 1000 grains, grain to head ratio, total dry weight, grain yield and number of grains per head. Source: Kassam & Andrews (1975).

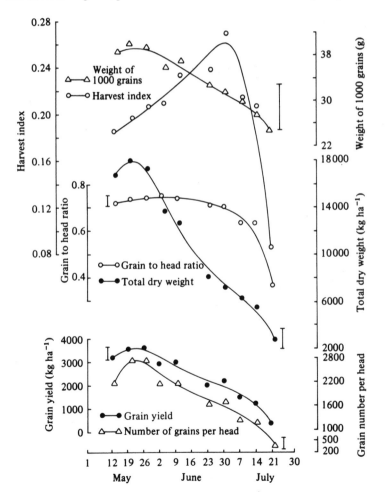

assimilate supply, estimated by the ratio of change in top dry matter during grain-filling to number of grains at maturity. Growth in excess of
grain requirements accumulated in other plant organs, including roots.
High grain yields (12–16 t ha^{-1}) therefore depend, after anthesis, on negligible vegetative growth (as in temperate sorghums: Vanderlip & Arkin,
1977) and high photosynthetic rates (populations in excess of 500 000
plants ha^{-1}: Fischer & Wilson, 1975a). However, in traditional tropical
sorghums only half of the dry weight gain after anthesis goes into grain
(Goldsworthy, 1970c) and probably less than this is channelled to grain
in late-season crops, which exhibit low grain weight and particularly low
harvest index (Fig. 7.4). Thus in tropical sorghums, low grain number
and low dry matter partitioning to grain both contribute to harvest
indices that are appreciably lower than for temperate sorghums growing
in tropical locations (Goldsworthy, 1970b).

7.5 Crop/soil relations

7.5.1 *Soil physical properties*

The main soils used for sorghum production in the tropics are Vertisols
(mainly in India but also in Sudan and Ethiopia) and Alfisols (particularly in India and Nigeria, but also in Sudan, Upper Volta and
Ethiopia). Entisols and Inceptisols are of minor importance (Myers &
Asher, 1982). The prevalence of sorghum on Vertisols and Alfisols can
be related to the fact that these soils, particularly Vertisols, are sufficiently fertile and have the potential to store adequate water for crop
requirements (Fig. 13.5) in the wet-and-dry climates to which sorghum
is adapted. On a Haplustalf at Hyderabad, Vittal, Vijayalakshmi &
Rao (1990) found grain yield increased with increasing depth of topsoil
– a response of 105 kg ha^{-1} per cm of topsoil.

The soil texture of Vertisols and Alfisols ranges through heavy clays
to loamy sands. On self-mulching Vertisols soil friability and structure
may range from a favourable friable granular structure to large hard
clods capable of inhibiting emergence. While the surface soil of many
Alfisols may be friable, some form crusts (Krantz, Kampen & Russell,
1978, Myers, 1978a). On Ustolls, conservation tillage may minimise
crusting problems for sorghum emergence (Unger, 1984).

Growth and distribution of sorghum roots are affected by soil water
(Section 7.4), soil physical resistance (Hemsath & Mazurak, 1974;

Baligar *et al.*, 1981) and soil porosity; root density doubled over the porosity range 39–48% in Senegal (Nicou & Chopart, 1979).

7.5.2 *Soil chemical properties*

The nutritional disorders of grain sorghum are reviewed by Grundon *et al.*, (1987). Although Vertisols and some Alfisols are among the more fertile soils of the wet-and-dry regions, where most sorghum is grown, the availability of water may place a more frequent and severe constraint on yield than the availability of nutrients. Where fertilisers are applied under low rainfall conditions, maximum grain yield may be obtained at moderate levels of application because, at high levels, early vegetative growth is excessive and leads to water stress; e.g., with nitrogen (Myers, 1978*a*) and phosphorus (Olsen *et al.*, 1962). Soil drying retards nutrient uptake late in the season: under dryland conditions on an Ustochrept, Locke & Hons (1988) found that over 80% of total plant nitrogen was taken up before anthesis, whereas under irrigation on a Haplustalf approximately 60% of nitrogen and phosphorus was taken up afterwards (Roy & Wright, 1974). At least 84% of total root weight may be in the surface 15 cm under favourable conditions (Kaigama *et al.*, 1977; Bloodworth, Burleson & Cowley, 1958). However, when sorghum was grown under severe water stress on a Paleustalf, root elongation continued below the fertilised surface 10 cm and the subsoil was a major source of nutrients, allowing 46% of total nitrogen and 37% of total phosphorus to accumulate during grain-filling (Smith & Myers, 1978). Recovery of nitrate nitrogen in Vertisols may be from as deep as 0.6–1.6 m (Standley *et al.*, 1990). The significance of these patterns of uptake is that under dryland conditions it is necessary to maximise the rate of nutrient uptake, from whatever source and location, before anthesis, to ensure its availability for retranslocation to the grain. Studies at Hyderabad (ICRISAT, 1976) showed that, among 49 genotypes examined, between 58% and 87% of nitrogen present in the whole plant at anthesis was present in the grain at maturity.

Sorghum crops in wet-and-dry environments may obtain much of their nitrogen from soil organic matter, which mineralises in a flush at the onset of the wet season (Jones & Wild, 1975) (see Fig. 8.3). In the wet-and-dry environment of northern Australia, Myers (1978*a*) calculated that arable soils could supply up to 110 kg nitrogen ha^{-1} per season, sufficient for sorghum yields of up to 2200 kg grain ha^{-1}. Application of nitrogen in this environment increased grain yields to

7730 kg ha⁻¹. Sorghum is grown on stored soil moisture on about 6.5 m ha in India, equivalent to 40% of the total sorghum area. These crops typically undergo nitrogen stress in addition to terminal drought stress (ICRISAT, 1987). Inoculation of sorghum with nitrogen-fixing bacteria such as *Azospirillum* spp. have shown significant increases in grain yield (ICRISAT, 1985), dry matter yield (Sarig, Blum & Okon, 1988) and uptake of nitrogen, phosphorus and potassium (Okon & Kapulnik, 1986; Pereira *et al.*, 1988) (Fig. 7.5).

Fig. 7.5. Effects of *Azospirillum brasilense* inoculation on dry-matter accumulation in *Sorghum bicolor*. Squares, panicles; circles, stems; triangles, leaves. Filled symbols, inoculated; open symbols, non-inoculated; vertical lines are standard errors. Source: Sarig, Blum & Okon (1988).

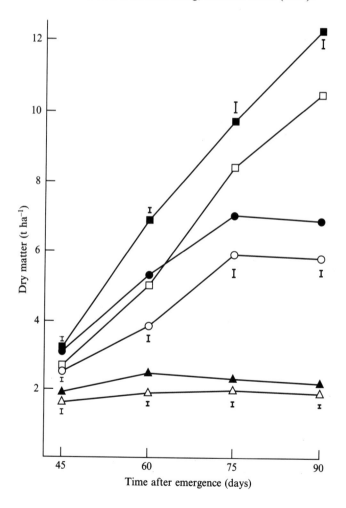

Phosphorus nutrition of sorghum is often inadequate on many of the weathered soils of the tropics, including Oxisols, Ultisols and Alfisols. Mokwunye (1979) reported responses of sorghum to phosphorus on 92% of sites in Nigerian savanna soils. Improved phosphorus nutrition may accelerate the development of a sorghum crop (Olsen *et al.*, 1962), a reduction of 7–8 days between emergence and anthesis being observed over two seasons by Myers (1978*b*). Late rains may ameliorate phosphorus deficiency by extending the growing season (e.g., Myers, 1978b). Cultivars differ in the efficiency with which phosphorus is taken up and utilised (Wieneke, 1990). Lavy & Eastin (1969) found that maximum uptake of ^{32}P occurs in the surface 15 cm and within a 20 cm radius of the plant. Root hairs are probably more important for phosphorus uptake on Oxisols and Ultisols than on Mollisols (Bidin & Barber, 1985).

On an Alfisol rock phosphate showed much greater sustained availability than superphosphate to seven successive crops of sorghum, the stubble of which was removed from the field (Fig. 7.6). After 7 years the residual value of rock phosphate was 60–70% of the initial value but that of superphosphate was only 8% (Arndt & McIntyre, 1963).

Fig. 7.6. Grain yield of sorghum in successive years of cropping after initial applications of three different levels (kg ha^{-1}) of (*a*) superphosphate and (*b*) rock phosphate to an Alfisol at Katherine, north Australia (latitude 14° S). *Note*: each year's yields were corrected for seasonal effects by multiplying them by the ratio of the mean of controls for all years to the mean of controls in that year. Source: Arndt & McIntyre (1963).

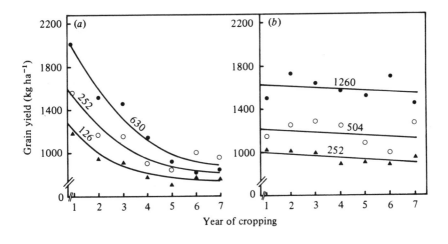

While inoculation with mycorrhizal fungi has been shown to improve phosphorus uptake (ICRISAT, 1987, 1988), the benefits of inoculation with an effective endophyte species varies with soil type, phosphorus level and sorghum genotype (e.g., Bethlenfalvay, Ulrich & Brown, 1985; Raju *et al.*, 1990).

Potassium deficiency is incipient in many West African savanna soils (Wild, 1971) and likely to constrain the yield of sorghum on acid tropical soils. Vertisols are likely to have adequate reserves of potassium. Responses to applied potassium may even be negative if phosphorus levels are low; sorghum grown as forage declined in yield from 25 to 18 t ha^{-1} with applications of up to 240 kg ha^{-1} of potassium where phosphorus was not applied (Reneau, Jones & Friedericks, 1983).

Sorghum is susceptible to aluminium toxicity (Shuman, Ramseur & Duncan, 1990): acid soils with aluminium saturation near or above 50% are usually toxic (Borgonovi, Shaffert & Pitta, 1987) and sorghum is regarded as less tolerant of aluminium than maize (Brenes & Pearson, 1973; Sanchez, 1976). However there are substantial genotype differences in tolerance to aluminium (Baligar *et al.*, 1989; Flores, Clark & Gourley, 1988); some genotypes may not attain maximum yield unless aluminium saturation is below 10% (Salinas, 1978). At low aluminium stress, toxicity effects may be ameliorated by phosphorus (Tan & Keltjens, 1990).

The response of sorghum to lime on highly weathered soils may sometimes be due to improved calcium supply and also to magnesium if dolomite is used. Sorghum has a higher requirement for calcium than rice and maize (Islam, 1981). Plants on calcareous soils are susceptible to iron deficiency (e.g., Bowen & Rodgers, 1987) as has been reported on Vertisols in India (Patil & Patil, 1981).

Sorghum is relatively tolerant of salinity (Table 11.3), which reduces calcium availability, but tolerance varies with genotype (Grieve & Maas, 1988). Salinity may also induce magnesium deficiency (Patel, Wallace & Wallihan, 1975). The grain yield of sorghum receiving saline water on a silty clay Torrifluvent was unaffected by a soil salinity of 6.8 dS m^{-1} in the saturation extract, but declined at the rate of 16% per dS m^{-1} above this, indicating moderate tolerance (Francois, Donovon & Maas, 1984). Tolerance at germination was higher, for while germination was delayed above 8.2 dS m^{-1}, final germination percentage was not significantly reduced by salinity as high as 22.1 dS m^{-1}.

7.6 Place in cropping systems

In the climatic spectrum of tropical rainfed cropping systems based on maize, sorghum and pearl millet, those based on sorghum are intermediate between those based on the other two crops. This is seen most clearly in West Africa (see Fig. 1.3 and Table 6.6). In Africa, cropping systems in which sorghum is an important component are characteristic of zones of 600–1000 mm annual rainfall (Kowal & Kassam, 1978). In the semi-arid tropics of India (400–1000 mm annual rainfall), sorghum is by far the most common rainfed crop. It occupies 22.9% of the total crop area; the next most widespread crops are pearl millet (11.6%) and groundnuts (9.2%) (Krantz *et al.*, 1974).

Under rainfed conditions, sorghum is frequently grown in mixed cropping patterns. In Nigeria, about 80% of the sorghum area is mixed, and in Uganda 46% (Okigbo & Greenland, 1976). In north Nigeria, sorghum figured in five of the seven most frequently recorded crop mixtures (Norman, 1974). These five, in order of frequency, are:

1. Millet/sorghum.
2. Millet/sorghum/groundnuts/cowpeas.
3. Millet/sorghum/groundnuts.
4. Millet/sorghum/cowpeas.
5. Sorghum/groundnuts.

Of these cropping patterns, millet/sorghum/groundnuts/cowpeas gave the highest net return per hectare, but millet/sorghum yielded the highest food energy output per man-hour of labour (Ruthenberg, 1980). It should also be noted that in West Africa sorghum may be sown in another type of 'mixed cropping': a mixture of lines of differing maturity.

In Ethiopia, *durra* sorghums are largely grown at altitudes of 1600–2000 m, either as a sole crop or in association with maize, groundnuts and common beans (Westphal, 1975). The most common associate crops in India are pearl millet, pigeonpea, groundnuts and chickpea.

Sorghum in crop sequences affects the yield of subsequent crops, including that of sorghum itself. Phillips (1959) found that the yield of sorghum, cotton and groundnuts was often lowest where sorghum had been the preceding crop. This was attributed to nitrogen immobilisation in the sorghum stubble (Norman, 1966), which can be very high in sucrose (Conrad, 1938). Similarly, in central Sudan, Gerakis &

Tsangarakis (1969) obtained poorer sorghum yields following sorghum than other crops or fallows.

Figure 4.1 illustrates the stability of income from mixed sorghum and pigeonpea compared with sole crops. This intercrop pattern has been studied in detail at ICRISAT, Hyderabad (ICRISAT, 1979). It is a good example of the efficient use of soil water in a growing season of limited duration (about 5 months). Sorghum and pigeonpea are planted together at the start of the wet season; sorghum is harvested after about 100 days, while the pigeonpea continues to grow, utilising end-of-season rains and stored soil water, for a further 60 days. Land equivalent ratios (LERs) of up to 1.76 have been recorded.

Rainfed cropping systems based on sorghum do not normally involve sequential cropping since the length of the period of available soil water is rarely adequate for a second crop, though they may involve relay cropping. However, on deep Vertisols in semi-arid India with very good soil water relations, some areas are sequentially cropped with safflower and pulses in early dry season following wet-season sorghum (Krantz *et al.*, 1974). Double cropping with sorghum using quick-maturing daylength-insensitive cultivars (90–120 days) has become feasible in the Guinea Savanna of West Africa (Kowal & Kassam, 1978). A typical relay crop pattern is millet/sorghum/cowpea on the Benue Plateau in Nigeria (Netting, 1968). Quick-maturing millet and sorghum are sown together in April–May and cowpeas are planted into the mixture in June–July; millet is harvested July–August and the sorghum and cow-peas in October–November.

One unusual but locally important sorghum cropping pattern on Vertisols in semi-arid India, particularly in Maharashtra (Chowdhury, 1974), involves a wet-season fallow followed by sorghum or a sorghum/oilseed mixture in early dry season. The reasons the land is not cropped in the main growing season period include difficulties in land preparation with draft animals, heavy weed infestation, and the likelihood of waterlogging. However, with increasing population pressure the area of land under wet-season fallow is declining (Krantz *et al.*, 1974).

Broadly speaking, in areas of the tropics where rainfed sorghum is a major crop human pressure on arable land is moderately heavy: e.g., the savanna zone of Africa north of the tropic, the Ethiopian highlands, semi-arid India. As a consequence, the crop is more characteristic of semi-intensive and intensive rainfed cropping systems – those with a cultivation frequency of 70–100% – than of shifting cultivation systems with a long fallow period. Typical rotations on Vertisols in semi-arid India are sorghum/pigeonpea followed by groundnuts, cotton or a 1-year fallow, or a 3-year rotation of sorghum/pigeonpea–groundnuts–cotton (Krantz *et al.*, 1974). Where sorghum is grown in systems that include a fallow, it is more commonly found in the early rather than the late phase of the cropping period (Miracle, 1967) owing to its relatively high fertility demand.

All the above refers to rainfed cropping. Sorghum is also important in some irrigated cropping systems within the tropics. The best-known example is the Gezira scheme in the Sudan (annual rainfall 200–400 mm), where it is grown in sequences that include cotton, wheat, groundnuts, lubia bean (*Lablab purpureus*) and fallow.

In the Gezira water supply and drainage are very closely controlled and yields are of the order of 2 t ha^{-1}. In contrast, irrigated sorghum in the Yemen is grown with irregular water supply dependent on stream floods from current rainfall in the catchments (Ruthenberg, 1980). The cropping system is extensive and labour input and yields are low, only a little higher than for Nigeria (Table 7.1). As an example of intermediate water control and level of input, sorghum is grown in basin irrigation systems in north Sudan, where water diverted from the Nile into adjacent depressions is allowed to stand for up to 30 days and then drained off. The crop is then sown and grows to maturity on stored soil water.

Since cultivated sorghums are not true annuals, they have the capacity for continued tillering after grain harvest if water is available in the root zone. In consequence, sorghum stubbles are frequently grazed by ruminant livestock. In areas where forage is in short supply, as in densely populated rainfed cropping areas of semi-arid India, the stover is carefully harvested and stored for subsequent dry-season feeding, and may even be sold for consumption by transport cattle or buffalo or urban dairy stock.

Further reading

Doggett, H. (1988). *Sorghum*, 2nd edn. Harlow, UK: Longman.

8

*Pearl millet (*Pennisetum glaucum*)*

8.1 Taxonomy

The genus *Pennisetum,* of the subfamily Panicoideae, tribe Paniceae, comprises over 140 species in both the Old and New Worlds. The section *Pennisetum* of the genus includes two species: a perennial tetraploid forage plant, and an annual diploid species (2n=2x=14) representing both wild and cultivated forms of pearl millet (Brunken, 1977).

Brunken divided the diploid species, which at the time was known as *P. americanum,* into three subspecies: ssp. *americanum,* cultivated pearl millet; ssp. *monodii,* its wild relatives; and ssp. *stenostachyum,* weedy types intermediate between the other two species, called *shibras* in West Africa. Clayton & Renvoize (1986) showed that P. *americanum* was not a valid name for pearl millet, which is now termed P. *glaucum* (L.) R. Brown.

Cultivated pearl millet is morphologically and phenologically very diverse. Brunken, de Wet & Harlan (1977) recognised four races, characterised by the size and shape of the caryopsis and of the inflorescence as a whole:

1. *typhoides,* found across the whole range of cultivation in Africa and India.
2. *nigritarum,* dominant in the eastern Sahel.
3. *globosum,* dominant in the western Sahel.
4. *leonis,* dominant in coastal West Africa.

Phenology has also been used as a criterion for broad grouping into early, medium and late types; in West Africa the boundaries between these two groups are 95 and 130 days to maturity, in India 100 and 180 days (Purseglove, 1972).

The tetraploid forage plant P. *purpureum,* napier or elephant grass, may be crossed with pearl millet, and the resultant hybrid is in use as a forage (Muldoon & Pearson, 1979).

8.2 Origin, evolution and dispersal

Wild taxa of *P. glaucum* are native along the southern fringes of the Sahara (de Wet, Bidinger & Peacock, 1992). Archaeological evidence indicates its use in this area, both in cultivation and as a wild cereal, from about 3000 BP. Cultivated pearl millet, initially of race *typhoides,* became widely distributed across the African semi-arid tropics and probably reached India by 2500 BP. Pearl millet in India is all of race *typhoides* and wild relatives and *shibras* are not found there.

Pearl millet was introduced to the USA in the 1850s, but it probably reached tropical America from southern Europe at an earlier date (Rachie & Majmudar, 1980). In the USA it is grown only as a forage, and as a grain crop it is unimportant in the New World tropics.

Crop statistics for pearl millet are not readily available, since many countries report data for 'millets' that include other species such as finger millet (*Eleusine coracana*). In the figures given below for the six major producers, FAO data for millets have been adjusted using information from Rachie & Majmudar (1980) on the proportion of pearl millet produced, which in the West African countries is over 97% but which in India is estimated at 56% of total millet production. Since the estimates are merely approximate, only relative production is given (India = 100; about 6 m t):

India	100	Mali	13
Nigeria	48	Chad	10
Niger	17	Senegal	9

8.3 Crop development pattern

Pearl millet is grown largely in the African Sahel and India within 200–800 mm average annual rainfall (AAR) and it particularly dominates cropping systems within 400–600 mm AAR (Bidinger *et al.*, 1981). As a consequence of this range in duration of growing season, genotypes vary in maturity from 55 to 180 days or more. In short-season cultivars, development is divided into three approximately equal periods: floral initiation occurs at about development index (DI) = 0.3 and anthesis at DI = 0.6–0.7 (Bidinger *et al.*, 1981). Floral initiation in most cultivars is stimulated by short daylengths (e.g., Begg & Burton, 1971; Patil, Reddy & Gill, 1978). This affects time to maturity, as demonstrated by serial plantings at high latitude (Ferraris, Norman & Andrews, 1973). Variation in development within long-season cultivars is mainly related to the timing of floral initiation and the period between initiation and anthesis. Long-season and short-season cultivars may take 20 days to proceed from a differentiated inflorescence to anthesis (Powers *et al.*, 1980); thus inflorescence development before anthesis may occupy one-sixth of the life span of a long-season type compared with one-third of the span of a short-season cultivar. Genetic variation in the duration of the grain-filling period is small relative to that in, for example, maize (Fussell & Pearson, 1978).

Developmental events of floral initiation, apex differentiation and spike development to anthesis have been described by electronmicrophotography (Powers *et al.*, 1980). Grain development and anatomical changes associated with black region formation at the cessation of grain growth are described by Fussell & Pearson (1978) and Fussell & Dwarte (1980).

Tillering is very important in pearl millet because crops are commonly grown at low populations under semi-arid rainfed conditions. Basal tillering occurs between the thirteenth and fortieth day after sowing and up to 40 tillers may be produced (Raymond, 1968). Axillary tillering from upper nodes occurs in flushes throughout grain development under intermittent drought. These tillers form two or three leaves and an inflorescence within 10–20 days (Bidinger *et al.*, 1981); they may contribute up to 50% of the total yield of pearl millet growing under natural rainfall (Rachie & Majmudar, 1980).

8.4 Crop/climate relations

The caryopsis of pearl millet is small (3–4 mm; 5–12 mg), which accounts for much of the field problem of establishment, especially in crusting soil. Seed viability and seedling vigour are not affected if seed development on the parent plant ends between mid-grain-filling and black region appearance (Fussell & Pearson, 1980); this is presumably an adaptation to ensure survival when early termination of the wet season interrupts grain development. High temperatures increase speed of germination and reduce variation about the modal date of both germination and emergence (Pearson, 1975), although high soil temperatures (>50 °C) following sowing cause seedling death (Soman *et al.*, 1986). The temperature requirements for pearl millet to complete various developmental events are given in Table 8.1.

Different levels of available soil water between 50% and 75% of field capacity do not influence germination, whereas soil held at field capacity may kill seedlings (Fawusi & Agboola, 1980). Dry soil at sowing and during the early seedling stage is a widespread limitation on growth: farmers adapt the seeding rate by varying the space between 'hills' (groups of seeds) and the intensity of thinning (usually leaving 3–6 seedlings per hill): optimum populations, because of restricted soil water, may be only 6000 ha^{-1} in the Sahel (Fussell, 1992).

Leaf photosynthetic and anatomical characteristics associated with a C4 photosynthetic pathway lead to high potential growth rates and sensitivity to low temperature (reviewed by Pearson, 1994). Lamina expansion increases linearly with increasing temperature (Monteith *et al.*, 1981) but, because of low plant populations, radiation interception is the primary constraint during vegetative growth. Crop growth rate is linearly related to intercepted radiation. In India maximum leaf area indexes (LAIs) are only in the region of 1.7–3, corresponding to 40–70% interception of radiation (ICRISAT, 1978). Complete light interception was attained at an LAI of 5–8 in north Australia (Begg *et al.*, 1964; Begg, 1965). Maximum apparent above-ground growth rates are in excess of 500 kg ha^{-1} d^{-1}: 580 kg ha^{-1} d^{-1} was reported by Phillips & Norman (1967) and 540 kg ha^{-1} d^{-1} by Begg (1965), the latter at an LAI of 9. At supraoptimal LAI values, higher photosynthetic rates may be attained but net growth may decline (McLeod, 1966) owing to changes in dry matter partitioning within the plant and to increased respiration and rate of senescence.

Table 8.1. *Development in a short-season pearl millet hybrid (BK560): thermal durations for 50% of the population*

Development	Thermal duration in day-degrees
Between successive leaves and root axes	26
Sowing to first tiller	200
Between successive tillers	80
Expansion of tiller leaves	270
Increase in mass of tiller leaves	420
Between successive spikelets	0.23
For all spikelets	190
Floral initiation to anthesis	460
Expansion and growth of main stem	360
Increase in weight of panicle	550
Increase of weight of grain	290

From Squire (1990).
The base temperature was about 10 °C for all processes.

Leaf area development, water use and energy balance have been described throughout the life of an 85-day crop at latitude 11° N in northern Nigeria (Kassam & Kowal, 1975) and for crops representative of African and Indian morphological types at latitude 17° N in India (Craufurd & Bidinger, 1988). In both African and Indian experiments, crops reached maximum LAI of 3.5–4.5. Maximum LAI was reached at inflorescence emergence and leaf area declined throughout grain development (e.g., Fig. 8.1). The seasonal growth rate of the Nigerian crop was 26.5 kg ha^{-1} d^{-1}, which represented an average of 5% conversion of photosynthetically active radiation (400–700 nm). The Indian crops averaged 17.6 kg ha^{-1} d^{-1} with efficiencies (E) of 1.9–2.4 g per MJ photosynthetically active radiation (PAR) (Craufurd & Bidinger, 1988). For the Nigerian crop the ratio of evaporation to pan evaporation (E/E_p) averaged 0.82 and was unity from floral initiation (at development index (DI) = 0.2) to anthesis (DI = 0.6) (Fig. 8.1). Crop water use was relatively efficient: 300 mm or 148 g water per gram top weight compared with 253 g g^{-1} for maize. However, because of low partitioning of dry matter into grain, water use per gram of grain weight was 863 g g^{-1} in contrast to 747 g g^{-1} for maize. The Indian study found that the cultivars representing African and Indian phenotypes (the African having a longer vegetative phase until floral initiation, being taller and having a lower harvest index) achieved equal yields under good conditions, although the extended period of stem growth resulted in greater competition between stem and panicle in the African type.

Despite its ability to survive drought and its reputation for being more productive than sorghum and maize at low soil water, there are few descriptions of the effects of water deficit on components of yield of pearl millet. Water deficits of 4 kPa may reduce leaf area and light interception by 40% (data collated in Squire, 1990) while the impact of water deficit on yield is less if the stress is imposed on vegetative, rather than reproductive, plants (Mahalakshmi & Bidinger, 1985). Stress-induced loss of yield components, e.g., grain on the main stem, can be

---

Fig. 8.1. Water use by pearl millet. (*a*) Relative evaporation (E/E_p ratio) and leaf area index (LAI) of rainfed pearl millet at Samaru, Nigeria, latitude 11° N. Development events are coded as in Fig. 2.6; HE, head emergence; (*b*) and (*c*) E/E_p ratio, dry weight (DW) and leaf area accumulation for rainfed and irrigated pearl millet sown in October at Hyderabad, India, latitude 17° N. Sources: (*a*) Kassam & Kowal (1975); (*b*) and (*c*) Hsiao, O'Toole & Tomar (1980).

compensated for by increased tillering (Mahalakshmi & Bidinger, 1986) to a greater extent than in sorghum and maize.

Studies of dry weight partitioning to roots indicate that under rainfed compared with irrigated conditions there are fewer root axes, root

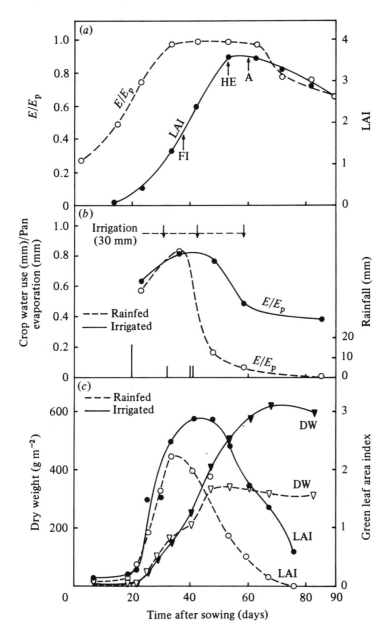

weight is lower, and a lower percentage of roots is found in the topsoil (Gregory & Squire, 1979). Nonetheless, pearl millet has a reputation for drought tolerance owing to its rapid and deep root penetration: e.g., to 1 m depth within 33 days after sowing (Gregory & Squire, 1979). Root penetration has been recorded to 3.6 m in north Australia (Begg *et al.*, 1964), although values of 1.2–1.5 m are probably more common (Wetselaar & Norman, 1960). As in other cereals, most of the roots and the widest lateral spread occur in the top 30 cm of the profile (Williamson, Willey & Gray, 1969; Gregory & Squire, 1979). Root length reaches a maximum at about anthesis: in an experiment in India, maximum root length was 3500 m m^{-2} ground area, compared with 2500 m m^{-2} ground area for groundnut (Gregory & Reddy, 1982). Desiccation of roots in dry upper soil is minimised by a well-developed exodermis and sclerenchymatous cells encircling the vascular bundles (Ratnaswamy, 1960).

Little is known of the effects of environment, other than the qualitative effect of daylength, on floral initiation and inflorescence development before anthesis. In glasshouse studies, spikelet number increased by 25% with increasing temperature from 19 to 31 °C and time from floral initiation to heading declined with increasing temperature (Monteith, 1980). In the field, grain number and grain size contribute equally to variation in yield (45% and 42% respectively for 50 genotypes at Hyderabad, India; Alagarswamy, Maiti & Bidinger, 1977). Grain number is reduced by abnormal temperature and by high vapour pressure deficit at flowering (Mahalakṣhmi, Bidinger & Raju, 1991). Avoiding high temperature and water deficit, maximum flowering occurs at night and flowering is weakest between 15.00 and 18.00 hours (Ayyangar, Vijiaraghavan & Pillai, 1933; Bhatnagar & Kumar, 1960).

Following anthesis, the 'lag phase' of grain development lasts between 0 and 10 days and decreases by 2 days per 3 °C increase in temperature (Fussell, Pearson & Norman, 1980). Rate of grain growth during the actual grain filling period (AGFP) varies substantially between genotypes (Fussell & Pearson, 1978). However, rate of grain growth appears to be constant over a range of temperatures (Fussell *et al.*, 1980); duration of AGFP accounts for all the temperature-dependent variation in grain size shown in Fig. 8.2. Soil water stress during grain-filling reduces individual grain weight (Mahalakshmi *et al.*, 1991) althought it is not known whether rate or duration of AGFP is more sensitive to stress: rate of grain-filling is more sensitive to nitrogen stress than is EGFP (Table 8.2)

Table 8.2. *Effect of nitrogen stress on components of yield and on grain growth of pearl millet*

	N3	N2	N1	N1 to N3	LSD at $P=0.05$
Change in total top weight during grain-filling (g per plant)	1.7	11.4	29.8	2.5	7.0
Grain dry weight (g per plant)	1.3	8.8	14.5	0.58	1.21
Grain N content (mg per plant)	34	184	392	17	33.6
Individual grain weight (mg)	6.1	7.7	10.1	6.7	0.7
EGFP (d)	36	32	27	–	4.1
Rate of filling (mg DW per grain d^{-1})	0.17	0.24	0.38	–	0.01

After Coaldrake, Pearson & Saffigna (1987).
EGFP, effective grain-filling period; DW, dry weight.
N1, N2, N3 were pearl millet cv Mx001 supplied continuously with 1, 4 and 12 mol nitrate nitrogen m^{-3}; N1 to N3 plants had supply increased at anthesis.

   The sensitivity of duration of grain-filling to temperature results in a relatively low temperature optimum for individual grain size (Fig. 8.2). It follows that where millet is maturing seed under a high-stress situation, e.g., with maxima of 33–45 °C and evaporation of 8–12 mm d⁻¹, high yields will depend on having both a high tiller population and adequate soil water because individual grain size will be small. In more typical situations of inadequate water from about anthesis onwards, axillary tillers fail to develop grain, reduced evaporation increases crop temperature, which reduces AGFP, and thereby grain size is again small. Experimental yields of 4 t ha⁻¹ are reported but national averages are commonly about 0.7 t ha⁻¹ (Table 4.1).

Fig. 8.2. Effect of temperature on rate of development, fertility and grain growth of pearl millet: (*a*) Sowing to anthesis; (*b*) anthesis to end of grain-filling; (*c*) spikelet fertility; (*d*) single grain weight. Different lines represent various experiments with cv. Tift 23DB × Tift 71. Bars show LSD at *P*= 0.05. Source: Fussell, Pearson & Norman (1980).

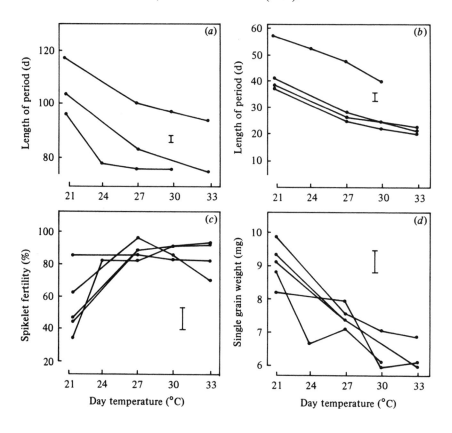

8.5 Crop/soil relations

8.5.1 *Soil physical properties*

The main soils on which pearl millet is grown for grain occur in the wet-and-dry tropics. In West Africa they include Aridisols, Alfisols and Entisols. As classified in the D'Hoore (1964) legend of the soil map of Africa, the main millet soils of north Nigeria are Juvenile soils, approximately three-fifths of which are aeolian sands, the remainder being riverine and lacustrine alluvium, also generally sandy (Klinkenberg & Higgins, 1968). In India pearl millet soils include Aridisols, Alfisols and Vertisols.

Pearl millet is preferred to sorghum on sandy soils (Norman & Begg, 1968; Ferraris, 1973), but grows best on light loams (Kowal & Kassam, 1978). It is one of the few crops adapted to the deep sands of the Sahelian region and similar areas such as the Rajasthan desert and extreme Western Punjab. The rainy season of the Sahelian region is generally preceded by dust storms with winds that can exceed 100 km h^{-1}. This contributes to erosion and can damage pearl millet seedlings by abrasion or burying, sometimes necessitating complete resowing (ICRISAT, 1990). Sandy Aridisols low in organic matter may develop surface crusts which inhibit seedling emergence (e.g., Camborthids of Hissar, India; Agrawal & Sharma 1984; Joshi 1987). Large seed is therefore a favoured quality (Lawan *et al.*, 1985) where crusts occur. Where heavier soils (e.g., Vertisols) are found in wet-and-dry environments, sorghum is preferred to pearl millet.

Pearl millet is also grown extensively as a grain crop on shallow mixed black and red light-coloured upland gravelly soils of the Deccan and South India (ICAR, 1961, quoted by Rachie & Majmudar, 1980). Soil depth is an important determinant of yield: over 4 years topsoil depth accounted for 76–85% of the variation in grain yield on a Haplustalf at Hyderabad; the relation was 77 kg ha^{-1}grain cm^{-1} topsoil (Vittal *et al.*, 1990).

Pearl millet does not tolerate waterlogging (Kowal & Kassam, 1978). Flooding a crop in sodic soils (Aquic Natrustalfs) for 2–6 days, at either the tillering or flowering stage, reduced grain yield by 15–27% (Sharma & Swarup, 1989). Studies in tanks by Williamson *et al.* (1969) showed more severe effects: 2 days of flooding 4 weeks before the first harvest reduced first harvest yield by 40%, although it did not affect the second harvest. Table 8.3 gives a comparison with other crops of the effect of depth to water table.

8.5.2 *Soil chemical properties*

A notable feature of pearl millet is its capacity to grow on soils of low chemical fertility. As a staple cereal it has been grown on sands (as mentioned above), on marginal soils, and on other soils which have declined severely in fertility (Ahn, 1970). Smith & Clark (1968) found that pearl millet was able to extract greater quantities of nitrogen, phosphorus and potassium than sudan grass (*Sorghum sudanense*) in an acid sandy loam. The capacity of pearl millet to grow on infertile soils may be related to its ability to root deeply and rapidly (see Section 8.4).

Mineral uptake proceeds throughout vegetative growth (Theodorides & Pearson, 1981) and was shown by Gregory (1979) to continue in an irrigated crop until about 2 weeks after root growth ceased at about anthesis. However, the pattern of nutrient uptake can be greatly modified by climate, nutrient availability, particularly of nitrogen, and availability of water. Gregory (1979) found that irrigation almost doubled the uptake of nitrogen, phosphorus and potassium on an Alfisol at Hyderabad when the crop was harvested 82 days after sowing. In the unirrigated crop, nutrient uptake ceased at about 40 days after sowing but continued for a further 7–21 days in the irrigated crop. The shorter period of nutrient uptake in the rainfed crop probably explains why split fertiliser applications of nitrogen, phosphorus and potassium did

Table 8.3. *Relative yields of crops at varying water table depth: highest yield within each crop = 100*

Crop	Soil texture	Water table depth (cm)						
		15	30	40–50	60	75	80–90	100
Maize	Silty clay loam	45	55	67	70	—	100	—
Sorghum	Clay[a]	73	86	93	100	93	—	—
Pearl millet	Loam	41	69	80	87	98	100	93
Common bean[b]	Clay	—	—	79	84	—	90	—

After Williamson & Kriz (1970).
[a] No surface water applied.
[b] Maximum yield when water table below 100 cm.

not increase yield under rainfed conditions at Samaru, Nigeria (Egharevba, 1978).

Although pearl millet is adapted to soils of low fertility, it is capable of taking up large amounts of nutrients when fertilised or grown on more fertile soils. Olsen & Santos (1976), for instance, obtained 35.3 t ha^{-1} dry matter of tops on a fertilised Alfisol in Brazil. An improved cultivar yielding 3.1 t ha^{-1} of grain in the West African savanna is reported to have removed per hectare 132 kg nitrogen, 28 kg phosphorus, 65 kg potassium and 56 kg calcium, which is greater than for an equivalent yield of maize (IRAT, 1972, quoted by Kowal & Kassam, 1978).

Where soil water is not limiting, pearl millet yield responses (dry matter or nitrogen) to fertiliser nitrogen are generally linear up to a high level of nitrogen. For example, in Brazil:

$$Y = 0.68 + 4.49F$$

where Y is t protein ha^{-1} and F is kg applied nitrogen up to 300 kg nitrogen ha^{-1} (Medeiros, de Saibro & Jacques, 1978). Coaldrake (1985) and Coaldrake & Pearson (1985*a,b*) describe the effects of nitrogen on growth parameters of well-watered plants.

However, in wet-and-dry climates where rainfed crops are grown for grain, rates of nitrogen are usually quite low; vigorous early growth promoted by fertiliser nitrogen may consume water required for later crop development and grain maturation. Etasse (1977), on the basis of West African experience, warned against nitrogen applications above 60 kg nitrogen ha^{-1}, which may promote vegetative growth but depress grain yield. Yield depression above 50 kg N ha^{-1} has been recorded in the Nigerian savanna (Egharevba, 1978). Under these dryland conditions uptake from fertiliser nitrogen may be quite low (20–37%) and losses severe (25–53%), as found in Niger by Christianson *et al.* (1990*b*). Nevertheless, pearl millet can be very efficient in recovering mineral nitrogen from subsoil. Wetselaar & Norman (1960) found that the crop could deplete soil nitrogen more efficiently than sorghum or sudan grass in the surface 150 cm of an Alfisol in north Australia (Fig. 8.3).

In soils of wet-and-dry climates, responses to nitrogen fertiliser may be modified by such factors as rate of nitrate leaching (particularly in sandy soils), nitrogen inputs from mineralisation, rainfall (Jones & Bromfield, 1970) and surface algal crusts (ICRISAT, 1978; Loftis & Kurtz, 1980). Yield increases have also been obtained by inoculation with *Azospirillum* spp. (e.g., 33.5%; ICRISAT, 1985) which appear to improve nitrogen uptake through the bacterial production of growth-promoting substances (Wani *et al.*, 1988) or nitrate reductase (Boddey & Dobereiner, 1988). ICRISAT (1988) now considers the contribution by nitrogen-fixing bacteria to pearl millet nutrition at Patancheru, India, not worth further examination.

Responses of pearl millet to phosphorus are not uncommon (Ferraris, 1973), but the phosphorus requirement of pearl millet does not appear to be high. While high responses to phosphorus have been obtained in Africa (e.g., in Niger; Christianson *et al.*, 1990*a*), in the African savanna, rates above 66 kg phosphorus ha^{-1} can depress yields if soil

Fig. 8.3. Nitrate-nitrogen in the surface 150 cm of an Alfisol before and after cropping on a fallow site at Katherine, north Australia, latitude of 14° S. A, September 1958.; B, May 1959, after no crop; C, after sorghum; D, after sudan grass (*Sorghum sudanense*); E, after pearl millet. Source: Wetselaar & Norman (1960).

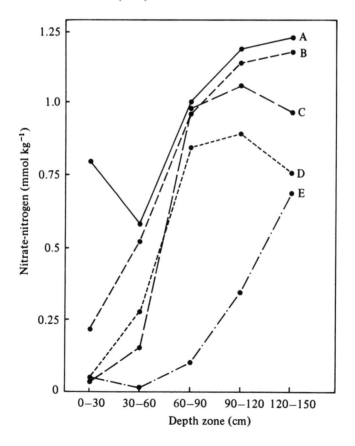

moisture is insufficient to mature the heavier crop (Egharevba, 1978). Surface soil appears to be an important location for phosphorus uptake: Shriniwas (1980) found that the 0–15 cm zone contributed 44–60% of total phosphorus taken up by five pearl millet hybrids. Mycorrhizal fungi can significantly increase phosphorus uptake (ICRISAT, 1985) and both colonisation and phosphorus uptake have shown heterosis. The percentage heterosis calculated over the better parent ranged from 61% to 260% for phosphorus uptake among the six crosses tested (ICRISAT, 1987). Rainfall and species of fungus used also affect the benefit of the infection (ICRISAT, 1985).

Pearl millet has a high requirement for potassium. In a total dry matter yield at the soft dough stage of 6749 kg ha^{-1}, Mehta & Shah (1958) recorded 121 kg ha^{-1} of potassium, in comparison with amounts for nitrogen, phosphorus, calcium and magnesium of 63, 28, 21 and 10 kg ha^{-1}, respectively.

Pearl millet appears to be very tolerant of acid soil constraints, including high (>50%) aluminium saturation (Flores, Clark & Gourley, 1991) and is generally more tolerant of acid soils than sorghum (Walker, Marchant & Ethredge, 1975; Flores *et al.*, 1991). In spite of this evident tolerance of pearl millet, a combination of soil acidity factors including low soil pH (<4.5), high aluminium saturation and decreased nutrient levels, appears responsible for the extreme variability in yield observed in the sandy soils (Psammetic Paleustalfs) at Niamey, Niger. Similar effects have been observed in other poorly buffered sandy Ustalfs, Ustults, Psamments and Tropepts of the semi-arid Sahelian environment of Niger (Scott-Wendt, Chase & Hossner, 1988).

Seedlings are moderately tolerant of salinity (Ashraf & McNeilly, 1987). Established plants have been grown to maturity on dilute (1%) seawater to give grain yields as high as with normal irrigation (Kurian, 1976). On aquic Natrustalfs, representative of sodic soils of the Indo-Gangetic Plains, yields were halved by an increase in ESP (exchangeable sodium percentage) from 28 to 45 (Singh, Sharma & Chillar, 1988).

8.6 Place in cropping systems

Pearl millet is an important component, and sometimes the only cereal component, of cropping systems at the arid limits to agriculture in wet-and-dry tropical climates. Because of limits to the growing season imposed by water availability, there is usually only one cropping period in the year. An account of cropping of millet in Africa is given by Fussell (1992).

In common with other upland tropical cereals, millet is frequently grown in mixed cropping patterns. In Nigeria, 90% of the area of millet is in association with other crops (Okigbo & Greenland, 1976). The most important are groundnuts, cowpeas and sorghum. In a survey of crop combinations in north Nigeria, millet was a component of four of the seven most frequently recorded crop mixtures (Norman, 1974; see Section 7.6).

In West Africa, however, the cropping systems are often more complex than they would appear because two distinct types of millet may be grown together: short-season non-photoperiodic 'Gero' millets maturing in 55–100 days and long-season photoperiodic 'Maiwa' types maturing in 120 days or longer. The former are more frequently sown. Common examples in north Nigeria (Norman, 1972, quoted by Kowal & Kassam, 1978) are: Gero millet/Maiwa millet/sorghum, and Gero millet/Maiwa millet/sorghum/cowpea. The quick-maturing Gero millets are particularly important in the food supply of subsistence farmers, which is at its lowest in early wet season, and they are generally sown first. They ripen while the wet season is at its height, but have a degree of resistance to grain mould diseases (Kowal & Kassam, 1978).

In India, rainfed pearl millet is grown, as in Africa, in association with sorghum, groundnuts and cowpeas, but pigeonpea is also a common associate crop (Krantz *et al.*, 1974). The relations between millet and pigeonpea are similar to those between sorghum and pigeonpea: a quick-maturing cereal in combination with a late-maturing crop that utilises end-of-season rainfall and stored soil water. Other legumes grown in association with millet in India include hyacinth or lubia bean (*Lablab purpureus*), horse gram (*Dolichos biflorus*), black gram (*Vigna mungo*), green gram (*Vigna radiata*) and guar (*Cyamopsis tetragonoloba*) (Rachie & Majmudar, 1980).

Yield advantages due to intercropping are commonly 20–30% but may be as much as 100% (Table 8.4; see Reddy & Willey (1980) for Indian examples). These arise from differences in growth habit of the co-habiting crops. Furthermore, the capacity of millet to compensate for water stress makes it adapted to late-season production under water deficits and thus complements short-season companion crops.

For the above examples of rainfed cropping patterns there is only one cropping period each year. However, in areas where water is available in quantity for a longer period but where millet is still a desired crop, for example on deep Vertisols or alluvials in semi-arid India, it is possible to grow two sequential rainfed crops in one season. Common

Table 8.4. *Intercropping effects from factorial experiments with pearl millet in West Africa.*

Year	Type of experiment	Rainfall (mm)	Cowpea hay/maize grain (kg ha⁻¹)	Millet grain (kg ha⁻¹)	Land equivalent ratios		
					Cowpea	Millet	Total
Millet/cowpea intercrop, Niger							
1982	Density × variety × intercrop proportion	372	318	277	0.58	0.69	1.27
1983	Intercrop proportion × rotations	599	768	385	0.48	0.69	1.17
1984	Intercrop proportion × rotations	216	22	435	0.09	1.16	1.25
1985	Intercrop proportion × rotations	495	734	648	0.48	0.72	1.20
1985	Variety × intercrop	495	360	921	0.31	0.93	1.24
			(261)ᵇ		(0.50)		(1.43)
Maize/millet intercrop, Mali							
1982	Intercropping study	579	2010	840	1.12	0.57	1.69
1982	Intercropping study	1046	1800	880	1.04	1.06	2.10
1983	Planting date × intercropping	765	590	1920	0.29	0.82	1.11

From Fussell (1992).
[a]These experiments included other management factors such as intercrop density, fertility, planting date and harvest time of intercrop.
[b]Cowpea grain yields.

patterns in regions of cool dry seasons are millet–wheat, millet–barley and millet–chickpea (Krantz *et al.*, 1974). Under similar soil water conditions but with a warmer dry season millet may be double-cropped. In Sudan millet is grown, on stored water only, in the early dry season after the subsidence of wet-season floods (Rachie & Majmudar, 1980).

Pearl millet figures in cropping systems varying in cultivation frequency, though it is more commonly recorded as a component of semi-intensive and intensive rainfed cropping systems than of shifting cultivation systems with a long fallow break. Typical rotations in semi-arid Africa that include millet (Charreau, 1974; Rachie & Majmudar, 1980) are:

Millet – groundnut – fallow
Millet – fallow – groundnut
Groundnut – groundnut – early millet – late millet – fallow
Groundnut – early millet/cowpeas – groundnut – late millet – fallow

In Senegal, the length of the fallow period is dependent on land pressure and may vary from 1 to 3 years. As population density increases, the fallow tends to be eliminated. A common sequence in Niger (Braun, 1974) is 5 years cropping/5 years bush fallow. The cropping phase is a succession of millet/legume intercrops, groundnuts being the most important legume. In north Nigeria, if fertiliser or manure is not used, 3 years of cropping with millet, groundnuts, sorghum and cowpeas in various combinations is usually followed by up to 6 years of fallow. Crop sequence effects of sorghum, groundnuts and millet are referred to in Section 10.6.

Crop rotations in India involving millet are summarised by Rachie & Majmudar (1980). On the lighter soils rainfed millet is grown in sequence with a wide range of crops, including finger millet, grams, sesame (*Sesamum indicum*), mustard, etc., either as a sole crop or in mixtures and with or without a short fallow. Rainfed rotations on Vertisols include various sequences of millet (with or without associated crops), sorghum and cotton. In India millet also finds a place in intensive irrigated systems: for example, with grams, forage sorghum and wheat in the north (Punjab) and with groundnuts, rice and sugarcane in the south (Tamil Nadu).

Further reading

Fussell, L.K. (1992) Semi-arid cereal and grazing systems of West Africa. In *Field Crop Ecosystems*, ed. C.J. Pearson, pp. 485–578. Amsterdam: Elsevier.

Pearson, C.J. (1984). Pennisetum millet. In *The Physiology of Tropical Field Crops*, ed. P.R. Goldsworthy & N.M. Fisher, pp. 281–304. Chichester: Wiley.

Rachie, K.O. & Majmudar, J.V. (1980). *Pearl Millet*. University Park, Pennsylvania: Pennsylvannia State University Press, 307 pp.

III

LEGUMES

9

Crop legumes in tropical agriculture

9.1 Introduction

This section of the book is concerned with the most important tropical food crops of the family Fabaceae: groundnut, common bean (*Phaseolus vulgaris*), soybean and chickpea. Other locally important tropical crop legumes include cowpea (*Vigna unguiculata*), concentrated largely in Nigeria; pigeonpea (*Cajanus cajan*) and the grams (*V. mungo* and *V. radiata*), grown mainly in India and southeast Asia; and hyacinth or lubia bean (*Lablab purpureus*), common in India and the Sudan. Adams & Pipolly (1980) give a useful general account of the biology of economic legumes.

9.2 Regional production

Table 9.1 gives the area, yield and production of tropical crop legumes. All pulses have been grouped together. Over half the world area sown to grain legumes is in tropical countries. Soybean and common bean (which is included among 'other pulses' in official statistics) are the most important tropical legumes, although groundnut is more specifically grown in tropical regions (Table 9.1). Tropical legumes contribute 35% to world production, which is higher than the comparable statistics for tropical cereals (30%, Table 4.1) but lower than for tropical root crops (40%, Table 14.1).

Production of crop legumes in tropical Asia equals that of Africa and America combined. The Asian figure would be further inflated if it were possible to include tropical soybeans in south China, which may amount to 7 m t of China's total soybean production of 9.8 m t per annum. Most of the world's groundnuts, chickpea and pigeonpea are grown in India. America dominates tropical soybean production, where high productivity per hectare reflects widespread cash cropping in the leading producer, Brazil. However, despite Brazil's production, 19 countries in tropical America import soybean and soybean products (Kueneman & Camacho, 1987). Production of crop legumes is lowest in tropical Africa where areas are small relative to cereals and root crops: a reason why some African systems are not sustainable (Pearson, 1994). In Africa, particularly West Africa, groundnuts are the main crop legume.

9.3 Role in tropical human nutrition

The energy and protein requirements of tropical peoples vary with weight and age. The average energy supply per head in developing countries is 9.2 MJ d^{-1} and the average protein supply 55 g d^{-1} (FAO, 1981). Although seed legumes are considered primarily as sources of protein, they are, of course, of direct value as energy food, particularly those with a high oil content. Thus cowpea provides 13.9, pigeonpea 14.8, soybean 18.0 and groundnut 22.9 MJ kg^{-1}, compared with an average of 15 MJ kg^{-1} for cereals.

Table 9.1. *Area, yield and production of crop legumes in the tropics, in 1991*

	Tropical Africa	Tropical America	Tropical Asia	Total Tropics	World	Tropics as % of world
Area (m ha)						
Soybean	0.47	10.32	4.81	15.60	55.37	28
Groundnut	5.80	0.32	10.00	16.12	20.36	79
Pulses	12.00	9.36	29.34	50.70	70.38	72
Total	18.27	20.00	44.15	82.42	146.11	56
Yield (t ha^{-1})						
Soybean	0.94	1.53	0.97	1.34	1.86	72
Groundnut	0.84	1.36	0.90	0.89	1.15	77
Pulses	0.55	0.59	0.60	0.59	0.85	69
Total	0.65	1.09	0.72	0.80	1.27	63
Production (m t)						
Soybean	0.44	15.78	4.66	20.88	103.07	20
Groundnut	4.89	0.44	9.02	14.35	23.37	61
Pulses	6.54	5.50	17.75	29.79	59.90	50
Total	11.87	21.72	31.43	66.02	186.33	35

After FAO (1992).

On a global basis, plants contribute about 70% and animals 30% to total protein consumption, but in some low-income developing countries in the tropics the contribution of animal protein may be as low as 10%. In the tropics, cereals on average account for about 68% of total plant protein consumption, legume seed 18.5%, tubers, nuts, fruit and vegetables 13.5% (Rachie, 1977). Legumes eaten as green vegetables contribute to the last category.

However, according to the spectrum of crops grown and the importance of animals in the rural economy, there is wide variation between regions in the contribution of legumes to human energy and protein intake. The percentage contribution of legumes (plus nuts and oilseeds) to total energy intake in three developing tropical countries, Uganda, India and Nigeria, was estimated to be 15.9%, 16.1% and 7.3% respectively (Rachie, 1977); protein intake was 36.2%, 36.7% and 14.2% respectively.

The protein content of grain legumes varies from 17% to 40%. Average protein content and yield for the crops of this section are given in Table 4.3. Phosphate amelioration may have positive effects on protein content, but other external influences appear to be small (Bressani & Elias, 1980). The legumes are all high in iron, fairly high in phosphorus but deficient in calcium.

The major nutritional limitation of legume seeds is a deficiency of sulphur amino acids: methionine and cystine. The compensation is a relatively high lysine content. Within any one genotype the amino acid spectrum is little influenced by environmental factors. The protein of some legumes is relatively indigestible: for example, the *in vitro* protein digestibility of raw common bean ranges from 36% to 56%, though figures for soybean and cowpea are much higher (70–80%). Cooking generally increases digestibility, though not, for example, in pigeonpea. Low digestibility in cooked legume seeds appears to be related to tannin content (Bressani & Elias, 1980). The processing of legumes for human consumption is considered in detail, crop by crop, by Kay (1979).

One important nutritional advantage is the complementarity of legume seeds and cereals, which is largely an effect of balancing out methionine and lysine content. Thus in mixtures of different proportions of rice and beans maximum protein efficiency is reached when about 80% of the total protein comes from rice and 20% from beans (Bressani, Flores & Elias, 1973). As mentioned earlier, the average contribution of cereals to protein intake in the tropics is 3–4 times that of legume seeds. Hence in subsistence economies where the main energy source is cereals, legumes play a secondary nutritional role. While it might be expected that in subsistence economies based on root and tuber crops, with their generally lower protein content than cereals, crop legumes would contribute more to protein intake, detailed information from Nigeria does not support this (FAO, 1966).

9.4 Place in tropical cropping systems

The place of legumes in tropical cropping systems has been reviewed regionally by Okigbo (1977*a*) for Africa, Pinchinat (1977) for Latin America, Moomaw, Park & Shanmugasundaram (1977) for Asia, Okigbo (1977*b*) for the humid tropics, Dart & Krantz (1977) for the semi-arid tropics and Gomez & Zandstra (1977) for the tropics in general. Important reviews devoted to individual crops include Shanmugasundaram, Kuo & Nalampang (1980) on soybeans, Steele & Mehra (1980) on cowpeas, Gibbons (1980) on groundnuts and Francis *et al.* (1976), largely on *Phaseolus*. There are two main situations to consider: intercropping patterns and crop sequences. The intercrop situation is listed first since in developing countries it is the norm rather than the exception.

Gutierrez, Infante & Pinchinat (1975) report that in Colombia 90% of the *Phaseolus* is intercropped, largely with maize, and Pinchinat (1977), summarising data from six Latin American countries, recorded that in five out of the six the proportion of *Phaseolus* intercropped was over 50%. In West Africa perhaps 90% of cowpeas are interplanted with sorghum and pearl millet (Okigbo & Greenland, 1976; Steele & Mehra, 1980) and the same pattern is common for cowpeas in semi-arid India, though less so in higher-rainfall regions or under irrigation (Steele & Mehra, 1980). In Nigeria and Uganda 95% and 56% respectively of groundnuts are intercropped (Okigbo & Greenland, 1976). Pigeonpea in India is largely sown in combination with sorghum, pearl millet or cotton (Dart & Krantz, 1977), and in southeast Asia soybeans are commonly intercropped with maize (Shanmugasundaram *et al.*, 1980).

There are numerous reasons why farmers intercrop, but the important one in the context of this chapter is that low-growing or climbing legumes may be integrated into the architecture of tall erect cereals – maize, sorghum or pearl millet – without a major reduction in cereal yield and, in many circumstances, with an increase in total yield per hectare compared with sole cropping (Willey, 1979) and a significant increase in total protein yield per hectare. The main role of the tall-growing pigeonpea as an intercrop component in wet-and-dry climates is to utilise stored soil water after the rains have ended and the short-season cereals have been harvested (Dart & Krantz, 1977). Examples of the direct biological advantages of legume intercropping include the weed-smothering effect of cowpeas in millet, sorghum and maize in West Africa (Steele & Mehra, 1980) and a reduction in insect attack relative to pure stands in maize–bean intercrops in Latin America (Altieri *et al.*, 1978).

When legumes are grown as an intercrop with non-legumes of comparable crop duration, it is unlikely that the companion non-legume benefits directly from nitrogen fixed by the legume and subsequently made available through nitrogen excretion or through the decay of roots, nodules or fallen leaves (though Eaglesham *et al.* (1981), using ^{15}N, obtained results indicating that nitrogen excreted from cowpeas was being taken up by intercropped maize). However, such internal nitrogen cycling probably occurs when a short-season legume is grown with a long-season crop such as cassava, and would certainly occur when legumes are grown in the early phases of a perennial crop such as sugar-cane. Furthermore, there is likely to be in all legume/non-legume intercrops a benefit to the non-legume companion in that competition for available soil nitrogen is less than if both crops were non-legumes. Residual effects of legume/non-legume intercrops on subsequent non-legume crops have received little attention, but Searle *et al.* (1981) have shown that nitrogen uptake by wheat following maize–soybean and maize–groundnut intercrops not receiving nitrogen fertiliser was about twice as great as that following maize alone without nitrogen and equivalent to that following maize alone receiving 100 kg nitrogen ha^{-1}.

The mention of residual effects leads us naturally on to consider the general situation of crop legumes grown in sequence with non-legumes, specifically fixation of nitrogen by the legume and its effects on subsequent crops (Peoples & Crasswell, 1992; Peoples & Herridge, 1991). Estimates of total nitrogen and the proportion derived from fixation in the four crop legumes dealt with in this book are given in Table 3.7. However, it is also important to draw attention to the role of nitrogen-fixing trees and shrubs, predominantly legumes, in the nitrogen economy of tropical cropping systems. Many of these have minor, but increasing, roles as the perennial components of intercrops with annual species; some also contribute nitrogen through transference when leaves, litter, animal dung, etc., are taken from tree-lots and crop surrounds to the crop itself. Table 9.2 provides estimates of the nitrogen fixation of the more important trees and shrubs.

In assessing the effect or value of tropical crop legumes on the nitro-gen economy of a cropping sequence, there are factors in addition to nitrogen fixation to be taken into consideration. First is the removal of nitrogen from the field at harvest: by maturity the major proportion of total plant nitrogen will be in the seed, and residues are frequently not returned nor consumed *in situ* by stock but are removed and utilised or disposed of elsewhere. Pate & Minchin (1980) report nitrogen harvest index values of 0.61, 0.75, 0.80 and 0.73 for cowpea, soybean, ground-nut and chickpea respectively (seed nitrogen/total plant nitrogen). Bunting & Anderson (1960) found that even when only the nuts of groundnut crops were harvested 60–70% of total plant nitrogen was removed from the field, and that when the whole crop was taken

Table 9.2. *Nitrogen fixation by tree and shrubs*

Species	Nitrogen fixation (kg N ha^{-1} yr^{-1})
Acacia albida	20
Acacia mearnsii	200
Allocasuarina littoralis	220
Casuarina equisetifolia	60–110
Coriaria arborea	190
Erythrina poeppigiana	60
Gliricidia sepium	100 (3 months)
Inga jinicuil	35–40
Inga jinicuil	50
Inga jinicuil	35
Leucaena leucocephala	100–500
Leucaena leucocephala	100–130 (6 months)
Prosopis glandulosa	25–30
Prosopis glandulosa	40–50
Prosopis tamarugo	200
Aeschynomene afrospera	140 (2 months)
Calliandra colothyssus	11 (3 months)
Sesbania species	125–140
	119–188 (2 months)
	140–290 (2 months)
Cajanus cajan	110–220

Data from various sources: Peoples & Craswell (1992), Nutman (1976). Values are per growing season or per year unless the number of months is given in brackets.

(remembering that at harvest the root system is dug up) over 90% was removed (see also Section 3.1.4).

The second important factor is the role of available soil nitrogen. When a legume crop is grown, the operations of ploughing and cultivation encourage the mineralisation of soil nitrogen, and a proportion of nitrogen made available may, in well-drained soils, be leached out of the crop root zone or, in poorly drained soils, be denitrified. For example, at Katherine, north Australia, 17 years of continuous groundnuts on a well-drained Alfisol actually reduced total nitrogen in the root zone, though leached nitrate accumulated in the subsoil (Norman, 1979). The proportion of total crop nitrogen derived from soil nitrogen uptake varies with the available nitrogen status of the soil (Table 3.7).

Wetselaar (1967*b*) in north Australia compared the nitrogen balance of four legumes – the self-seeding annual forage Townsville stylo (*Stylosanthes humilis*), groundnuts, cowpeas and guar (*Cyamopsis tetragonoloba*) – each grown successively for 3 years (Table 9.3). Only with Townsville stylo, where the soil remained undisturbed for 3 years and leaching was of no significance, and with guar, which involved a major nitrogen return in crop residue (largely fallen leaf), was the original topsoil nitrogen status – organic plus mineral nitrogen – maintained. Three successive groundnut or cowpea crops, the bulk of their plant nitrogen removed from the field and with opportunities for leaching during the land preparation phase, resulted in a significant fall in total topsoil nitrogen.

The results of this experiment clearly indicate that it is hazardous to generalise on the contribution of annual tropical legumes to the nitrogen balance of cropping systems: it depends on the crop species, the disposal of the components of the crop, and on the transformations and

Table 9.3. *Approximate removal and return of nitrogen (kg ha⁻¹) by annual legume crops at Katherine, north Australia (averages of 3 years)*

	Removed	Returned
Townsville stylo (cut for hay)	200	Small amount[a]
Groundnuts (nuts only removed)	150	50
Cowpeas (cut for hay)	400	–
Guar (grain only removed)	50	200

[a]Aftermath allowed to die back on field.

movement of soil nitrogen as influenced by the conditions of cropping.

The foregoing discussion of legumes in crop sequences has been largely within the context of intensive upland cropping patterns. There are, in conclusion, two other situations to consider: semi-intensive upland cropping including a period under fallow, and wet rice cropping systems.

In semi-intensive patterns the amount of nitrogen available to crops declines during the cropping phase. Hence nitrogen-demanding non-legumes tend to predominate early in the crop sequence and nitrogen-independent legumes to be grown later. (There are confounding factors: e.g., declining available phosphate, increasing weeds, and whether crops are grown for cash or subsistence.) For example, from a survey of bush fallow systems in Zaïre, Miracle (1967) found that millet and sorghum were grown twice as frequently as the first crop than as the last crop, whereas legumes were grown 50% more frequently as the last crop than as the first.

In wet rice cropping systems in Asian wet-and-dry climates, there is often some water available after the main wet-season rice crop, from extended rainfall or from irrigation, though not enough for a second rice crop. In these circumstances legumes – groundnuts, soybeans or grams – may be grown as an early dry-season crop. Gomez & Zandstra (1977) discuss such sequences. The main difficulty is the unsuitability of the flat puddled ricefield for upland legumes; for good yields it is usually necessary to create raised beds or ridges.

Further Reading

Gigou, J., Ganry, F. & Pichot, J. (1985). Nitrogen balance in some tropical ecosystems. In *Nitrogen Management in Farming Systems in Humid and Subhumid Tropics*, ed. B.T. Kang & J. Vander Heide, pp. 247–68. Ibadan, Nigeria: Institute for Soil Fertility/International Institute for Tropical Agriculture.

Mulongoy, K., Gueye, M. & Spencer, D.S.C (1992). *Biological Nitrogen Fixation and Sustainability of Tropical Agriculture.* Chichester: Wiley.

Peoples, M.B. & Crasswell, E.T. (1992). Biological nitrogen fixation: investments, expectations and actual contributions to agriculture. *Plant and Soil*, **141**, 13–39.

Smartt, J. (1990). *Grain Legumes: Evolution and Genetic Resources.* Cambridge: Cambridge University Press, 379 pp.

10

Groundnut (Arachis hypogaea)

10.1 Taxonomy

The genus *Arachis*, the taxonomy of which is described by Gregory & Gregory (1976), Gregory, Krapovickas & Gregory (1980) and Smartt (1990), includes 37 named species and a number of undescribed species. The groundnut, *A. hypogaea*, is within the section *Arachis*, one of the seven into which the genus has been divided. The section *Arachis* comprises annual and perennial diploids (2n = 2x = 20) and two annual tetraploids (2n = 4x = 40), one of which is the cultivated *A. hypogaea*.

There have been numerous past attempts to classify genotypes within the species on the basis of agronomic characters (see Gregory *et al.*, 1973), but it has become recognised that branching and floral axis patterns are the primary discriminating characters. The grouping of Krapovickas (1973), which is generally accepted, is given below:

Subspecies *hypogaea*	No floral axis on main stem, alternating pairs of vegetative and floral axes on laterals
var. *hypogaea*	Less hairy, branches short (Virginia type)
var. *hirsuta*	More hairy, branches long (Peruvian Runner type)
Subspecies *fastigiata*	Floral axes on main stem, continuous runs of floral axes on laterals
var. *fastigiata*	Little branched (Valencia type)
var. *vulgaris*	More branched (Spanish type)

10.2 Origin, evolution and dispersal

The genus *Arachis* is South American in origin, and all species are located east of the Andes, south of the Amazon and north of the River Plate (Krapovickas, 1968; Gregory *et al.*, 1980).

A. hypogaea is cross-compatible with all diploid species within the section *Arachis*, forming infertile or partially fertile triploids. Gregory *et al.* (1980) believe that it originated as a wild allotetraploid of two species in the section *Arachis*, somewhere along the eastern front of the Andes. 'The present candidates for this parentage', to quote their tentative conclusion, are the quasi-annual *A. batizocoi* and the perennial *A. cardenasii.*

The earliest archaeological records of groundnuts in cultivation are from Peru, 4000–5000 BP (Hammons, 1973). Cultivated groundnuts were widely dispersed through South and Central America by the time Europeans reached the continent, probably by the Arawak Indians, and there is archaeological evidence in Mexico, 1300–2200 BP (Krapovickas, 1968).

It is therefore not surprising that, after European contact, more than one major genotype was dispersed around the world. The Peruvian runner type was taken to the western Pacific, China, southeast Asia and Madagascar (now Malagasy Republic). The Virginia type was probably introduced to Mexico (and thence across the Pacific via the Philippines) by the Spanish in the sixteenth century. It was then taken to Africa, and later India, via Brazil by the Portuguese. Virginia types apparently reached southeast USA with the slave trade. The Spanish type was introduced to the Old World by the Portuguese in the eighteenth century (Krapovickas, 1968). Gibbons, Bunting & Smartt (1972) noted substantial secondary diversity in Africa and Asia; the types they found and their location generally support the above conjectures regarding dispersal.

The groundnut is now cultivated throughout the world from 36° N to 36° S in a range of temperature regimes from warm temperate to equatorial. Countries producing more than 500 000 t nuts-in-shell are shown in Table 10.1. All major producers except USA are tropical countries; in the case of China, production is located within the tropics. Total production in China and Nigeria has doubled in the past decade owing to increases in both productivity per hectare and area, while production increases and productivity remain modest in other tropical countries. Total production in Sudan, about 850 m t in the 1980s, has fallen by 86% during the past decade owing to war and drought.

10.3 Crop development pattern

Both subspecies of groundnut feature indeterminate axillary branch and flower formation and the production of subterranean fruit. Subspecies *hypogaea* and *fastigiata* range in maturity from 120 to 160 and 90 to 105 days respectively. The timing of development of *hypogaea* is summarised in Table 10.2. There are relatively few comparisons of *hypogaea* with *fastigiata*: those in Nigeria and Argentina suggest that flowering begins at the same time relative to maturity in both subspecies (development index (DI) = 0.2–0.28, depending on cultivar and season) and reaches a distinct peak at DI = 0.6, or about two-thirds of the time between first flowering and harvest. After anthesis (syngamy) the ovary elongates and forms a peg or carpophore (Smith, 1950). The peg grows most rapidly 5–10 days after anthesis; there is then a period of rapid cell division in the embryo and the peg enters the soil at 8–16 days after anthesis (Smith, 1956). Elongation of the peg stops after it has penetrated to about 5 cm; the apical region then swells, bends through 90° and enlarges into a pod which differentiates into seed (nuts) and shell.

Indeterminacy and subterranean fruit production give rise to an extended period of seed formation and to low reproductive efficiency. The percentage of pegs that bear pods ranges from 20% (Smith, 1954) to 70% (Kowal & Kassam, 1978; Choudhari, Udaykumar & Sastry, 1985; Chapman *et al.*, 1993*c*) and the mature seeds may represent only 10–20% of the flowers that are produced (Smith, 1954; Donovan, 1963).

Table 10.1. *Major groundnut-producing countries: area, yield and production of nuts-in-shell in 1991 of countries producing more than 500 000 t per annum*

Country	Area (m ha)	Yield (t ha^{-1})	Production (m t)
India	8.26	0.85	7.00
China	3.05	1.98	6.06
USA	0.81	2.76	2.24
Nigeria	1.05	1.16	1.22
Indonesia	0.63	1.46	0.92
Senegal	0.80	0.87	0.70
Myanmar	0.56	0.91	0.51

After FAO (1992).

10.4 Crop/climate relations

The seedbed requirements of groundnuts are considered in Section 10.5.1. Speed of emergence increases with increasing temperature to 33 °C (Bolhuis & de Groot, 1959). Seedling dry weight increments are highest at 27–28 °C, at least in controlled environments (de Beer, 1963; Cox, 1979); measurable growth ceases at about 11–14 °C (e.g., Bagnall & King, 1991a). Gas exchange measurements from single leaves confirm that although potential photosynthetic rates are high (140 ng CO_2 cm^{-2} s^{-1}; Pallas & Samish, 1974), the plant has a C3 (Calvin cycle) photosynthetic pathway which exhibits lower rates of net photosynthesis at 30–35 °C than at 20–25 °C (Pallas, Samish & Willmer, 1974). Dry weight of tops increases curvilinearly with irradiance, the rate reaching a plateau at 23 Einstein (E) m^{-2} d^{-1}. A relationship between top dry weight (W, g per plant), days after emergence (D) and E m^{-2} d^{-1} (E) is given by Cox (1978):

$$W = 0.44 \ e^{c(D-8)}$$

where $c = 0.13(1 - 3^{-0.12E})$. Crop dry weight (including pods but excluding roots) increases linearly after canopy closure and interception of 95% of incident radiation, which occurs at a leaf area index (LAI) of 3 at approximately 50 days after emergence (Duncan *et al.*, 1978). Differences between genotypes and seasons in crop development, patterns of LAI and leaf area duration are described by Choudhari *et al.*, (1985). Williams, Dutta & Nambiar (1990) found that LAI (or frac-

Table 10.2. ⅃ *ime to various developmental events in var.* hypogaea *ground-nuts*[a]

Developmental events	Days
Imbibition to emergence	4(36 °C)–13(21 °C)
Emergence to first flower	27–37
Emergence to maximum number of flowers	65–110
Emergence to first pegging	40–60
Emergence to end of pegging	95–150
Emergence to maximum dry weight	90–150
Emergence to maximum leaf area	75–150
Emergence to maximum nitrogen yield	85–140
Emergence to maturity	100–160

[a] Data from Argentina, north Australia, Florida, Nigeria, Zimbabwe and controlled environments.

tional interception of radiation) explained 90% of variation in rhizobial nitrogen fixation.

Maximum growth rate is broadly coincident with maximum LAI, and the time at which it is reached is accelerated by temperature (e.g., Fig. 10.1, where temperature varies inversely with altitude). Peak rates of 280 kg ha^{-1} d^{-1} (Williams, Wilson & Bate, 1975) are consistent with the relatively high gas exchange but C3 characteristics of single leaves. Crop photosynthetic efficiency (E) is of the order of 1 g dry matter MJ^{-1} (Chapman *et al.*, 1993*a*).

Dry weight partitioning between tops and roots during vegetative growth has not received much attention. Data indicating that roots account for 9–14% of total dry weight at 28 days, declining to 2–5% at maturity (Yayock, 1979), seem so low as to question the method of root extraction. Roots penetrate to 30 cm within 12 days and reach 60 cm at 21 days after sowing (Hall *et al.*, 1953). Some (Inforzato & de Tella, 1960, cited by Sigafus, 1972; Robertson *et al.*, 1980) report shallow root profiles with 30% of roots in the upper 10 cm and little penetration

Fig. 10.1. Trends with time in growth of groundnut var. *hypogaea* cv. Makula Red at 900 m, mean temperature 23 °C (continuous line), and 1600 m, mean temperature 18 °C (broken line) in Zimbabwe. LAI, leaf area index; GR, whole-plant (excluding root) growth rate; DW, whole plant dry weight; K, kernel weight. Dates of formation of first pegs are shown by arrows on the abscissa. Source: Williams *et al.* (1975).

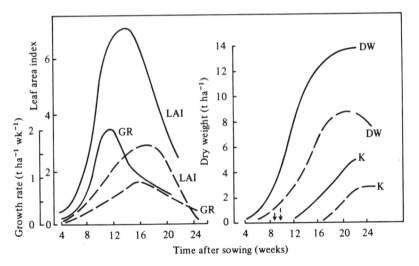

below 90 cm, but Prabowo, Prastowo & Wright (1990) found dryland crops extracted water from at least 1.2 m. Nutrient uptake presumably follows dry weight increment closely and reproductive development is associated with increased rates of nitrogen accumulation (to 3.8 kg nitrogen ha^{-1} d^{-1}; Williams, 1979).

Time to first flower appearance does not depend much on daylength (eg., Leong & Ong, 1983; Bagnall & King, 1991a). It occurs earliest at high temperature: Bunting & Elston (1980) cite times to maturity for cultivar Natal Common of 90, 105 and 120 days in Sudan, Tanzania and Transvaal, where mean temperatures during the growing season were 25, 22 and 19 °C respectively. High radiation also accelerates flowering (Bagnall & King, 1991b). In contrast, maximum flowering and highest peg and pod numbers occur under short days (Bagnall & King, 1991b) and at moderate temperatures (24–27 °C; e.g., Cox, 1979) although there are cultivar differences in flowering response to temperature (Bolhuis & de Groot, 1959). Pollen viability is also highest at moderate temperature (de Beer, 1963), as is whole-plant growth rate during flowering (Wood, 1968a). Final pod size is greatest at or below 24 °C (Cox, 1979; Williams *et al.*, 1975).

The primary branches contribute the majority of pods (90% in Choudhari *et al.*, 1985). Pod formation results in the channelling of between 40% and virtually 100% of current photosynthate into reproductive parts (Duncan *et al.*, 1978). In Florida, within a range of cultivars representative of those grown commercially since 1943, genotypic differences in pod growth rate and final yield were closely related to variation in the percentage of photosynthate partitioned to the pods (Duncan *et al.*, 1978). The corollary is that pod formation is particularly sensitive to variation in photosynthate supply: when supply is reduced by 35–65% by medium-to-high infestation of leaf spot (*Cercospora arachidicola* and *C. personatum*) the number of pods is affected more than individual pod weight (Boote *et al.*, 1980; Kowal & Kassam, 1978). Likewise, increasing plant population causes a reduction in all yield components per plant, but number of pods and seed weight per pod are reduced more than individual pod weight. Defoliation experiments (R. Santos, personal communication) also show that, once a pod has reached some critical size, it will continue to develop and reach maturity at the expense of photosynthate supply to younger pods, which abort under stress. This results in discrete populations or cohorts of pods reaching maturity and the abortion of complete pod age-classes. Pod growth rates in the field are about 0.02 g per pod d^{-1}; they range from 0.026 at 34/30 °C to a maximum of 0.047 at 26/22 °C in controlled environments at 680 μE m^{-2} s^{-1} (Cox, 1979).

The coincidence of a moderate (20–30 °C) temperature optimum for net photosynthesis, flower formation and pod growth, and the absence of daylength sensitivity for the onset of flowering, results in groundnut yields being highest outside the hot tropics: Williams *et al.* (1975) report kernel yields of 6, 5 and 3 t ha^{-1} at 1300, 900 and 1600 m elevation respectively at 18° S latitude, whereas 2–3 t ha^{-1} is a common maximum in lowland African savanna (Kowal & Kassam, 1978) and lowland India and Indonesia.

Water is the major constraint to yield in groundnut-growing areas (Virmani & Singh, 1986). In controlled environments soil water deficits reduce root length and pod numbers (Ike, 1986). By contrast, in sandy soil in the field, moderate water deficits do not necessarily affect root growth (Robertson *et al.*,1980) or yield (Wright, 1989); presumably the lack of sensitivity to drought in the field is due to exploitation of subsoil moisture. Water deficits may halve crop photosynthetic efficiency (E) e.g. from 1.12 to 0.63 g MJ^{-1} at 49–70 days after sowing (Chapman *et al.*, 1993*a*). Drought in the early reproductive stage reduces flowering and peg initiation; flowering and peg growth may begin again after re-watering (Fig. 10.2).

The crop appears to be adapted to high relative humidity (Fortanier, 1957; Lee, Ketring & Powell, 1972) and has relatively poor water use efficiency (1.8 kg dry weight per tonne of water used throughout the growing season; Pallas & Stansell, 1978). Consequently, yields often reflect the pattern of soil water availability, and there are numerous reported yield responses to irrigation. Unfortunately, these have rarely been related to soil water potential. In one exception, a 4-year experiment in Oklahoma, nut-in-shell yields under no irrigation and when soil water potential was maintained at –0.6 and –0.2 MPa were respectively 0.5, 0.8 and 0.9 of the yield at –0.1 MPa (Sturkie & Buchanan, 1973). Yield responses such as these are the cumulative effect of water deficit on flowering and peg formation (Fig. 10.2) and pod development. Drought during pod-filling causes abortion: e.g., up to 45% loss of yield through the death of the youngest pods (Chapman *et al.*, 1993*c*).

10.5 Crop/soil relations

10.5.1 *Soil physical properties*

Groundnuts grow best in well-drained soils of a loose to friable consistence. Texturally many important groundnut soils are sandy and occur within the orders Alfisols, Entisols, Inceptisols and Ultisols. In the USA, F.R. Cox & S.W. Buol (personal communication) estimate that most groundnut production is on Udults (64%), Ustalfs (32%) and Entisols (deep sands) (4%). The most prevalent groundnut soils in the

Fig. 10.2. Effect of water deficits on flowering (*a*), pegging and pod numbers (*b*); lower graphs show peg elongation in an irrigated crop (*c*) or under water deficit (*d*). Well-watered plants are coded CO in (*a*) and (*b*) and periods of water deficit S1 and S2 are shown as horizontal bars, where applicable. Source: Chapman *et al.* (1993).

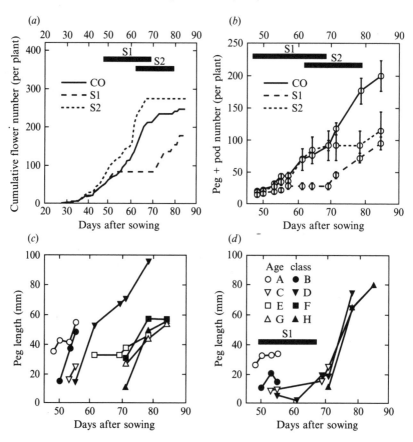

Sahel region of West Africa are coarse-textured, containing more than 65% sand; those in Niger exceed 80% sand (ICRISAT, 1986). However, heavy-textured soils, well-aggregated, friable and well-drained, such as Oxisols and oxidic Ultisols, are also quite suitable. The main groundnut soils in Queensland, Australia, are oxidic.

Light-textured soils allow ease of sowing and establishment, although Arndt (1965) showed that the relatively large stem diameter of the groundnut seedling ensured emergence from most surface crusts. However, the soil below the surface should be loose and porous as even small reductions in porosity can severely affect growth: a decrease in soil porosity in the cultivated layer from 44% to 38% reduced root yields in the 10–20 cm horizon from 1000 to 100 kg ha^{-1} (Nicou & Chopart, 1979). More importantly, loose and friable soils allow peg penetration and development and ease of lifting at harvest. Underwood, Taylor & Hoveland (1971) found that though pegs were capable of exerting great force, a surface crust of 1.5 cm could adversely affect peg penetration and pod development. Deep sowing practised by Indian farmers results in reduced yield, even though the intention is to utilise residual soil mosture (Nambiar & Rao, 1987); pod yields of two varieties averaged 4.1 t ha^{-1} when sown at 4–6 cm but yielded only 2.8 t ha^{-1} when sown at 8–10 cm (ICRISAT, 1985).

Heavy clay soils make harvesting difficult: yields are reduced through peg fracture and pods may be stained by adhering clay. Light-textured soils are also less liable to waterlogging, to which the groundnut seedling is sensitive after periods as short as 24 h (Hack, 1970). Waterlogging affects the growth of rhizobia of young plants when nitrogen is in high demand (Reid & Cox, 1973). Some groundnuts are grown successfully on heavier soils (e.g., in India, Thailand and Taiwan), though usually on raised beds.

10.5.2 *Soil chemical properties*

Groundnuts are particularly sensitive to low levels of available calcium even though they are tolerant of aluminium (Rogers, 1948; Adams & Pearson, 1970) and manganese (Morris & Pierre, 1949). Calcium deficiency is revealed by 'pops'; that is, pods containing aborted or shrivelled kernels. Low calcium levels are associated with increased vegetative growth that remains green later in the season, with greater, but infertile, flower production and with reduced kernel yield (Nicholaides & Cox, 1970; Wolt & Adams, 1979).

Calcium for normal pod and kernel development is very dependent on direct uptake by the pod from the soil (Wolt & Adams, 1979). This may be facilitated by pod hairs (Wright, 1989). Groundnuts growing in dry surface soil but with roots in moist subsoil (such as in post-harvest rice paddy fields) may therefore exhibit poor pod development and kernel abortion (Wright, 1989). This mode of calcium uptake and distribution explains why there may be no response to currently applied lime if it is not incorporated in the pegging zone (7–9 cm) (Hartzog & Adams, 1973) and why drought may induce calcium deficiency (Rajendrudu & Williams, 1987).

As runner types disperse their pods in the soil more than do bunch types they are less susceptible to the development of 'pops' resulting from calcium deficiency (ICRISAT, 1990). Of the bunch types, small-seeded Spanish types are less responsive to liming and have fewer 'pops' than large-seeded Virginia Bunch types (Hobman, 1985). This is consistent with Boote *et al.*'s (1982) observation that large-seeded varieties have a lower surface to volume ratio than small-seeded varieties. Other pod characteristics also influence calcium movement to the seed: thin, light hulls and long pod maturation allow calcium absorbed in the pod walls to move on into the seed (Kvien *et al.*, 1988).

Nye & Greenland (1960) observed that groundnuts were among the most consistent of all food crops in their response to lime in the savanna zones of Africa, but this was due to molybdenum deficiency rather than a requirement for lime (R.W. Pearson, 1975). In a glasshouse experiment Chong *et al.*, (1987) found that while maximum root growth of groundnut occurred at pH 7.3, shoot growth, nodulation and nitrogen fixation were best at pH 5.9–6.3. Nevertheless, the groundnut is considered tolerant of acid soil (see Fig. 11.4 and Table 3.9; Munns & Fox, 1977). In calcareous soils groundnut is very susceptible to chlorosis from lime-induced iron deficiency, e.g., in Barbados (Graham, 1986).

High potassium fertilisation can antagonise calcium uptake, leading to a higher requirement for this element (Chesney, 1975). This antagonism appears to have occurred on a Paleudult in Peru (Wade, 1978, cited by Ritchey, 1979), where shelling percentages above 60% were obtained only when calcium : potassium ratios exceeded 0.53. On an Alfisol in the rainforest belt of Nigeria, Kayode (1987) found that while potassium application reduced calcium in the leaf, it increased calcium levels in the seed and shell, as well as increasing seed yield.

The response to nitrogen fertiliser is often small and erratic, even on nitrogen-deficient soils. This is in spite of the fact that at very high yield levels the nitrogen requirement of nodulated groundnuts cannot be wholly met from symbiotic nitrogen (Williams, 1979). In Ghana, Ofori (1975) reported no response but earlier (1973) had reported responses to up to 15 kg nitrogen ha^{-1}, which were higher when there had previously been a grass fallow. Acuna & Sanchez (1969) found no response to nitrogen in Venezuela, but in Brazil de Tella, Canecchio & Da Rocha (1970) obtained yield responses in three out of five experiments. Responses are influenced by cultivar (Walker, Morris & Carter, 1974), the effectiveness of rhizobia and the sensitivity of rhizobia to soil nitrogen: sequentially branched cultivars may be more responsive to soil nitrogen than alternate-branching types (Cox, Adams & Tucker, 1982), while high levels of nitrogen absorbed through the pod wall suppressed translocation of nitrogen from root uptake and fixation (Inanaga *et al.*, 1990).

Groundnuts in the tropics are seldom inoculated because they nodulate effectively with naturally occurring cowpea cross-inoculation group *Bradyrhizobium* sp. (Gaur, Sen & Subba Rao, 1974). The median efficiency of 168 rhizobium isolates from South African soils was 70% of a commercial strain: the range was 38–96% (Staphorst, Strifdom & Otto, 1975). While groundnut may fix as much as 190 kg nitrogen ha^{-1} (Nambiar, Rego & Rao, 1986), inoculation with rhizobium does not always lead to a yield increase (Subba Rao, 1976; Hadad *et al.*, 1986). Part of the unpredictability in response may be attributed to the variable efficiency with which a particular strain fixes nitrogen in different groundnut genotypes (Graham & Donawa, 1982; Joshi, Kulkarni & Bhatt, 1990). Comparisons of nitrogen fixation by groundnut relative to other tropical grain legumes are given by Peoples & Herridge (1991), Peoples & Crasswell (1992) and in Table 3.7.

Sandy soils are liable to be low in phosphorus, but as phosphate fixation and crop removal on such soils are generally low, only low rates of phosphorus application are required. Because of higher rates of fixation, Oxisols and oxidic Ultisols require higher rates of phosphorus application. Mycorrhizal fungi can improve the uptake of phosphorus and other elements, e.g., zinc, even at high levels of applied phosphorus (Bell, Middleton & Thompson, 1989). Mycorrhizal colonisation of the roots varies, being 20–50% for Spanish, 17–31% for Valencia, 22–50% for Virginia bunch, and 21–36% for runner types in India (ICRISAT, 1986).

Groundnut is susceptible to excess exchangeable sodium found in sodic soils. ESP values above 15 delay germination and flower emergence, while yield was reduced to 50% of controls at an ESP of 20 (Singh & Abrol, 1985).

10.6 Place in cropping systems

Detailed information on the place of groundnuts in tropical farming systems may be found scattered in the regional literature: for West Africa, FAO (1974), Okigbo & Greenland (1976), Rachie & Silvestre (1977), Kowal & Kassam (1978), Haen & Runge-Metzger (1989) and others; for India, ICAR (1972); for south and southeast Asia, Harwood & Price (1976) and McIntosh *et al.* (1977). A more general account is given by Gibbons (1980).

Within the tropics the groundnut, as a species with a range of maturity types and a moderately high temperature requirement for rapid growth, finds its main place as a rainfed summer crop in wet-and-dry climates, particularly in West Africa and India. Thus in a transect across West Africa, Cochemé & Franquin (1967) established four main cropping zones on the basis of rainfall and length of growing season: moving from low to high rainfall, groundnuts of early maturity first appear as a main crop in zone B, with a mean annual rainfall of 550 mm and mean growing season of 100 days, and are grown on suitable soils through to zone D, with a mean annual rainfall of 1000 mm and a growing season of 145 days.

However, groundnuts develop satisfactorily, though more slowly, under conditions somewhat cooler than the lowland wet-and-dry climate summer, and are daylength-insensitive. Hence they are also grown in areas of the wet-and-dry tropics in the dry season under irrigation if temperature permits. In continental north India, however, the dry seasons are too cold for groundnuts. In the southern Guinea and derived savanna zones of West Africa, with extended rainfall, two rainfed groundnut crops are often grown sequentially in one year (Kowal & Kassam, 1978).

In general, the more rapidly maturing types are grown in the tropics, for crushing or for local consumption. However, within quite restricted geographical areas different ecotypes of groundnuts are cultivated under different conditions and for different markets. In Sudan, for example, late-maturing types are grown under irrigation for export, whereas small farmers grow, under rainfed conditions, quick-maturing types for local sale or subsistence. In Malawi, late-maturing confectionery cultivars for export are grown on the plateaux, cultivars for crushing are grown on the lakeshore, and short-season cultivars for local consumption are grown in the drier, hotter valleys (Gibbons, 1979).

With their adaptation to well-drained soils, groundnuts are normally grown within the rainfed cropping systems of regions dominated by sands or loams and are less frequently encountered on Vertisols. The crop is often grown on beds or ridges, particularly where drainage is poor. A special situation is the paddy field of south and southeast Asia, where groundnuts may be a wet-season substitute for rice or, more commonly, a dry-season crop following rice, often under irrigation. In these cases, the crop is grown on raised beds. One exception is when groundnuts are grown as a quick catch crop after summer rice on a declining water regime, as in East Java, where to save time (and to save money if the crop fails) the crop is dibbled directly into rice stubble.

As a legume, the groundnut occupies an important position in cropping patterns based on cereals or root crops: e.g., with pearl millet and sorghum in West Africa and India, or cassava and rice (upland or lowland) in southeast Asia (Braun, 1974; ICAR, 1972; Okigbo & Greenland, 1976; McIntosh *et al.*, 1977; Kowal & Kassam, 1978). However, as noted in Section 9.4, it cannot be assumed that groundnuts are significantly restorative of topsoil nitrogen.

On the other hand, groundnuts may have substantial beneficial effects, relative to non-legume crops, on immediately subsequent non-legume crops, which may be ascribed to soil nitrogen differences. For example, at Katherine, north Australia, sorghum grain and stover yields were respectively 77% and 56% higher after groundnuts than after sorghum, and grain nitrogen yield was almost double (Phillips & Norman, 1961). Total nitrogen yield of pearl millet was about 50% higher after groundnuts than after sorghum (Phillips & Norman, 1962). In north Nigeria, the average yield over 6 years of sorghum immediately following sorghum, cotton or groundnuts was respectively 2037, 2553 and 2861 kg ha^{-1}, and the average yield of maize after 7 years continuous sorghum, cotton or groundnuts was respectively 2503, 3568 and 4478 kg ha^{-1} (Lombin, 1981).

Groundnuts are cultivated both as a sole crop and as a component of intercrop mixtures (see Reddy & Willey, 1980, for a detailed account of the competitive relations of groundnuts as an intercrop). In the northern Guinea savanna zone of West Africa about 70% of the crop is grown in mixtures with two to four crops, including pearl millet, sorghum and cowpea (Norman, 1972, quoted by Kowal & Kassam, 1978), though in Senegal a higher proportion of sole crop is found. Okigbo & Greenland (1976) report the proportion of groundnuts in mixed crops in Nigeria and Uganda as 96% and 78% respectively of the total area sown to the crop in each country. Whereas in short-season wet-and-dry climates

Table 10.3. *Land equivalent ratios of contrasting groundnut intercrop systems, according to soil type, weed control, and plant protection, from 4 years' experimentation in India*

System	Soil type[a]		Weed control[b]		Plant protection[c]	
	Black	Red	Complete	Limited	Protected	No protection
Groundnut/pigeonpea						
Groundnut	0.59	0.60	0.46	0.49	0.57	0.73
Pigeonpea	1.05	0.84	1.07	1.20	0.90	0.77
Total	1.64	1.44	1.53	1.69	1.47	1.50
Groundnut/pearl millet						
Groundnut	0.61	0.72	0.60	0.82	0.61	0.68
Pearl millet	0.51	0.52	0.58	0.46	0.55	0.49
Total	1.12	1.24	1.18	1.28	1.16	1.17

From Rao & Singh (1990).
[a]Soil type: four sites on black soil and five sites on red soil, over 4 years.
[b]Weed control: two sites representing the two soil types for 4 years for each weeding system.
[c]Plant protection: five sites across soil types that received some protection in comparison with two sites with no protection over 4 years. Weed control was similar for both situations.

groundnuts are sown in mixtures with other short-season crops, as indicated above, in more extended rainfall regimes they are sown with both short- and long-season crops. For example, in the South Cameroon, groundnuts are a component in intercrop mixtures with short-season maize and long-season tubers and plantains (Mutsaers, 1978). In India, groundnuts may be intercropped with pigeonpea and with cotton (Gibbons, 1980).

Land equivalent ratios for contrasting groundnut intercropping systems are shown in Table 10.3. The presence of groundnuts appears to confer only a small yield advantage over the yield of sole crops when the duration of growth of the crops is similar, e.g., of groundnut and pearl millet, while intercropping gives a large yield increase when growth duration differs by up to 3 months, as in the groundnut/pigeonpea intercrop. As might be expected, land equivalent ratios were somewhat higher under conditions of poor weed control.

Further reading

Gibbons, R.W. (1980). Adaptation and utilization of groundnuts in different environments and farming systems. In *Advances in Legume Science,* ed. R.J. Summerfield & A.H. Bunting, pp. 483–93. Kew: Royal Botanic Gardens.

Pattee, H.E. & Young, C.T. (eds.) (1982) *Peanut Science and Technology.* Yoakum, Texas: American Peanut Research and Education Society.

Virmani, S.M. & Singh, P. (1986). Agroclimatological characteristics of the groundnut-growing regions in the semi-arid tropics. In *Agrometeorology of Groundnut*, pp. 3–46. Patancheru, India: International Crops Research Institute for the Semi-Arid Tropics.

11

Common bean (Phaseolus vulgaris)

11.1 Taxonomy

The taxonomy of *Phaseolus* has been reviewed by Debouck (1991). The genus, of New World origin, is in the tribe Phaseoleae and is tentatively regarded as including 55 species, of which five are cultivated:

P. *vulgaris* L., common bean

P. *coccineus* L., scarlet runner bean

P. *lunatus* L., Lima bean

P. *acutifolius* A. Gray, Tepary bean

P. *polyanthus* Greenman, year-bean

All are diploid (2n = 2x = 22). There are some genetic affinities between the species: crossing experiments suggest that P. *vulgaris*, P. *coccineus* and P. *polyanthus* are closely related, but there is little compatibility between P. *lunatus* and other species (Debouck, 1991).

Classification of types with P. *vulgaris* is based on determinacy of the main axis, growth habit, crop duration and seed characteristics (Voysest & Dessert, 1991). Plants may be erect ('bush' beans), semi-climbing or climbing; crop duration may vary from 75 days for an early bush type to 270 days for a late climbing type in a cool climate.

The seeds of the common bean are highly variable and consumer preferences are very refined, particularly in Latin America (Voysest & Dessert, 1991). The CIAT classification is based primarily on colour (nine groups), secondarily on size (three groups), and also on shape (round, elongated or kidney). Voysest & Dessert list 66 individually named market classes for seed, distinguished by colour, size and shape.

11.2 Origin, evolution and dispersal

The origin, evolution and dispersal of the common bean have been reviewed by Gepts & Debouck (1991). The archaeological record indicates a very early domestication. Remains found in Peru (Kaplan, Lynch & Smith, 1973) are dated 8000–10 000 BP, in Mexico 6000–7000 BP (Kaplan, 1965) and in Argentina 6000–9000 BP. All the material appears to be of fully domesticated plants.

Wild common beans are to be found throughout Latin America. The Andean populations are *P. vulgaris* var. *aborigineus* (Berglund-Brücher & Brücher, 1976) and those in Central America *P. vulgaris* var. *mexicanus* (Delgado Salinas, Bonet & Gepts, 1988). They grow from 33° S to 13° N latitude and at altitudes of 500–1800 m. Hybridisation trials between wild and cultivated beans have yielded fertile F_1 and F_2 generations.

Gepts *et al.* (1986), on the evidence of phaseolin-protein variation, suggest two main centres of domestication: in Central America and in the southern Andes. The former led to small-seeded, the latter to large-seeded types. There may have been a third centre in the northern Andes producing small-seeded types. Evans (1976) and Gepts & Debouck (1991) have reviewed dispersal: in general, both Central America and Andean types spread to the same regions of the world, but the former became predominant in lowland South America and the latter in Africa. Dissemination routes are uncertain, however: the spread of common beans to Africa may have been direct or via the Iberian peninsula. They were probably taken to Asia via the Philippines.

Brazil and Mexico are the largest producers of common bean (Table 11.1). While the statistics in Table 11.1 are only approximate as they may also include cowpea, it appears that productivity per hectare of common bean is low relative to other legumes or to temperate areas such as Argentina.

Table 11.1. *Area, yield and production of dry beans: all species of Phaseolus[a] in 1991*

Country	Area (m ha)	Yield (t ha^{-1})	Production (m t)
Brazil	5.51	0.50	2.75
Mexico	2.04	0.71	1.45
Uganda	0.50	0.80	0.40
Tanzania	0.44	0.61	0.27
Rwanda	0.26	0.88	0.23
Argentina	0.19	1.16	0.22

[a]FAO statistics for dry beans show production in 1991 was highest for India (4.05 m t), but this includes mainly *Vigna* spp. Similarly, African production (Uganda, Tanzania, Rwanda) would include species other than *Phaseolus*.

11.3 Crop development pattern

In developmental terms there are two types of common bean: determinate, in which the main axis terminates in an inflorescence and produces no vegetative nodes after flowering, and indeterminate. CIAT (1980*a*) has classified the world bean collection into four main growth habit types on the basis of determinacy, node production after flowering and growth habit (height and climbing tendency) (Table 11.2). Debouck (1991) gives a detailed analysis of these various morphologies and of the development of leaves and flowers in common bean.

The determinate type (Table 11.2) is short, self-supporting or bushy and of short growth duration. Debouck (1991) recognises at least two types of determinate habit (in contrast to CIAT, Table 11.2): a 'few-noded type' having three to seven trifoliate leaves on the main stem below a terminal double raceme, which is known as a 'bush' or 'dwarf' type and is found in the North Andes, and a 'many-noded' type with 15–25 leaves on the main stem, from the south Andes. Indeterminate

Fig. 11.1. Growth attributes of common bean, Type II, cv. Porillo Sintetico, at Palmira, Colombia (latitude 3° N, mean temperature during the growing season 24 °C). LAI, leaf area index; DW, whole plant (excluding root) dry weight. Source: Laing *et al.* (1984).

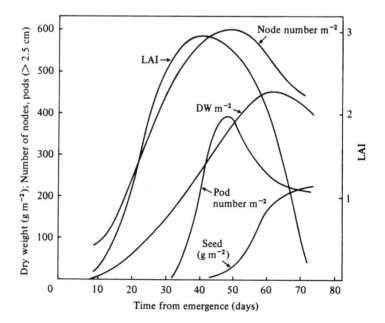

Table 11.2. *Morphological and phenological characters for the four growth habit types of* P. vulgaris. *Data show means, with standard deviations in parentheses, from evaluation of* 9.5×10^3 *germplasm collections at CIAT-Palmira (3° N, 1000 m altitude, mean temperature 23.8 °C), at a density of 20–25 plants per* m^2.

Attributes	I Determinate bush	II Indeterminate bush	III Indeterminate semi-climbing	IV Indeterminate climbing
Total collections evaluated (%)	23.7	23.2	33.3	19.8
Plant height (cm)	44 (18)	92 (44)	103 (28)	160 (50)
Nodes[a] at flowering	8.3 (1.5)	13.3 (1.9)	14.3 (2.1)	14.8 (4.0)
Nodes[a] at maturity	8.6 (1.4)	17.0 (2.0)	18.9 (2.6)	22.7 (4.0)
Seed weight (mg per seed)	336 (100)	236 (80)	257 (89)	294 (107)
Days to flowering	34 (3)	39 (4)	38 (5)	40 (6)
Duration of flowering (days)	22 (6)	25 (5)	28 (6)	29 (7)
Leaf area duration (days)[b]	114	130	164	261
Yield (t ha^{-1})[b]	2.4	2.7	3.2	3.7
Yield (kg) per leaf area duration[b]	2.1	2.1	2.0	1.4

After Laing *et al.* (1984).

[a] Number of mainstem nodes.

[b] From 'representative' cultivars.

genotypes show a wide range of node number on the main stem, climbing tendency and growth duration. As Laing, Jones & Davis (1984) point out, an indeterminate genotype may change in growth habit category with a change in temperature/daylength combination. All types bear flowers on short, lax, axillary racemes; the flowers are self-fertilised and develop slender pods which usually carry four to six seeds.

The development pattern of bush types is predictable (Fig. 11.1). At least half of the determinate bush types (type I) are not sensitive to daylength (CIAT, 1978*a*). In these, growth duration depends primarily on temperature and we may expect the relativity between various developmental events to remain fairly constant under a wide range of thermal environments. Flowering begins at about the time of maximum leaf area index (LAI) at development index (DI) = 0.5, pod number reaches a peak at about DI = 0.61–0.7 and above-ground dry weight at DI $\geq$ 0.7. Since bush types are of short growth duration and have only four to ten nodes on the main stem at maturity, their yield is usually lower than that of other growth habit types (Table 11.2). Tayo (1986) reports a detailed study of flower and pod development and the relationship between pod and vegetative growth, albeit in a temperate climate. It emphasises the importance of the terminal and upper three inflorescences to yield in determinate beans, and that most (90%) of the pods which reached maturity were formed from early flowers.

Indeterminate climbing or 'pole' beans are either insensitive to daylength or sensitive to long days. Daylength-sensitive beans do not have a juvenile phase (Zehni, Saad & Morgan, 1970). There is, however, interaction between daylength and temperature in promotion of flowering: increasing temperature accelerates flowering under short, but not necessarily under long, daylengths (Wallace & Enriquez, 1980). There are not enough data to indicate whether promotion of flowering by both temperature and daylength results from multiplicative (Laing *et al.*, 1984) or additive (Clarkson & Russell, 1979) temperature–daylength effects. Such interactions are probably of little importance in the lowland tropics, nor for intermediate bush genotypes (type II, Fig. 11.2). However, they are important in intermontane valleys, where shadows influence effective photoperiod (Bryant & Humphries, 1976) and for the morphologically more variable climbing types. In the climbing types morphological variation may occur independently of daylength effects on flowering because the climbing tendency is under phytochrome control (Kretchmer *et al.*, 1977). Indeterminate types have relatively high yield potential, though seed yield per day or per unit leaf area duration may be similar to or lower than that of determinate bush types (Table 11.2). In contrast to the development of yield in determinate types (Tayo, 1986), most pods and yield in indeterminate types are located at low or intermediate positions on the mainstem and branches (CIAT, 1978, cited by Laing *et al.*, 1984).

Full maturity, that is, to dry seed, is reached from 45 to 150 days after emergence, depending on growth habit type and location. For example, types I and II take about 45–60 days to maturity in lowland Hawaii, 70–75 days in lowland Colombia and 90 days in East Africa (Kay, 1979). Alternatively, green pods may be harvested as a vegetable at about DI = 0.65.

11.4 Crop/climate relations

As Meiners & Elden (1980) put it: 'Of all edible legumes, *Phaseolus* spp., and particularly *Phaseolus vulgaris*, have the widest geographical distribution. They are grown whenever temperatures between 10 and 35 °C prevail and insects and diseases are always present to a greater or lesser

Fig. 11.2. Effect of mean temperature during the growing season at five locations in Colombia on seed yield of five genotypes of common bean. Locations A to E were Pasto, Popayan, Palmira, Patia and Sante Fé respectively, ranging from 350 to 2700 m altitude. Source: Laing *et al.* (1984).

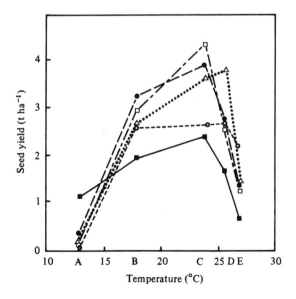

extent'. Diseases, particularly bean common mosaic virus, *Anthracnose*, blights and leaf spots, are responsible for part of the 30% lower yield found in the tropics compared with temperate locations (CIAT, 1978*a*) and, of course, contribute to yield variation within the tropics.

There is variability between bean genotypes in thermal tolerance. However, P.G. Jones's analysis (CIAT, 1979*a*) of South American bean-growing areas showed that nearly all are found where temperatures at flowering lie between 17.5 and 25 °C; 55% of the areas have tempera-tures at flowering between 20 and 22.5 °C. According to Laing *et al.* (1984), where seasonal temperatures vary substantially planting time is usually adjusted to make the crop flower when temperatures are within 21 ± 2 °C. The 21 °C optimum corresponds to 1250 m altitude in the tropics. The constraint on bean yields associated with narrow tempera-ture tolerance is shown in Fig. 11.2. Yield reductions below or above the optimum are related to plant mortality (which is very substantial at high temperatures; Ishag & Agoub, 1974), reduced photosynthesis (e.g., Cookston *et al.*, 1974) and failure of flowers to produce mature pods. Abscission of flowers and failure to set and fill pods are probably the greatest constraints to bean yields. Failure rates are commonly 50–70% of opened flowers and the proportion increases above 30/25 °C day/night (Kay, 1979).

The second climatic constraint to yield is water. Bean-growing areas have an annual rainfall of 500–1500 mm. Production in Colombia and East Africa is most successful where rainfall during the growing period is 300–400 mm and seed maturation occurs in dry weather (Kay, 1979). Water stress is common in major bean-producing areas (Laing *et al.*, 1984). However, the crop is not particularly tolerant: stomates close at moderate leaf deficit (–0.5 MPa) (Walton, Galston & Harris, 1977). Stress during flowering, when the crop is most sensitive, reduces yield through increased flower failure and to a lesser extent by reducing the number of seeds per pod (Stoker, 1974). At the other extreme, heavy rain or irrigation produces a humid crop micro-climate conducive to diseases; e.g., white mould (*Sclerotinia sclerotiorum*; Weiss *et al.*, 1980).

Water use follows leaf area patterns (e.g., Fig. 11.1). For a 75-day crop in Brazil, total evaporation was 220 mm; crop evaporation was 3.2, 3.2 and 1.7 mm d^{-1}, or 0.62, 0.77 and 0.38 of E_p, from germination to flowering, flowering to pod development and pod development to matu-rity respectively (de Silveira & Stone, 1979).

Maximum leaf photosynthetic rates are 0.6–1.1 mg m^{-2} s^{-1} (Davis & McCree, 1978; Catsky & Ticha, 1980; Laing *et al.*, 1984). Canopy rates of 0.8–1.1 mg m^{-2} s^{-1} were recorded at LAIs from 4.5 to 7.3 in a high-radiation temperate environment (35° S) (Sale, 1975). Calculated canopy net assimilation rates ranged from 30 to 60 mg m^{-2} d^{-1} at Campinas, Brazil (23° S) (Magalhaes, Montojos & Miyasaka, 1971; Montojos & Magalhaes, 1971). Leaf area development is sensitive also to phosphorus supply, largely through nutritional effects on branching (and hence, node numbers) and rate of leaf appearance; mycorrhizae, in making phosphorus available to the host plant at low soil concentrations, substitute for phosphorus fertiliser (Lynch, Lauchli & Epstein, 1991). Peak LAIs are usually 2–4 (Fig. 11.1) but under high fertility, as with split application of nitrogen, they may rise higher owing to increased leaf area duration and higher crop growth rates (Montojos & Magalhaes, 1971). Since an LAI about 4 intercepts essentially all radiation (Aguilar, Fisher & Kohashi, 1977), it is not surprising that additional leaf area does not necessarily increase grain yield (Magalhaes *et al.*, 1971).

Partitioning of dry matter during vegetative growth has received little attention; a single study of roots suggests that 40% of root dry weight is in the tap root and its branches, 50% in the basal root system and only 10% in adventitious roots (Stofella *et al.*, 1979). Rhizobial activity has been studied more fully; nitrogenase activity is highest at the beginning of seed development (DI = 0.7 in the cultivar studied) and may contribute up to 90 kg nitrogen ha^{-1} during the life of the crop (Westermann *et al.*, 1981).

The course of reproductive growth is shown in Fig. 11.1. Seed yield is closely correlated with number of pods per plant and number of plants surviving to maturity (Adams, 1973); likewise, the yield profile of a single plant reflects the number of pods at each node (Laing *et al.*, 1984). Again, emphasis on pod number relates to large flower losses (Fig. 11.1), which in a single plant are highest at low, shaded nodes (cf. Section 12.4).

Varying the plant population or intercropping may affect yield, depending on cultivar mortality and sensitivity in pod formation. Indeterminate cultivars are sensitive to intercropping because competition increases mortality and reduces the number of pods per plant in both climbing and non-climbing genotypes (Francis, Prager & Tejado, 1982). Bush beans are less affected by intercropping because of earliness, which allows them to escape severe competition at flowering, but their yields are still below those of climbing beans (Leakey, 1972; Willey & Osiru, 1972; Westermann & Crothers, 1977; Francis, Prager & Laing, 1978*b*).

The seed develops in a bi-sigmoidal pattern with an intermediate period of 3 days when growth is slow (Carr & Skene, 1961; Hsu, 1979). Low temperatures produce large seed (Siddique & Goodwin, 1980), presumably by prolonging the actual grain-filling period (see Section 8.4). Seed yield is not necessarily related to photosynthetic rate (Peet *et al.*, 1977). However, yield depends on photosynthate production during the reproductive period (Wein, Standsted & Wallace, 1973) and there is a general correlation between yield and post-anthesis leaf area duration (e.g., Montojos & Magalhaes, 1971).

The elaboration of seed yield and its sensitivity to water and temperature stress is summarised, at least at a gross level, by the stress-degree-day (SDD) concept applied in the field in California (38° N). A SDD is defined as the elevation of canopy temperature above air temperature (deg C) when measured about 1 h after solar noon. Yield was related to SDD (Walker & Hatfield, 1979) by:

$$Y = 297 - 0.04 \left(\sum_{E}^{F} SDD \right) - 0.71 \left(\sum_{F}^{M} SDD \right), \ (r^2 = 0.83);$$

where E, F and M are emergence, flowering and maturity. Note that the coefficient is very low before flowering but appreciable during reproductive development because of the sensitivity of flower failure and photosynthate production to climatic stress. In the absence of stress, seed yields of 4–5 t ha^{-1} can be attained. Such yields are reported from the Constanza valley in the Dominican Republic (Kay, 1979), which probably has Papadakis's 1.3 type climate of marine savanna with cool dry winters.

11.5 Crop/soil relations

11.5.1 *Soil physical properties*

Common bean can be grown on soils varying from light sands to heavy clays (Wallace, 1980), including peats (Purseglove, 1968), but preferably the soils should be friable, and where rains are erratic, also deep (Kay, 1979). If the soil is liable to crusting, there may be problems with seedling emergence (Wallace, 1980). Field (Tu & Tan, 1991) and pot studies (Schumacher & Smucker, 1981) show that soil compaction affects plant growth; at the same degree of compaction plants yielded more in clay loam than in sandy loam (Tu & Tan, 1991).

The fact that common bean can be damaged while emerging from wet soil (Orphanos, 1977), and does not grow well under waterlogged conditions (Williamson & Kriz, 1970; Table 8.3), contributes to its low yields on poorly drained soils in the wet tropics (de Londono, 1977; CIAT, 1978*a*). Even transient submergence, after high-intensity rain, adversely affects established plants (Forsythe, Victor & Gomez, 1979). Pot studies reveal that waterlogging effects operate through reduced soil air content (Dasberg & Bakker, 1970) and reduced nodule development and nitrogen fixation efficiency (Sprent, 1976).

11.5.2 *Soil chemical properties*

Common bean is nodulated by the nitrogen-fixing *Rhizobium leguminosarum* biovar *phaseoli*. Acetylene reduction studies show that generally the determinate early type I cultivars fix less nitrogen symbiotically than do the strongly indeterminate type III and IV cultivars (Graham & Rosas, 1977), but the common bean is generally considered a poor nitrogen-fixing plant (Graham & Halliday, 1977) and less effective than soybean (Piha & Munns, 1987; Isoi & Yoshida, 1991). Comparisons of nitrogen fixation in common bean with other tropical legumes are reviewed by Peoples & Herridge (1991) and Peoples & Crasswell (1992) (see Table 3.7).

Inoculation with elite rhizobia has increased yield by up to 145% (Habish & Ishag, 1974). However, few bean crops are inoculated in the tropics (e.g., 1% in Brazil; Graham, 1978) and the response to inoculation is unreliable (Graham, 1981; Redden, Diatloff & Usher, 1990) and sometimes related to the capacity of the cultivar to nodulate in the presence of high soil nitrate levels (Park & Buttery, 1989).

Nitrogen fertiliser rates required to overcome the poor contribution from rhizobia and to maximise yields can be high: from 80 kg nitrogen ha^{-1} (Lugo-Lopez, Badillo-Feliciano & Calduch, 1977) to 600 kg nitrogen ha^{-1} (Bazan, 1975). Applied nitrogen usually suppresses nodulation and nitrogen fixation (e.g., Rigaud, 1976) and yield responses can be variable (Franco, 1977*a*).

Phosphorus is probably the factor most commonly limiting nitrogen fixation in common bean in developing countries (Graham, 1981); around 50% of beans grown in Latin America are in low phosphorus soils (CIAT, 1987). It is vital to the development and function of the nodules (Graham & Rosas, 1979; Pereira & Bliss, 1989). Genotypes vary in their response to phosphorus (Schettini, Gabelman & Gerloff, 1987; Fagaria, 1989); black-seeded varieties are more tolerant of deficiency

than those of other colours (CIAT, 1979*a*). Plants that are responsive
and efficient are not necessarily tolerant of the toxic levels of aluminium
and manganese often associated with low levels of available phosphorus
in tropical soils (CIAT, 1980*a*). However, varieties efficient in using
phosphorus appear to be efficient also in using soil nitrogen
(CIAT, 1980*a*). While mycorrhizal fungi may enhance phosphorus
uptake (Pacovsky *et al.*, 1991), root phosphatase activity may also be
important where a high proportion of labile phosphorus is organic
(Helal, 1990).

Responses to phosphorus in common bean on acid tropical soils are
not uncommon: e.g., in Kenya (Ssali & Keya, 1986) and Brazil
(Lathwell, 1979*a*). Besides a response to phosphorus, Awan (1964) in
Honduras also obtained a response to lime which was partly explained
by increased mineralisation of soil organic phosphorus.

Numerous responses to potassium in common bean have been
obtained in East Africa (Anderson, 1974*a*), but responses in tropical
America are infrequent (e.g. Cox, 1973*b*; Bazan, 1975). Even negative
responses have been obtained (e.g., Anderson, 1974*b*; da Eira *et al.*,
1974) and cultivars may vary in their potassium requirement (Sale &
Campbell, 1987).

Common bean is normally grown in the tropics on Oxisols, Ultisols,
Entisols and Inceptisols – all acid soils (Thung, 1991). Soil acidity
affects all stages of the legume–rhizobium symbiosis, from strain survival in soil and on the seed, to root-hair infection, nodule initiation
and nitrogen fixation (Graham *et al.*, 1982; Vargas & Graham, 1988).
Soil pH should preferably be between 6.0 and 6.8 but critical problems
seldom develop unless it falls below 5.2 or rises above 7.0 (Kay, 1979).
This is consistent with the data of Abruña, Pearson & Perez-Escolar
(1975), who found that snap beans in Puerto Rico were highly sensitive
to aluminium toxicity (Fig. 11.3) but gave maximum yield when pH was
raised by liming to about 5.3. Similarly, when Buerkert *et al.*, (1990)
limed Ustults, Tropepts and Ustolls in Mexico with an initial soil pH
range of 4.6–5.0, pH increased by 0.4 to 1.3 pH units and aluminium
saturation decreased by 13% to 38%; the result was yield increases of
76–313%.

Manganese toxicity was implicated in the Puerto Rico study of
Abruña *et al.* (1975). Solution culture studies of Asher & Edwards
(1978) show that common bean is far more susceptible to solution manganese concentration than is soybean (see also Table 15.6).

Liming Oxisols can greatly improve the growth of common bean in the presence of a basal fertiliser application, maximum yields being attained at pH 7.1 (saturation soil paste) (Fig. 11.4). Improvement in growth was probably due to elimination of manganese toxicity below about pH 5.5 and increased calcium availability above this (Munns & Fox, 1977). Thus common bean has a greater need for lime than soybean, groundnut and cowpea. The study also showed that, in common bean, nodule number per plant was depressed at pH values below 5.7, but that individual nodule weight was depressed by an increase in pH over the full range. In acid soils, responses by common bean to molybdenum may not be obtained until the pH is raised above 5.5 (Franco & Day, 1980). The benefits of liming may also be expressed through improved survival of rhizobia in acid soils (Danso, 1977). Improved plant growth and nodulation in acid soil may be shown when either the

Fig. 11.3. Effect of aluminium saturation on relative yield of common bean on five Utisols and one Oxisol in Puerto Rico. Filled circles, Humatas Clay; open circles, Corozal Clay; crosses, Corozal Clay (eroded phase); open squares, Corozal Clay (level-phase); open diamonds Los Guineos Clay; filled diamonds, Coto Clay. $Y = 98.21 - 0.53x - 0.0072x^2$, $r = 0.93^{**}$. *Note*: Soil pH values for unlimed soils ranged from 3.9 to 4.6. Adapted from Abruña *et al.* (1975).

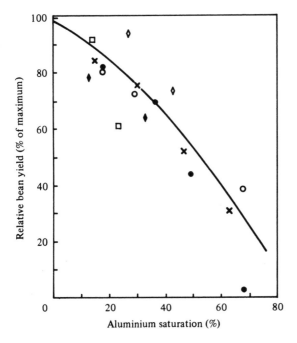

cultivar or the *Rhizobium* strain is tolerant of acidity (Vargas & Graham, 1988).

Studies on micronutrients in common bean, as for other legumes, must take into account the requirements of both the plant and the rhizobia. The nodule–rhizobia complex of most legumes requires molybdenum, manganese, cobalt, iron, zinc, boron and copper (Andrew, 1977; Franco, 1977b). As mycorrhizal fungi may improve uptake of iron, zinc and copper of infected plants (in addition to improved phosphorus uptake, Kucey & Janzen, 1987), they may be quite important in relation to legume sensitivity to zinc nutrition (Moraghan, 1984). Common bean

Fig. 11.4. Effect of soil pH (of saturation paste), as affected by liming a nitrogen-deficient Oxisol (Wahiawa silty clay), on the relative yield of common bean, soybean (cvs. Kahala and Kanrich), groundnut and cowpea (*Vigna unguiculata*). Adapted from Munns, Fox & Koch (1977).

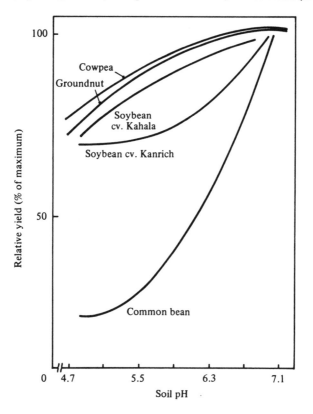

is quite susceptible to salinity (Table 11.3); black-seeded varieties are more tolerant than varieties of other colours (Leon & Medina, 1977). Salinity affects shoot growth more than root growth (Wignarajah, 1990). Sodium chloride is more detrimental than sodium sulphate to the translocation of photosynthetic assimilates (Bhivare & Chavan, 1987).

11.6 Place in cropping systems

Cropping systems including the common bean have been reviewed by Laing *et al.* (1984), Davis, Woolley and Moreno (1989) and Woolley *et al.* (1991). *Phaseolus* is not an important component of tropical Asian or West African cropping systems, though it is probably the most important grain legume in East Africa. In fact, Rwanda has the highest per capita bean consumption in the world (CIAT, 1981). African cropping systems have been reviewed by Allen *et al.* (1989); in this section attention is largely confined to tropical America.

About 80% of the bean area in Latin America is intercropped, nearly all with maize. Some aspects of the maize/bean intercrop have already been discussed in Chapter 6. Dietetic complementarity of maize and beans is referred to in Chapter 9.

Woolley *et al.* (1991) have grouped the world's bean cropping systems in six classes:

Table 11.3. *Relative salt tolerance of food crops*

Crop	Specific electrical conductivity ($dS\ m^{-1}$ at 25 °C) at which yield reduced by [a]		
	10%	25%	50%
Sorghum	6	9	12
Soybean	5.5	7	9
Rice[b]	5	6	8
Maize	5	6	7
Sweet potato	2.5	3.5	6
Common bean	1.5	2	3.5

After Bernstein (1974).

[a] In gypsiferous soils, conductivities causing equivalent yield reductions will be about 2 $dS\ m^{-1}$ greater than the listed values.

[b] Less tolerant during seedling stage. Salinity at this stage should not exceed 4 or 5 $dS\ m^{-1}$.

1. Sole crops of bush beans.
2. Relay cropping of bush or semi-climbing beans with maize. Systems in which the beans are planted at physiological maturity of the maize are most common.
3. More or less simultaneous row cropping of bush or semi-climbing beans with maize.
4. Mixed (i.e., non-row) intercropping of semi-climbing or climbing beans with maize.
5. Intercropping, in rows or not, with crops other than maize; e.g., cassava and sweet potatoes.
6. Relay cropping of maize and climbing beans.

Figure 11.5 is a schematic representation of Latin American maize/bean cropping systems in relation to rainfall pattern.

Research at CIAT on the yield advantages of maize/bean intercropping over sole cropping were considered in Chapter 6 (see Table 6.7). The agronomy of the intercrop has been reviewed by Woolley & Davis (1991). Other aspects of intercropping research of relevance include stability of income, insect attack and nitrogen fixation.

An economic analysis was made of 20 trials comparing sole climbing beans (with artificial support), maize/climbing bean intercrops and sole maize (Francis & Sanders, 1978). Although the yield of intercropped beans was only about 40% that of sole beans, it was obtained with virtually no reduction in maize yield compared with sole maize. Yield variability between seasons was lower for intercrop than for sole beans, giving greater stability of income.

Beans grown in association with maize suffer less insect attack than sole beans (Altieri *et al.*, 1978). Populations of the leafhopper *Empoasca kraemeri* and the chrysomelid *Diabrotica balteata* were respectively 26% and 45% lower in the intercrop, when maize and beans were planted at the same time, than in the sole crop situation. Relay cropping was particularly effective in reducing leafhopper populations: maize planted 20 and 30 days before beans reduced leafhoppers on beans by 66% compared with simultaneous planting.

With respect to nitrogen fixation, although rapid leakage of fixed ^{15}N from beans has been demonstrated, Graham (1981) is of the opinion that fixation does not directly benefit the associated maize crop. Graham & Rosas (1978) found that although the dry matter yield of climbing beans grown in association with maize was reduced relative to sole beans, this reduction did not occur until active nitrogen fixation had ceased. In experiments with bush beans, post-flowering fixation was

Fig. 11.5. Examples of common bean/maize cropping systems and rainfall patterns. Source: Woolley *et al.* (1991).

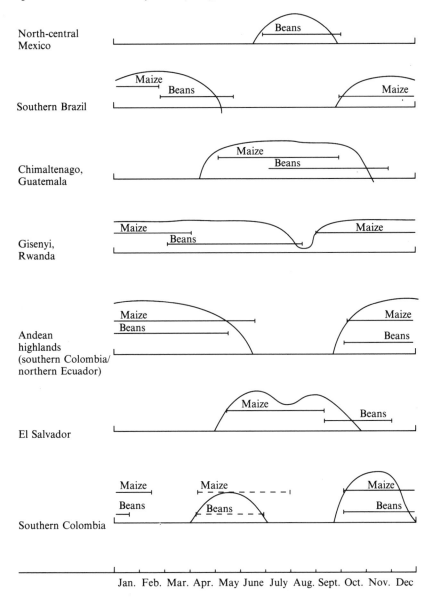

Jan. Feb. Mar. Apr. May June July Aug. Sept. Oct. Nov. Dec

significantly reduced in intercrop beans compared with sole beans. The adverse effects were greater if bean planting was delayed.

Other crops grown in association or sequence with beans and maize/bean intercrops include squash, cassava and sweet potato, and to a lesser extent sorghum. In a major experiment carried out at Turrialba, Costa Rica, reported by Pinchinat *et al.* (1976), 54 cropping patterns based on beans, maize, cassava and sweet potato were compared. The three most productive sequences that included beans were: (1) sole beans followed by two successive crops of sole maize, (2) bean/maize intercrop followed by sole maize and (3) bean/cassava intercrop followed by sole maize.

Further reading

Laing, D.R., Jones, P.G. & Davis, J.H.C. (1984). Common bean (*Phaseolus vulgaris* L.). In *The Physiology of Tropical Field Crops*, ed. P.R. Goldsworthy & N.M. Fisher, pp. 305–52. Chichester: Wiley.

Schoonhoven, A. van & Voysest, O. (eds.) (1991). *Common Beans: Research for Crop Improvement*. Wallingford, UK: CAB International/Centro Internacional de Agricultura Tropical.

12

*Soybean (*Glycine max*)*

12.1 Taxonomy

In the genus *Glycine* there are two subgenera: *Glycine* and *Soja*. The subgenus *Glycine* comprises six perennial Australasian species (2n = 40 or 80), none of which is used in agriculture. The subgenus *Soja* comprises two east Asian species: *G. soja*, the wild soybean, and *G. max*, the cultivated soybean (Singh & Hymowitz, 1987). Both are annuals (2n = 40). No hybrids between the subgenera *Glycine* and *Soja* have been reported, but there are virtually no cytogenetic barriers to hybridisation between *G. soja* and *G. max*.

Soybeans in the North American continent are classed in 12 maturity groups based on daylength response. However, the US classification is of little value for discriminating between types within the general group of tropical soybeans. A classification for the tropics, shown in Fig. 12.1, has been proposed by Shanmugasundaram, Kuo & Nalampang (1980).

12.2 Origin, evolution and dispersal

G. soja is native to northern, northeastern and central China, adjacent regions of Kazakhstan, Korea, Japan and Taiwan. Evidence based on chromosome number and size, geographical distribution, and the electrophoretic banding pattern of seed proteins strongly suggests that *G. max*, which is not known in the wild state, is derived from *G. soja*. The linguistic, geographical and historical evidence indicates that *G. max* emerged as a domesticate in the eastern part of northern China about 3000 BP (Hymowitz, 1970). Changes during domestication include increased plant and seed size, modification from a twining to an erect habit, and reduced dehiscence of pods. A complex of intermediate weedy forms is found where the distribution of the two species overlaps.

For a crop of major importance in the tropics, the soybean is therefore unusual since it evolved in a temperate continental climate. It spread down to the south and southeast Asian tropics from about 1800 BP onward. Soybeans reached Europe in the early eighteenth century and were taken from France to the USA in the late eighteenth century. The great expansion of soybean growing in Latin America, particularly Brazil, has taken place only in the last 35 years.

Data on area, yield and production are given in Table 12.1. Only four of the eleven countries producing more than 500 000 t — Brazil, Indonesia, Paraguay and Mexico — are in the tropics. Total production in the four major tropical countries has increased by 20–40% during the past decade owing to increases in both productivity and area. However, of the tropical producers, only in Mexico is productivity per hectare comparable to that in temperate regions.

12.3 Crop development pattern

Soybean is a small erect branching annual. Its anatomy and morphology have been reviewed by Carlson (1973) and Wilcox & Frankenberger (1987), and the growth pattern of temperate soybeans by Shibles, Anderson & Gibson (1975) and Wilcox & Frankenberger (1987). In the

Fig. 12.1. Proposed classification of tropical soybean cultivars. Source: Shanmugasundaram *et al.* (1980).

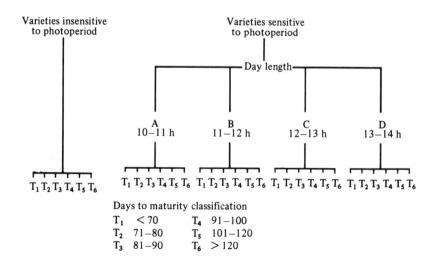

tropics the growth duration of adapted cultivars and landraces is commonly 90–110 days and up to 140 days (Osafo, 1977; Nangju, 1979; Pearson, Masduki & Moenandir, 1980). This relatively short growth duration is due primarily to sensitivity to short daylength. 'Numerous studies have shown [that] the induction of flowering, the duration and extent of vegetative growth, the production of viable pollen, the length of the flowering and pod-filling periods, and maturity characteristics, are all subject to modification by photoperiod' (Lawn & Byth, 1979). Time to flower initiation shows a complex daylength–temperature interaction (e.g., Jones & Laing, 1978) and there are marked differences between genotypes in sensitivity to short days or high temperatures (e.g., Criswell & Hume, 1972; Lawn & Byth, 1973, 1974). Temperature alone is a poor predictor of time to flowering or total growth duration (e.g., Major *et al.*, 1975). At relatively high latitudes within the tropics there is a linear decline in number of days to flowering and in total plant height with delay in sowing after midsummer (Kununurra, north Australia, 16° S; D.F. Beech, personal communication).

Flowers first appear on lower mainstem nodes and form progressively towards the tip of the mainstem and along axillary branches. There are determinate and indeterminate genotypes. The flowers are normally closely self-pollinated. The flowering period lasts 15–30 days in determinate soybeans. Many flowers may abort, the most common stage of loss

Table 12.1. *Major soybean-producing countries: area, yield and production in 1991 of countries producing more than 500 000 t per annum*

Country	Area (m ha)	Yield (t ha^{-1})	Production (m t)
USA	23.45	2.30	54.04
Brazil	9.52	1.55	14.77
Argentina	4.86	2.31	11.25
China	7.95	1.23	9.81
India	2.65	0.79	2.10
Indonesia	1.37	1.13	1.55
Canada	0.58	2.44	1.41
Italy	0.41	3.23	1.32
Paraguay	0.70	1.86	1.30
Former USSR	0.81	0.94	0.76
Mexico	0.34	2.08	0.72
Thailand	0.49	1.22	0.61

After FAO (1992).

being in the first week after flowering (Carlson, 1973). Pod and seed development, reviewed by Howell (1963), Carlson (1973) and Carlson & Lersten (1987), takes 30–70 days (e.g., Osafo, 1977; Summerfield & Wein, 1980). Physiological maturity (maximum seed dry weight at a moisture content of 50–60%) coincides with the growth stage R7 of Fehr & Caviness (1977), i.e., when one pod on the mainstem has reached its mature colour (Te Krony, Egli & Henson, 1981). Delaying harvest after physiological maturity by as little as 2 weeks increases seed discoloration and damage (Nangju, 1979). As with other legumes in the tropics, seed quality may decline rapidly owing to pre-harvest weathering and during storage.

12.4 Crop/climate relations

The size of soybean seed ranges from 50 to 500 mg; seeds of tropical landraces usually weigh 70–120 mg. Seedlings developing from small (*c.* 80 mg) or large (>220 mg) seed may have reduced vigour (Burris, Wahab & Edge, 1971; Edwards & Hartwig, 1972; Burris, Edge & Wahab, 1973); within the normal range of seed size, genotype is more important than size in determining emergence (Johnson & Luedders, 1974). Soybeans are reputed to require relatively high soil water for germination. For germination to occur in 5–8 days at 20 °C, soil water potential should not be less than –0.66 MPa for soybean, compared with –1.25 MPa for maize (Hunter & Erickson, 1952).

The temperature minimum for germination and emergence is less than that for groundnuts and perhaps other tropical legumes: regression analysis suggests a threshold of 9.9 °C and 70.5 day-degrees to emergence compared with 13.3 °C and 76 day-degrees for groundnut (Angus *et al.*, 1980); other workers suggest thresholds between 2–4 °C and 13–15 °C. The optimum temperature is about 30 °C (Delouche, 1953; Inouye, 1953), which coincides with the optimum for relative growth rate (Trang & Giddens 1980). Emerging seedlings die above 42–44 °C (Inouye, 1953).

Rates of leaf production and expansion increase with increasing temperature, but individual leaf area and branching are greatest at <24/19 °C day/night, i.e., temperatures substantially below those found in the wet and wet-and-dry tropics (Shibles, 1980).

Pearson *et al.* (1980) provide one of the few descriptions of leaf area and dry weight accumulation within the tropics. In Java, at 7° S latitude, two genotypes reached maximum leaf area indices of 5.5 and 4 in high populations (1 million plants ha^{-1}) at development index

(DI) = 0.6. This was approximately the time of maximum growth rate of tops (160 kg ha^{-1} d^{-1}), whereas peak above-ground dry weight of about 6 t ha^{-1} was attained later, at DI = 0.80–0.85. Leaf carbon dioxide exchange rates of soybean, a C3 species, are low relative to those of tropical cereals; however, peak rates during pod filling may reach 1.7–1.9 mg m^{-2} s^{-1} (Zhailibaev & Khosenov, 1966; Hesketh *et al.*, 1981; Fig. 7.1). The thermal optimum for carbon dioxide exchange is about 35 °C (Hofstra & Hesketh, 1969). Hommertzheim (1979) gives an algebraic description of soybean canopies: their spherical structure and planophile leaf orientation ensure that most radiation is intercepted at the periphery of the canopy, to such an extent that opening or turning the canopy may increase yield (Pearson *et al.*, 1980).

Water use by closed canopies may be 3–7.5 mm d^{-1} (Whitt & van Bavel, 1955; Mason *et al.*, 1982). Soil water losses are approximately half of total evaporation (*E*) when the soil is wet but fall to 0.25–0.5 of *E* in the dry season (Peters & Johnson, 1960). Thus soybean shows high water use per unit dry matter (DM) produced relative to the tropical cereals (Fig. 7.1); water use efficiencies of 1.4–4.4 kg DM (root plus top) m^{-3} water were reported by Mason *et al.* (1982). Meyer and others (Meyer *et al.*, 1990; Dugas *et al.*, 1990; Meyer & Mateos, 1990) present a detailed description of growth and water use of soybeans in lysimeters in a high-latitude, high radiation climate. Crop water use differed by 30% between soil types and maximum rooting depths differed by 40 cm; in both soils 35–45% of crop evaporation came from water stored in the top 10 cm of soil.

Soybean leaves are relatively sensitive to water deficit: stomatal conductance begins to fall when leaf water potential is about –1 MPa (Meyer & Green, 1981) and 50% closure takes place at –1.6 MPa (Turner *et al.*, 1978). Partial stomatal closure in turn reduces evaporation (by 40–70%) and further increases canopy temperature (e.g., Reicosky, Deaton & Parsons, 1980). Leaf enlargement rates decline sharply at –0.4 MPa (Boyer, 1970).

The root system of soybeans consists of a primary root and laterals that arise in four rows along the top 10–15 cm of the tap root. Total root length per unit ground area reaches a maximum of 1000–4000 m m^{-2} at early pod-filling (Willatt & Olsson, 1982, and references therein). Likewise, there is a marked developmental pattern in nodule nitrogen fixation, with peak rates at early pod-filling and a sharp decline at mid-pod-filling (Thibodeau & Jaworski, 1975). There is a broad temperature optimum for root growth (Earley & Cartter, 1945), although several workers have shown that nitrogen fixation is more adversely affected by high temperature than is root growth: with increasing temperature, the growth of soybeans dependent on nitrogen fixation begins to fall before that of plants supplied with fertiliser nitrogen (Munevar & Wollum, 1981, and references therein). Similarly, rhizobial nitrogen fixation is more sensitive to intermittent drought than are processes, e.g., growth, in the host soybean (Kirda, Danso & Zapata, 1989).

It therefore appears that development constraints, particularly in rooting depth and nitrogen fixation, have most effect on growth and water relations of soybeans in the tropics. Most roots are in the upper 30 cm of soil but root penetration and water extraction take place to 2 or even 3 m depth (e.g., Willatt & Taylor, 1978; Mason *et al.*, 1982).

Flowering terminates vegetative growth of the mainstem of determinate soybeans. Thus plant population and population × time of planting interactions are critical in determining the optimum vegetative structure to support subsequent grain-filling. For small early-flowering plants, as found at high temperature (Shibles, 1980), the optimum population for highest yields may be dense: 0.5 million plants ha^{-1} is considered optimal for yield under irrigation in Indonesia and Brazil (Andrade & Sedijama, 1977; Pearson *et al.*, 1980). These early-flowering high-population tropical crops contrast with temperate soybeans which may show long growth duration and in which intermediate population may be optimal for grain yield (e.g., Dunphy, Hanway & Green, 1979; Gay, Egli & Reicosky, 1980; Blumenthal, Quach & Searle, 1988).

Flowering is tolerant of a wide temperature range: flower formation is inhibited at night temperatures of 10–13 °C or 15 °C (Parker & Borthwick, 1939; Roberts & Struckmeyer, 1939; Goto & Yamamoto, 1972; Hume & Jackson, 1981) and pod set is low above 40 °C (Mann & Jaworski, 1970). Once filling begins, grain growth rate is related to leaf area index (or, more likely, intercepted radiation) (Pearson *et al.*, 1980), and individual grain growth rate increases with temperature from 18/13 °C to a plateau at 27/22–33/28 °C (Egli & Wardlaw, 1980). The grain-filling period in controlled environments is constant up to 30/25 °C but reduced by 3 days at 33/28 °C (Egli & Wardlaw, 1980).

The sensitivity of any one component of vegetative growth (e.g., branching, leaf production) to water deficit is offset in part by compensation among yield components (Shaw & Laing, 1966; Momen *et al.*, 1979). For example, stress at early flowering causes flower and pod abortion at low node positions but compensatory pod formation later, at upper nodes, so that there is little net loss of pods (Fig. 12.2). Naturally the crop's ability to compensate diminishes during grain-filling so that end-of-season water deficits reduce all yield components (Fig. 12.2).

In conclusion, it seems that low yields (Table 12.1) are due to developmental constraints on growth and crop water relations. Such constraints are exacerbated by high temperature and low water availability, as are found particularly in the wet-and-dry tropics. If water is not limiting, high populations will give reasonable seed yields (e.g., 3 t ha^{-1}; Pearson *et al.*, 1980). Singh, Rachie & Dashiell (1987) review our knowledge of plant types which are adapted to tropical environments. However, even when grown with irrigation in the tropics, Muchow, Robertson & Pengelly (1993) have found soybean has a relatively low

Fig. 12.2. Changes in yield of soybean as a consequence of changes in yield components because of water stress. Stress periods were successive 1-week intervals during which relative water content of the upper leaves was at or below 85% for 4 days. Adapted from Shaw & Laing (1966) by Shibles *et al.* (1975).

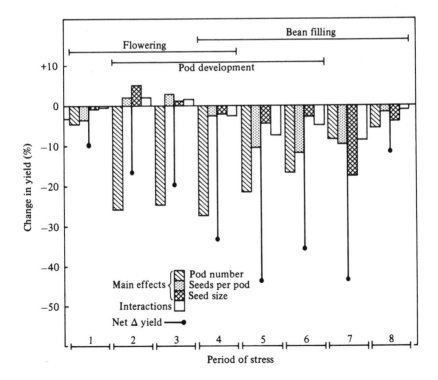

growth efficiency (0.85–0.90 g dry matter MJ^{-1} compared with approx. 1.1 for mungbean, *Vigna radiata*) which was insensitive to time of sowing and environment.

12.5 Crop/soil relations

12.5.1 *Soil physical properties*

Soybeans can be grown on soils of a wide range of textural classes (Purseglove, 1968); drained Histosols can be physically satisfactory (Wahab, 1979). Mineral soils that crust can reduce emergence and yield, as found on the Indo-Gangetic plains by Rathore, Ghildyal & Sachan (1981). While soil compaction due to the use of agricultural equipment may also reduce yield (by up to 27%; Johnson *et al.*, 1990) the capacity of seedlings to emerge through compacted soils varies with cultivar (Howle & Caviness, 1988). Soybean crops may also influence soil properties: soil aggregate size, stability and carbon content were lower following soybeans than following maize (Bathke & Blake, 1984).

The yield of soybeans following wet rice is often lower than that following upland rice (e.g., Floresca, 1968), owing to impaired soil/water relations. Soybean establishment after rice can be reduced by flooding where the crop is grown in slaking alluvial soils (Talsma *et al.*, 1977). In the Philippines, Herrera & Zandstra (1979) found that continuous flooding for 7 days at 30 days after sowing reduced the yield of mungbean, groundnut and soybean by 56%, 49% and 37% respectively. The relative effects of flooding soybeans and maize at two development stages are shown in Table 12.2. Determinate soybeans were more susceptible to prolonged flooding during early reproductive growth than during early vegetative growth, and when grown on a clay soil than on a silt loam (Scott *et al.*, 1989). Harwood (1975) rated the tolerance of upland crops to soils of high bulk density and poor aeration following wet rice: the order of increasing tolerance was maize and groundnuts, sorghum, soybean and mungbean, cowpea.

In spite of potential adverse effects of high soil moisture on soybeans, the crop may be grown by careful maintenance of the water table, after emergence, at 10–15 cm. Roots and nodules develop in the shallow layer of wet but aerated soil (Wright & Smith, 1987). This system has shown substantial increases in plant, grain and nodule yields over conventional irrigation (Troedson *et al.*, 1989).

The survival of rhizobia in soils can be affected by soil physical properties such as soil texture (survival being lower in sands and clay loams than in soils of intermediate texture; Mahler & Wollum, 1981) and soil moisture (Osa-Afiana & Alexander, 1979). Waterlogging of soils used for wet rice reduced soybean nodulation by 15% (Wu, Lee & Chiang, 1979).

12.5.2 *Soil chemical properties*

Where no soybean crop has been grown before it is usually necessary to inoculate with an efficient rhizobial strain to maximise yield in the absence of fertiliser nitrogen (e.g., Dadson & Acquaah, 1984), as fixation can contribute 25–75% of total soybean nitrogen (Deibert, Bijeriego & Olson, 1979). Matheny & Hunt (1983) obtained values of 55–60% in non-irrigated plots and 76–91% in irrigated plots. Favourable responses to inoculation have been obtained in most tropical countries (Ayanaba, 1977), yield increases being as high as six-fold (Bromfield & Ayanaba, 1980). In some soils where soybeans have been grown previously, continued inoculation may still be necessary (Rao *et al.*, 1985), apparently because of poor survival of introduced rhizobia. Introduced inoculant strains must exhibit both saprophytic and symbiotic superiority, relative to indigenous strains, if they are to maintain soybean yields in the absence of continued inoculation (Fuhrmann & Wollum, 1989).

Most rhizobia that fix nitrogen in soybean are slow-growing *Bradyrhizobium japonicum*, but fast-growing soybean rhizobia also exist

Table 12.2. *The effects of crop growth stage and duration of flooding on yield (t ha^{-1}) of maize and soybean at Los Baños, Philippines*[a]

Flooding duration (d)	Maize		Soybean	
	At 15 DAS[b]	At grain-filling	At 15 DAS[b]	At pod-filling
0	4.68 a	4.32 a	1.15 a	1.23 a
4	3.91 ab	3.03 a	1.06 ab	1.27 a
8	3.04 b	3.44 a	0.81 b	1.07 ab
12	1.95 b	3.84 a	0.84 b	0.88 b

After Herrera & Zandstra (1979).
[a] Column means followed by a common letter are not significantly different at the 5% level.
[b] DAS, days after sowing.

(Keyser *et al.*, 1982). Some indigenous rhizobia may be efficient in fixing nitrogen and have the advantage of being adapted to the local environment (Awai, 1981). However, introduced soybean cultivars may not always be nodulated effectively by indigenous rhizobia (La Favre *et al.*, 1991). Thus, in Nigeria, Nangju (1980) found that while three southeast Asian soybean cultivars nodulated quite well with indigenous strains, three American cultivars nodulated poorly. Level of inoculation (Smith, Ellis & Smith, 1981), seedling vigour (Smith & Ellis, 1981) and soil nitrogen status (Lopes, 1977) may also affect nodulation.

The application of nitrogen to soybean crops in the tropics has given variable results, depending on such factors as rhizobial strain, the particular host–rhizobium association (Senaratne, Amornpimol & Hardarson, 1987), the levels of fertiliser nitrogen applied, soil nitrogen status (Herridge & Betts, 1988), soil acidity and other nutritional and environmental factors affecting both plant and rhizobial growth. Thus Kang (1975) in Nigeria found that an inoculated crop required 30 kg nitrogen (N) ha^{-1} and an uninoculated crop 60 kg N ha^{-1} to attain maximum yields (around 1.7 t ha^{-1}), while in Brazil yields of nodulated soybeans were not affected by applications up to 300 kg N ha^{-1}, the amount required to maximise yields (around 2 t ha^{-1}) of non-nodulating isolines at the same site (Pal & Saxena, 1975). Nitrogen applications usually depress nodulation (e.g., Singh & Saxena, 1972) but this may not always happen, even where nitrogen rates are high (e.g., 448 kg N ha^{-1}; Olsen, Hamilton & Elkins, 1975). Yield responses to nitrogen have been reported even where plants were well nodulated (Kang, Nangju & Ayanaba, 1977). Negative yield responses to nitrogen have also been obtained (e.g., Singh & Saxena, 1977).

Soybeans are usually grown as a sole crop in sequence with other crops, but are sometimes intercropped. In the Philippines intercropped soybeans and maize took up significantly more nitrogen than either of the crops grown alone (Fig. 12.3). However, when 180 kg N ha^{-1} was applied, nitrogen uptake of the intercrops and maize alone was increased much more than that of soybeans alone, presumably owing to adequate nitrogen fixation. The residual value of nitrogen from unfertilised intercropped maize and soybeans to a succeeding wheat crop can be equivalent to that from 100 kg N ha^{-1} applied to a prior sole maize crop (Searle *et al.*, 1981). The allelopathic effects of weeds on soybean growth and nodulation observed in temperate areas by Mallik & Tesfai (1988) are also likely to exist in the tropics.

Low phosphorus availability is liable to limit soybean yields on many highly weathered soils in the tropics, although the external phosphorus requirement of soybean is low compared with some other grain legumes (Table 12.3). Applications of phosphorus to correct deficiency should take into account differences in extractable phosphorus and clay content or surface area (Lins & Cox, 1989).

Low soil phosphorus may also contribute to the poor survival of some rhizobial strains (Cassman, Munns & Beck, 1981*a*) in the soil between crops. Improved dry matter and grain yield resulting from improved phosphorus nutrition of soybean are often accompanied by increases in nodule number and weight (e.g., Haque, Walker & Funnah, 1980; Singleton, AbdelMagid & Tavares, 1985). However, it appears that not only is phosphorus important to the plant but that, above a certain level of phosphorus supply, it is relatively more important to

Fig. 12.3. Effect of nitrogen fertiliser on nitrogen uptake of sole and inter-cropped soybeans and maize at Los Baños, Philippines. Open symbols, nil nitrogen; filled symbols, 180 kg nitrogen ha^{-1}. Adapted from IRRI (1976).

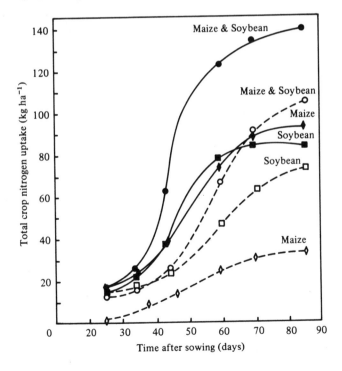

nitrogen fixation. For instance, Cassman, Whitney & Fox (1981*b*), showed that inoculated soybeans growing in a nitrogen- and phosphorus-deficient Tropohumult required 750 kg phosphorus ha^{-1} to attain maximum yield when dependent on nitrogen fixation, compared with only 370 kg phosphorus ha^{-1} to attain maximum yield when supplied with combined nitrogen. Soybeans appear also to have a higher phosphorus requirement than either groundnut or cowpea (Fox, 1978; Cassman *et al.*, 1981*b*; but see Table 12.3, in which soybean and cowpea requirements are the same). Phosphorus uptake can be reduced by drying in the surface soil, in spite of water available elsewhere in the profile (Kaspar, Zahler & Timmons, 1989). Phosphorus uptake is also increased by mycorrhizal infection (Yost & Fox, 1989; El-din & Moawad, 1988).

In a soybean/maize intercrop, inoculation with mycorrhizae (*Glomus intraradix*) increased the growth of both the host crops, appeared to the authors to increase the transfer of nitrogen from soybean to maize, and caused greater growth and thus competitiveness in the maize than in the soybean (Hamel & Smith, 1991).

The review of R.W. Pearson (1975) indicates that soybean responses to lime on acid soils in the tropics are common (see Fig. 11.4) and sometimes high, particularly on Oxisols. Often it is not possible to determine from the literature whether the responses were due to the elimination of aluminium or manganese toxicity or to improved calcium, magnesium or molybdenum availability, nor whether the effects were primarily on the plant or on the nodule–rhizobia complex. Abruña (1980) attributed the benefit of liming an Oxisol in Puerto Rico to elimi-

Table 12.3. *External phosphorus requirements of some tropical grain legumes: the external phosphorus requirement in the solution concentration at which the greatest change in slope of the response curve occurs*

Species	External phosphorus requirement (µmol)
Soybean cv. Fitzroy	0.8
Cowpea cv. Vita 4	0.8
Pigeonpea cv. Royes	1.0
Mungbean cv. Regur	2.0
Guar cv. Brooks	3.0

Adapted from Fist, Smith & Edwards (1987).
Plants were harvested at 16 days, except for guar which was harvested at 20 days.

nation of manganese toxicity, to which some soybean cultivars are very susceptible (Heenan & Campbell, 1980). On two Ultisols and five Oxisols in Puerto Rico and southern Brazil, Abruña (1980) and Martini *et al.* (1974) found that aluminium saturation had to be reduced to 1–5% (pH 5.2–5.7) to maximise yields, although improved calcium and magnesium nutrition may also have been involved. Improved nodulation has also been shown with a reduction in toxic levels of aluminium by liming: 0 and 75 nodules per plant at 56% and 3% aluminium saturation respectively (Abruña, 1980). In soils with toxic levels of subsoil aluminium, aluminium-tolerant species are also more likely to tolerate drought (Goldman, Carter & Patterson, 1989).

Solution culture studies of Alva *et al.* (1986*a–c*) and Noble, Fey & Sumner (1988) have shown that the activities of monomeric aluminium species and calcium account for reduction in growth of tap roots of soybean. Fulvic acid, a humic compound commonly found in acid soils, can alleviate aluminium toxicity (Suthipradit, Edwards & Asher, 1990). This could explain the effect of organic matter incorporation, as an alternative to lime, in reducing aluminium toxicity (e.g., Ahmad & Tan, 1986). In acid soils with no aluminium toxicity it may be calcium deficiency which restricts plant growth. In 20 acid Queensland soils (including subsoils) with no aluminium toxicity, Bruce *et al.*, (1988) showed that relative root length of soybean cv. Forrest was reduced to less than 90% of maximum when the calcium activity ratio of the soil solution and the soil calcium saturation were lower than 0.05 and 11% respectively. While it appears that soybeans are generally quite sensitive to soil acidity, reasonable grain yields of 2–2.5 t ha^{-1} and quite satisfactory nodulation are sometimes obtained with tolerant varieties on soils with pH values in the range 4.1–4.6 and aluminium saturation in the range 36–55%, as found in Guyana (Wahab, 1979), Nigeria (Bromfield & Ayanaba, 1980) and Puerto Rico (Abruña, 1980). Some studies show that where soybean growth is poor on acid soils it may be due to effects on the plant rather than on the degree of nodulation or effectiveness in nitrogen fixation (e.g., Munns *et al.*, 1981). A contrasting result was found by Cline & Kaul (1990), who attributed the reduction in nodule number and nodule weight per plant in acidified soils to hydrogen ion toxicity and not to aluminium or manganese toxicity. In addition, the survival of rhizobia between crops in acid infertile soils may be adversely affected by high exchangeable aluminium (Ayanaba, Asanuma & Munns, 1983), low soil pH, and low phosphorus and potassium (Hiltbold, Patterson & Reed, 1985).

Soybeans are comparable to rice in salt tolerance (Table 11.3). To grow soybeans in saline soils it is important to select both salt-tolerant cultivars and *Bradyrhizobium* strains (Velagaleti & Marsh, 1989). Saline soils may also increase the inherent sensitivity of some soybean cultivars to high levels of soil phosphorus, owing to the chlorine (Grattan & Maas, 1988).

12.6 Place in cropping systems

As indicated in Tables 9.1 and 12.1, soybeans are of minor importance in tropical Africa, and they receive little attention in the literature of African agriculture. In tropical America there are two distinct types of soybean culture: sole cropping and intercropping with maize (Knight, 1971). Over large areas of the Rio Grande do Sul in Brazil soybeans are grown as a sole crop in summer, often in sequence with winter wheat in a double cropping system (see Lanzar, Paris & Williams, 1981). A substantial proportion of the crop is mechanised. The maize/soybean inter-crop system is associated with less developed agricultural regions and the crop is all hand-harvested.

In southeast Asia, soybeans have an important role as a secondary crop in intensive cropping systems based on wet rice, since they are fairly tolerant of poorly aerated soil (see Section 12.5). Shanmugasundaram *et al.* (1980) present data from Taiwan, Indonesia and Thailand to show that, through the use of cultivars of appropriate daylength response, soybeans may be grown at virtually any time of the year in lowland locations in southeast Asia. The crop is sometimes harvested in the green pod stage as a vegetable.

The most intensive cropping systems involving wet rice and soybeans are found in Taiwan. Kung (1969) describes irrigated systems with three or four crops a year based on two rice crops: a spring–early summer crop (February–June) and a late summer–early autumn crop (August–November). In the three- and four-crop systems, soybeans (or other short-season upland crops such as maize, peanuts and vegetables) are grown between rice crops, and are normally relay-planted into the preceding rice crop during the period from heading to a week before harvest.

In Asia soybeans are grown as a sole crop in dryland, upland crop-ping systems. Most are grown early in the wet season or in the late wet or early dry season (AVRDC, 1986). They are also one of the short-term upland crops included in 3-year rotations of wet rice and sugarcane: Ruthenberg (1980) gives an example from Taiwan in which sugar-cane (16 months duration) is followed in the ensuing 20 months by green manure–wet rice–sweet potatoes/wheat–soybeans. In East Java (Norman, 1979), 15–18 months of sugarcane is followed by wet rice–soybeans–wet rice before the next cane planting. An example of a less intensive soybean–wet rice cropping system is from northern Thailand, where in this wet-and-dry tropical region summer-grown rice is followed by irrigated dry-season soybeans. AVRDC (1986) review a range of soybean intercropping systems in Asia, including those with cassava and with tree crops such as coconut, rubber and oil palm.

In southeast Asia there is a variety of cultural systems for growing soybeans after wet rice, governed by the intensity of the cropping pattern, soil drainage characteristics and the reliability of expectation of rainfall or irrigation water. The relay-planting method of Taiwan has already been mentioned; it permits the compression of three, four or even five crops into a year. In northern Thailand, to save time and thus to make full use of a somewhat limited supply of irrigation water, soybean seed is dibbled directly into the stubble after rice harvest. In East Java, for the same reasons, seed is dibbled or broadcast into a mulch of rice straw. While minimum tillage for soybean following other crops is widespread, where time permits and the expected returns make it worth while, the stubble from the previous crop, e.g., rice, may be ploughed and the soil formed into beds or ridges to improve drainage before sowing the soybean.

In conclusion, it should be noted that in all the examples of southeast Asian cropping systems quoted above the soybean is only one of a range of short-season upland crops that may be grown, and that it may not necessarily be grown as a sole crop. For example, maize/soybean intercrops may be found in sequence with wet rice.

Further reading

AVRDC (1986). *Soybean in Tropical and Subtropical Cropping Systems.* Shanhua, Taiwan: AVRDC (Asian Vegetable Research and Development Center), 471 pp.

Singh, S.R., Rachie, K.O. & Dashiell, K.E. (eds.) (1987). *Soybeans for the Tropics: Research, Production and Utilization.* Chichester: Wiley, 230 pp.

Summerfield, R.J. & Bunting, A.H. (eds.) (1980). *Advances in Legume Science.* Kew: Royal Botanic Gardens.

Wilcox, J.R. (ed.) (1987). *Soybeans: Improvement, Production and Uses*, 2nd edn. Madison, Wisconsin: American Society of Agronomy, 888 pp.

13

*Chickpea (*Cicer arietinum*)*

13.1 Taxonomy

The genus *Cicer*, of the monogeneric tribe Cicereae, includes about 40 species, distributed throughout central and western Asia. *C. arietinum*, an annual, is not known in the wild state. Most members of the genus, including chickpea, have $2n = 2x = 16$, though $2n = 14$ has also been reported for chickpea and two other species (Ramanujam, 1976). Only one wild species, *C. reticulatum*, crosses readily with cultivated chickpea, and may be its wild progenitor (Ladizinsky & Adler, 1976). The taxonomy of chickpea is summarised by Smartt (1990), who also reviews the evidence of domestication and evolution.

Agronomically, two main types, kabuli and desi, are recognised. The desi type is the one most widely grown in the tropics; it is of short stature and relatively short growth duration, with small, dark-coloured wrinkled seed. The kabuli type is characteristic of the Mediterranean region; it is taller and of longer growth duration, with large, light-coloured and relatively unwrinkled seed.

13.2 Origin, evolution and dispersal

The main centre of diversity of the genus *Cicer* is the Caucasus and Asia Minor (van der Maesen, 1972, 1984). The forerunner of the present domesticate spread westward along the Mediterranean and eastward to India (Ramanujam, 1976). The earliest record, 6400 BP, is from Turkey. There is evidence of chickpea cultivation in the Mediterranean from 4000–6000 BP. A secondary centre of diversity is found in Ethiopia; chickpeas are likely to have reached this region from the Mediterranean. Vishnu-Mittre (1974) has summarised the archaeological evidence from India: the oldest record is thought to be about 4000 BP, from Uttar Pradesh.

The divergent spread of chickpeas east and west was associated with an evolutionary divergence in morphology, giving rise to the kabuli type in the west and the desi type in the east. In the eighteenth century the kabuli type was introduced to India and a range of desi × kabuli crosses resulted (Ramanujam, 1976).

Chickpeas were taken from the western Mediterranean to South and Central America by the Spanish and Portuguese in the sixteenth century. Expatriate Indians introduced it to East and South Africa and the West Indies (Ramanujam, 1976).

Area, yield and production data are given in Table 13.1. India continues to produce most of the world's chickpeas: 67% of world production from 69% of the chickpea area.

13.3 Crop development pattern

It is impossible to generalise about development patterns in chickpea because three-quarters of the main area of production, India, is sown to unselected local cultivars, mostly of the desi type, about which we know little.

Both the widely grown, small-seeded desi type and the kabuli type are long-day plants (Sethi *et al.*, 1981, and references therein). They are indeterminate, the floral primordium forming behind the vegetative shoot apex (Moncur, 1980). Self-pollination usually occurs before the flowers open (Biderbost *et al.*, 1974). Timing of development events has received little attention; we may expect, given daylength sensitivity, that floral initiation

Table 13.1. *Major chickpea-producing countries: area, yield and production in 1991 of countries producing more than 100 000 t per annum*

Country	Area (m ha)	Yield (t ha^{-1})	Production (m t)
India	7.41	0.70	5.20
Turkey	0.88	0.98	0.86
Pakistan	1.06	0.55	0.58
Australia	0.17	1.17	0.20
Mexico	0.14	1.07	0.15
Ethiopia	0.13	0.96	0.13
Myanmar	0.16	0.74	0.12

After FAO (1992).

is variable with respect to maturity, although 50% flowering commonly occurs at development index (DI) = 0.55–0.6 (Murty, 1975; ICRISAT, 1978). Growth duration is usually 90–110 days at Hyderabad (latitude 17° N) and 140–180 days at Delhi (latitude 28° N); the effect of latitude on time to anthesis is shown in Table 13.2. Moreover, because the crop is indeterminate, growth duration is markedly affected by irrigation: maturity of cv. Annigeri was delayed from 87 to 127 days by irrigation at Hyderabad (ICRISAT, 1980). When the green seeds are to be used as a vegetable, chickpeas are harvested prematurely, preferably 23 days after flowering (Hedge, Raghunatha & Narayanswamy, 1975).

13.4 Crop/climate relations

Chickpea is grown within areas having 600–1000 mm average anuual rainfall (AAR), although it tolerates 280–1500 mm (Duke, 1981; ICRISAT, 1987). In the tropics it is sown under declining temperatures at the start of the season of low sun angle; at Delhi the daily minimum

Table 13.2. *Effect of latitude on time to anthesis of chickpea in Australia*

Location	Latitude (deg. S)	Days to seedling emergence	Days to anthesis
Kununurra	16	8	62
Walkamin	17	8	77
Millaroo	20	8	85
Emerald	23	8	83
Gatton	28	9	89
Warwick	28	14	107
Trangie	32	17	101
Wagga Wagga	35	16	122
Rutherglen	36	16	149
Launceston	41	27	174

D.F. Beech, personal communication.

Fig. 13.1. Seasonal rainfall, temperature, open-pan evaporation and available soil moisture (*a*) on an Entisol (depth 100 cm) at Hisar (17° N), and (*b*) on a Vertisol (depth 150 cm) at Patancheru near Hyderabad (29° N), during the chickpea season in India. S.D., short duration; L.D., long duration. Source: Saxena (1987).

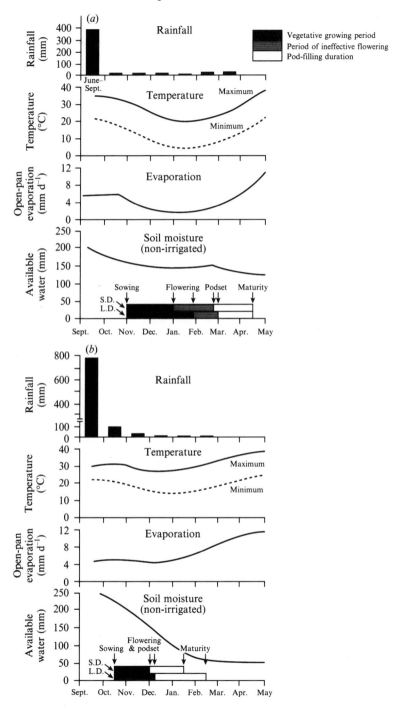

falls from 20 to 10 °C during the October-November sowing period. However, since sowing in India takes place after the end of the monsoon season, soil moisture at sowing may be marginal for good establishment and soil moisture declines throughout the growing season (Fig. 13.1). The onset of water stress is early and more severe in warmer regions, e.g., Patancheru in peninsular India (Fig 13.1*b*) and in west Asia, while stress is more mild at Hisar in northern India (Fig 13.1*a*), Pakistan or following winter sowing in West Asia (Saxena, 1987).

There is a very sharp optimum in sowing time, as shown by Alvarado (1972), and in Fig. 13.2. This optimum is related to thermal optima for germination, growth and rhizobial activity, to water relations, and to diseases. Germination and emergence are tolerant of a broad range of temperatures, from 10 to 35 °C. Emergence falls to 35% and 55% at 5 and 40 °C respectively (Singh & Dhalival, 1972). Nonetheless, over the temperature range in which emergence occurs readily, an increase in temperature of 1 °C will reduce the time to 50% emergence by 0.26–0.33 days (Roberts *et al.*, 1980). As in other crops, large seed size may confer advantages of rapid emergence and high percentage establishment (Townsend, 1972). Imbibition and germination appear relatively insensitive to water stress. This insensitivity is associated with a high seed water diffusivity: 0.1 cm^2 d^{-1}, compared with 0.008 for vetch at 40% v/v soil water (Hadas & Russo, 1974*a,b*). As in other crops, emergence is delayed by increasing sowing depth and by heavy (erosive) rainfall (Sivaprasad & Sundara Sarma, 1987).

Root development is slow below about 20 °C (Smoliak, Johnston & Hanna, 1972). Nitrogen fixation changes little from 15 to 25 °C but is low at 30 °C, owing partly to lower activity per nodule and partly to lower rate of nodule formation; seedlings in culture did not form nodules at 33 °C and died at 35 °C (Dart, Islam & Eaglesham, 1975).

The base temperature below which chickpea does not grow is about 8 °C (Huda & Virmani, 1987). The optimum temperature for seedling growth is 25–27 °C (Smoliak *et al.*, 1972; Dart *et al.*, 1975), which conforms with a broad plateau of photosynthetic rates from 19 to 28 °C (van der Maesen, 1972).

Fig. 13.2. Effect of sowing date and soil type on (*a*) plant mortality and (*b*) yield of chickpea at El-Damer, Sudan. Filled triangles, sandy clay loam; filled circles, loamy clay; open circles, clay. Source: Ageeb & Ayoub (1976).

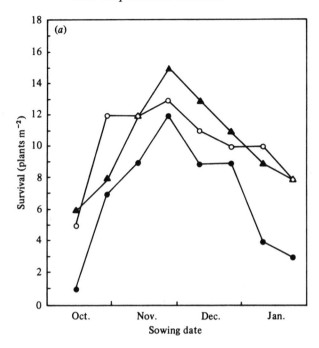

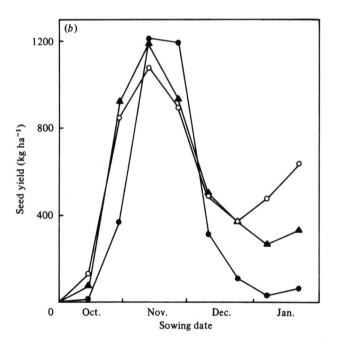

Photosynthetic rates of 1.1 mg CO_2 m^{-2} s^{-1} were recorded by van der Maesen (1972). These are high and comparable to maize. However reported maximum crop growth rates are only moderate (about 100 kg ha^{-1} d^{-1}) because of relatively low leaf area indices (to 3.1, Katiyar, 1980). Singh (1991) has demonstrated the range of leaf area index (LAI) which commonly occurs (from <1 to 3.2) by varying water availability; at peak LAI these crops intercepted between 60% and 100% of radiation.

Diseases are major contributors to the relatively low yields of chickpea and to the sharp optimum of sowing time (Fig. 13.2). Major diseases are wilts (*Rhizoctonia, Fusarium*), blight (*Ascochyta*) and rust (*Uromyces*) (Singh & Auckland, 1975); the main pest is *Heliothus armigera* in the Old World and *H. virescens* in the Americas (ICRISAT, 1978). It is largely to avoid these diseases and pests that chickpea is sown most often as an early dry-season crop in the wet-and-dry tropics. Humidity in itself has little effect on crop growth (van der Maesen 1972), but humidity and high temperatures together render disease infestation more likely. Plant death may be correlated with excess water and is linearly related to mean air temperature during early growth (Ageeb & Ayoub, 1976).

Sixteen per cent of Indian chickpeas are irrigated (van der Maesen, 1972). However, studies of chickpea water use are rare and not definitive (Gupta & Agrawal, 1976; Chatterjee & Sen, 1977; Nagarajrao, Mallick & Singh, 1980). At Hyderabad, unirrigated chickpeas use about 180 mm of water during the growing season; thrice-irrigated crops may use twice as much water (Saxena, Krishnamurthy & Sheldrake, 1982).

When crops are irrigated, time to 50% flowering is delayed, early formed flowers may fail to set seed and maturity may be delayed by 28–40 days compared with rainfed crops (ICRISAT, 1980; Saxena, Krishnamurthy & Sheldrake, 1981). Singh (1991) has quantified the converse effect, i.e., the extent to which water deficit accelerates chickpea development. On the basis of crops grown on a Pallustert at Patancheru, Singh found that the time from emergence to flowering, from flowering to the beginning of pod-filling, and from the beginning of pod-filling to maturity were reduced by 4.5, 3.1 and 3.8 day-degrees per mm kPa of evaporation deficit.

Thus at any one time, rainfed crops usually have a higher proportion of dry weight in pods than do irrigated crops (Fig. 13.3; also Singh, 1991). Whether the potential yield increment from irrigation is realised depends on good water relations and avoidance of disease by the longer-duration irrigated crop. When chickpeas are grown early in the dry season under high solar radiation, even irrigated crops experience leaf water stress in the middle of the day. Under such atmospheric conditions, early maturation may be advantageous: early-maturing rainfed chickpeas outyield genotypes having longer growth duration and higher yield potential (Fig. 13.4).

'Ineffective flowering', when early flowers form but fail to develop, is common in the cool tropics and may last for 48 days, from DI = 0.44 to 0.72 (ICRISAT, 1978). Temperatures below 15 °C cause low fruit set (N.P. Saxena, 1980; Savithri, Ganapathy & Sinha, 1980), whereas high temperatures increase flower and pod shedding, which in India account for of the order of 20% and 20–40% of the flowers and pods formed respectively (Varma & Kumari, 1978). In north India, maximum and minimum temperatures at flower formation are 25–31 and 10–14 °C

Fig. 13.3. Effect of irrigation on vegetative growth and yield of chickpea cv. Annigeri at Hyderabad, India. Irrigation was applied at vegetative stage (30 days after sowing) and at first flowering (38 days). Bar graphs show dry matter 67 days after sowing (at early pod-filling stage); numbers above bar graphs are final grain yield in t ha^{-1}. Source: Saxena *et al.* (1982).

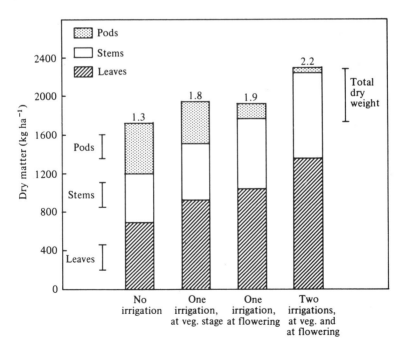

respectively and daylength is usually 11.5–12.5 h (van der Maesen, 1972). Owing to its daylength sensitivity, time to first flowering in chickpea may not be responsive to temperature, whereas duration of grain formation (from first perfect flower to maturity) decreases by 4–6 days per 1 °C increase in temperature from 15 to 25 °C (Roberts *et al.*, 1980). There is, however, no close relationship between duration of grain-filling and yield (Minchin *et al.*, 1980). Extended filling periods, as at high latitudes, may be associated with both longer effective grain-filling periods (EGFP) and higher grain growth rates: e.g., 50 kg grain ha^{-1} d^{-1} at Hissar (29° N) compared with 28 kg ha^{-1} d^{-1} at Hyderabad (17° N) (ICRISAT, 1978). On the other hand, longer filling, associated with flowering at high-order nodes, may be inefficient because yield components decline at successive apical nodes (ICRISAT, 1976; Sheldrake & Saxena, 1979), and long-season cultivars appear not to retranslocate dry matter from roots to fruit during grain-filling (Minchin *et al.*, 1980).

Chickpeas show a wide yield/population plateau and have the ability to compensate for flower removal (ICRISAT, 1978). Furthermore, being indeterminate, they produce appreciable dry matter after first flowering. Katiyar (1980) suggested that 75–85% of total top dry matter was accumulated after first flowering in early- and mid-season cultivars. It is unfortunate that current cultivars, through disease susceptibility, intoler-

Fig. 13.4. Relationship between days to flower of chickpea and (*a*) rainfed yield, (*b*) drought tolerance index (ratio of dryland to irrigated yield) on an Alfisol (open circles) and a Vertisol (filled circles) at Hyderabad, India. Sources: Saxena *et al.* (1981).

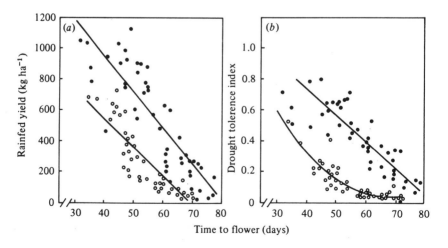

ance of high temperature, and ineffective partitioning of dry matter to reproductive growth, seldom yield above 1.5–1.7 t ha^{-1} in the tropics.

13.5 Crop/soil relations

13.5.1 *Soil physical properties*

Chickpeas are grown in a wide range of soil textural classes, but medium- to-heavy-textured soils are preferred (Singh & Manjhi, 1975). Heavy-textured soils (Vertisols and soils with vertic characteristics in other orders) are used in India (in Maharashta and on the Deccan Plateau), in Ethiopia (Bezuneh, 1975) and Tanzania (Clegg, 1947). In north India, chickpeas are grown in the moderately heavy grey and brown alluvial soils (mainly Inceptisols) of the upper Ganges Basin. Heavy soils are favoured in the wet-and-dry climate regions of India because of their high residual moisture content in the post-monsoon growing period. The available water content of Vertisols at Hyderabad, for instance, while affected by depth, is usually considerably greater than in adjacent coarser-textured Alfisols (Fig. 13.5), which are less commonly used for chickpeas. In the western parts of the Punjab, Haryana and Rajasthan, lighter soils, mostly sandy loams, are used (van der Maesen, 1972). Deep sowing (10 cm as against the normal 5 cm) by farmers in India, to place the seed near the moist soil layer in Vertisols, is liable to result in late emergence of plants which have fewer nodules and lower acetylene reduction activity (ICRISAT, 1986).

Soil compaction, heavy soil texture and excess water are liable to induce poor aeration, to which the crop is susceptible (Kay, 1979). Under these conditions the incidence of wilt diseases may increase. A coarse seedbed tilth is preferred (Westphal, 1974; Duke, 1981); a fine tilth may reduce germination (van der Maesen, 1972). However, the crop is sometimes also sown directly into moist uncultivated soil following wet rice (van der Maesen, 1972).

13.5.2 *Soil chemical properties*

Most chickpea cultivars are adapted to rainfed soils of only moderate fertility. Because they have been selected for yield stability rather than high yield, some consider that they are unresponsive to fertilisers (e.g., Cummings, 1976). At Hyderabad, responses of chickpeas to nitrogen and phosphorus have been inconsistent and generally low (ICRISAT, 1980). The ability of chickpeas to yield adequately on marginal soils is probably related to their extensive (Allen & Allen, 1981) and deep root system (over 1.2 m; Dart & Krantz, 1977) and their ability to fix nitrogen. Extremely high fertility leads to luxuriant vegetative growth and may depress grain yield through poor pod set (van der Maesen, 1972).

The nitrogen-fixing symbiont in chickpea (*Rhizobium loti*) is highly specific (Corbin, Brockwell & Gault, 1977; Dart *et al.*, 1975). The proportion of nitrogen in the plant derived from symbiotic sources varies greatly depending on the presence of chickpea rhizobia in the soil, their efficiency of nitrogen fixation in association with the particular chickpea genotype grown, and environmental conditions, including the nitrogen status of the soil. Somasegaran, Hoben & Gurgun (1988) showed that the kabuli type has a greater potential to fix nitrogen than the desi type. Strains of rhizobia compete differently according to soil type: in an

Fig. 13.5. Average weekly available water storage capacity (AWC) of Vertisols and Alfisols, as affected by soil depth, at Hyderabad, India. Data based on 1901–70 rainfall records. Low AWC (50 mm), shallow Alfisols, medium AWC (150 mm), deep Alfisols and medium-depth Vertisols; high AWC (300 mm), deep Vertisols. Adapted from Kanwar, Kampen & Virmani (1982).

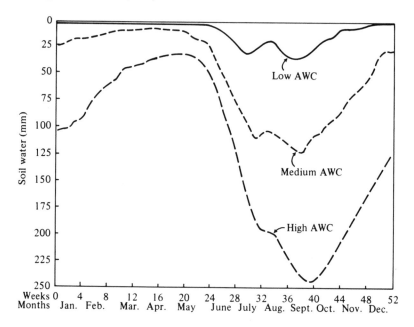

Ultisol there were significant strain differences and genotype × strain interactions, while in an Oxisol there was no strain competition (Somasegaran *et al.*, 1988). Chickpea genotype × rhizobia strain interaction has also been demonstrated on a Mollisol in India by Chandra & Pareek (1985).

Symbiotic nitrogen sometimes accounts for almost all the nitrogen in the crop (Saxena, 1979). However, in some areas of India, chickpeas are not nodulated, or poorly nodulated. This may reflect low rhizobial populations or poor soil moisture conditions (ICRISAT, 1976). Vertisols at Hyderabad appear to contain adequate numbers of rhizobia for nodulation, but numbers are reduced by wet rice cropping so that reinoculation is necessary (ICRISAT, 1981). Soils examined in Sudan lacked chickpea rhizobia (Salih, 1979).

Inoculation can increase yield (e.g., Vaishya & Gajendragadkar, 1982; Saxena & Singh, 1987; Namdeo *et al.*, 1989), have no effect (e.g., Dart *et al.*, 1975) – presumably where indigenous rhizobia are adequate or soil nitrogen is high – or may reduce yield (e.g., Subba Rao, 1976). The pattern of nodule development and period of effectiveness is influenced by factors such as growth duration of the cultivar, soil moisture and nutrient stress (e.g., iron deficiency) (ICRISAT, 1976, 1978, 1980). Nearly all nodules observed on plants at Hyderabad were formed in the first 2 or 3 weeks after sowing under rainfed cool-season conditions (ICRISAT, 1980). Irrigation resulted in an approximately 40-fold increase in nodule weight and number, associated with an increase in grain yield from 1.3 to 3 t ha^{-1} (ICRISAT, 1981). At Hyderabad, chickpeas on Alfisols tended to have higher nodule number and weight per plant than those on Vertisols receiving similar treatments (ICRISAT, 1978). On Vertisols at the same location about 90% of total nodules were formed in the top 15 cm of soil but in lighter soils, such as Entisols, substantial nodule development has been observed to a depth of at least 30 cm (ICRISAT, 1986). Soil properties may affect the distribution of nodules through their effects on the chemotaxis of rhizobia towards exudates from the seed coat (Gitte, Rai & Patil, 1978) and roots (Rai & Prasad, 1986).

Nitrogen fertiliser would be expected to increase yield where plants do not fix sufficient nitrogen and the soil is low in available nitrogen. Thus Islam (1978) reported an increase in yield of a desi-type chickpea in India of 62% from application of 150 kg nitrogen ha^{-1} in comparison with 65% from inoculation. Small starter doses (20 kg nitrogen ha^{-1}) have increased yield by 42% over the uninoculated control (Kadem *et al.*,

1977). In Sudan responses to nitrogen up to 120 kg ha^{-1} have been obtained under irrigation (Saxena, 1979). However, widespread field experiments in India showed yield depression due to nitrogen in 6 out of 16 experiments, even where nitrogen rates were as low as 15–25 kg nitrogen ha^{-1} (Subba Rao, 1976). While these results need to be interpreted in the light of information on soil nitrogen status and the adequacy of other nutrients and soil moisture for the resulting greater growth, the study of McNeil, Croft & Sandhu (1981) suggests that the response to fertiliser nitrogen may not be a reliable index of potential yield following inoculation by an effective strain of rhizobium. The findings of the Regional Pulse Improvement Project in India and Iran indicate that most regions supply sufficient rhizobia for local cultivars so that there is no widespread benefit from applying nitrogen (Saxena, 1979).

Responses of chickpea to phosphorus are reported in Indian literature (e.g., Chundawat, Sharma & Shekawat, 1976); they are found to be improved by irrigation (e.g., Singh & Sharma, 1980). The responses include improved root growth and nodulation (Chowdhury, Ram & Giri, 1975). Phosphorus uptake by chickpea is fairly small: e.g., 13.5 kg phosphorus ha^{-1} in a crop yielding a 3.0 t ha^{-1} of grain and 4.5 t ha^{-1} of straw (Saxena, 1979). Indian rates of fertiliser application are correspondingly low: about 12–25 kg phosphorus ha^{-1} (van der Maesen, 1972; Suryawanshi & Chaudhari, 1979). The general conclusion is that the crop is very efficient in taking up phosphorus from low-phosphorus soils (e.g., ICRISAT, 1978, 1980) but can benefit from phosphorus applications. Inoculation of chickpea with vesicular-arbuscular mycorrhizal (VAM) fungi has increased nodulation, grain yield, and nitrogen fixation and phosphorus uptake in the field (e.g., Singh & Tilak, 1989), while in the glasshouse inoculation with the 'phosphate solubilising' bacteria *Pseudomonas striata* and *Bacillus polymyxa* has had similar effects (Alagawadi & Gaur, 1988).

Responses to potassium are rarely reported; heavy-textured soils such as Vertisols are usually adequately supplied with potassium.

van der Maesen (1972) concluded that a soil pH range of 6–9 is favourable to the growth of chickpea. While there are no data to indicate a lower pH limit, Paliwal & Anjaneyulu (1967) found that the crop was badly affected by pH values above 9, which were all associated with salinity and to which the crop is known to be susceptible (Saxena *et al.*, 1982). Irrigation water with a conductivity of 10 dS m^{-1} can reduce yield by 55% (Saxena, 1979), while in pots, yield was reduced by 50% at 2.5 dS m^{-1} and seed set prevented at 4.5 dS m^{-1} (ICRISAT, 1986). Salinity directly affects germination as well as crop growth, although the affect varies with variety (Kumar *et al.*, 1980; Goel & Varshney, 1987). In arid and semi-arid areas of India where chickpea is grown, saline groundwater, dominated by either chloride or sulphate salts, is used as supplementary irrigation and can cause problems. Manchanda & Sharma (1989) showed that chickpea seed yield was more tolerant to sulphate than to chloride salts of sodium (Fig. 13.6). There was practically no seed development when the chlorine content of plants reached 5%.

While the main effect of salinity is on the plant rather than on the rhizobia (Balasubramanian & Sinha, 1976; Lauter, Munns & Clarkin, 1981) survival of chickpea rhizobia may be lower than those of soybean in salt-affected soils (Elsheikh & Wood, 1990). Calcareous, high-pH soils may induce iron deficiency (M.C. Saxena, 1980).

Fig. 13.6. Effects of chloride-dominated (open circles) and sulphate-dominated (filled circles) salinities on the average seed yield of the chickpea cultivars H355 and H208 grown in the glasshouse in a sandy soil (Typic Torripsamment) at five EC values ranging from 1.8 to 8.0 dS m⁻¹, dominated either by chloride (Cl : SO_4 = 7 : 3) or sulphate (Cl : SO_4 = 3 : 7) salts of sodium, calcium and magnesium (4 : 1 : 3). Source: Manchanda & Sharma (1989).

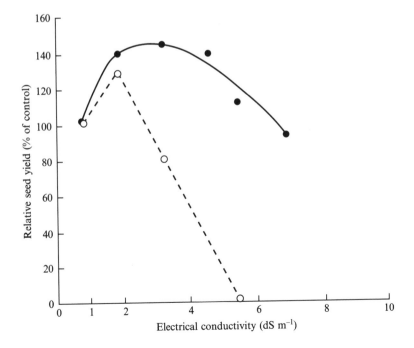

13.6 Place in cropping systems

As already indicated, chickpeas are normally grown as a rainfed crop in the early dry season of wet-and-dry climate regions, often almost wholly on residual soil water after the monsoon rains (Dart & Krantz, 1977; Fig. 13.1). In north India this supply is sometimes augmented by winter rain and to a small extent by irrigation (van der Maesen, 1972). Supplementary irrigation is rarely used on heavy soils since it impairs aeration (Saxena & Yadav, 1975).

Chickpea is grown on 15–25% of the arable area in central and north-west India, and on 5–15% of the arable area throughout the remainder of India and in Bangladesh and Pakistan (Reddy *et al.*, 1987). Cropping sequences in India in which chickpea figures as a component are considered by van der Maesen (1972), Krantz *et al.* (1974), Chowdhury (1974) and Saxena & Singh (1987). On Vertisols where cropping is difficult in the wet season, chickpeas, preceded by a 'monsoon fallow', may be the only crop grown in the year. This pattern provides the best possible conditions for soil water storage. On deep Vertisols with inherently high water storage potential (see Fig. 13.5), which, however, comprise less than 10% of the total Vertisol area of India, two crops a year are feasible, and chickpeas may be grown immediately after short-season summer crops such as pearl millet, sorghum or maize. On Alfisols there is rarely enough soil water storage capacity for early dry-season cropping.

Chickpeas may be grown as a sole crop, but in India they are often sown mixed with other crops (van der Maesen, 1972). The most important of these is wheat; others, in descending order of importance, are barley, mustard, linseed, safflower and vegetables. Where the summer crop is sorghum/pigeonpea (*Cajanus cajan*), chickpeas may be interplanted in the maturing pigeonpea after the sorghum has been harvested at the end of the wet season.

Further reading

ICRISAT (1987). *Adaptation of Chickpea and Pigeonpea to Abiotic Stresses.* Patancheru, India: International Crops Research Institute for the Semi-Arid Tropics.

Saxena, M.C. & Singh, K.B. (eds.) (1987). *The Chickpea.* Wallingford, UK: CAB International, 409 pp.

IV

NON-CEREAL ENERGY CROPS

14

Non-cereal energy crops in tropical agriculture

14.1 Introduction

The group of crops reviewed in this chapter, the four most important of which are considered in detail in Chapters 15–18, comprise species other than cereals that are grown in the tropics largely as human energy foods. Their somewhat clumsy designation as 'non-cereal energy crops' is the only properly inclusive and exclusive title that can be applied to such a diverse group. Thus cassava (*Manihot esculenta*, Euphorbiaceae) is a dicotyledonous short-lived perennial tuberous shrub grown as an annual or biennial; sweet potato (*Ipomoea batatas*, Convolvulaceae) is a dicotyledonous perennial tuberous herb grown as an annual; yams (*Dioscorea* spp., Dioscoreaceae) are monocotyledonous annual tuberous herbs; and bananas (*Musa* spp., Musaceae) are monocotyledonous perennial herbs with aerial fruit (Fig. 14.1).

Because in all instances the edible portion has a high water content and is therefore perishable, the crops are normally consumed locally in the fresh or home-processed state. There are two major exceptions: cassava, which although very important as a human food is also processed industrially for starch and liquid fuel and is exported in the dried form as stockfeed; and the sweet banana, the demand for which in high-income countries is such that it is exported fresh in large quantities.

There is a wide range of other less important non-cereal energy crops. They include taro and eddoe, varieties of *Colocasia esculenta*, grown throughout the Pacific and the West Indies; tannia, *Xanthosoma sagittifolium*, cultivated in the Pacific, the West Indies and West Africa* edible canna, *Canna edulis*, found mainly in Latin America; arrowroot, *Maranta arundinacea*, largely confined to the West Indies; sago,

*In West Africa, both *Colocasia* and *Xanthosoma* are known as cocoyam.

Metroxylon spp., and breadfruit, *Artocarpus altilis*, utilised in the South Pacific and southeast Asia. The white or Irish potato (*Solanum tuberosum*) is also grown in the tropics in cool highland zones. Summary accounts of these crops are given by Kay (1973) and

Fig. 14.1. Gross morphology of non-cereal energy crops: (*a*) yam, (*b*) sweet potato, (*c*) cassava, (*d*) banana. Approximate scales are ÷20, 10, 30 and 60 respectively.

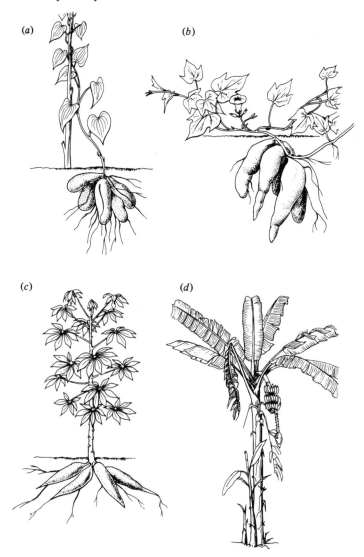

Purseglove (1968, 1972). Wilson (1977) has reviewed the physiology of 'root crops', which includes all crops mentioned above except bananas, sago and breadfruit.

14.2 Regional production

Table 14.1 gives area, yield and production of the non-cereal energy crops as far as FAO statistics permit. There is a sharp distinction between cassava and bananas on the one hand, which are grown almost wholly within the tropics, and sweet potatoes and other roots and tubers on the other hand, which are of far greater importance in temperate zones. About 85% of the world's sweet potato production is in China. A high proportion of other roots and tubers grown outside the tropics is contributed by the white or Irish potato.

Production of the tropical root and tuber crops, principally cassava, sweet potato and yams, is changing with time owing to changes in dietary preference and markets for livestock feed. Production continues to increase in Africa (where sweet potato production is now about 80% higher than in 1960) but generally root and tuber crops are less favoured and human consumption declines as discretionary income increases. The marked shifts in production and consumption are illustrated with data for sweet potato in Fig. 14.2.

Africa grows over half the world's cassava, one-quarter of the tropical sweet potato and most of the world's yams, the vast bulk of which

Fig. 14.2. Trends in production of sweet potato in Africa, Asia, Latin America and developed countries, principally Japan and USA. Source: Horton (1988).

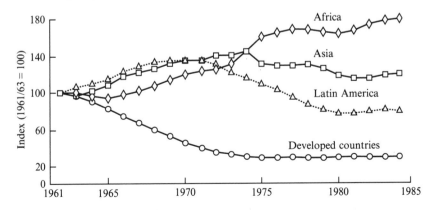

Table 14.1. *Area, yield and production of non-cereal energy crops in the tropics, in 1991*

Crop	Tropical Africa	Tropical America	Tropical Asia	Tropical Oceania	Total Tropics	World	Tropics as % of world
Area (m ha)							
Cassava	8.92	2.53	3.72	0.02	15.19	15.67	97
Sweet potato	1.32	0.25	0.94[a]	0.12	2.63	9.26	28
Other	4.35	1.07	1.82	0.09	7.33	22.25	33
Total roots and tubers	14.60	3.85	6.47	0.23	25.15	47.18	53
Yield (t ha^{-1})							
Cassava	7.73	11.50	12.90	11.23	9.63	9.81	98
Sweet potato	4.80	6.89	15.03	4.83	5.86	13.63	43
Other roots and tubers	7.36	10.72	12.71	10.44	9.66	13.25	73
Production (m t)							
Cassava	68.93	29.06	48.14	0.19	146.32	153.69	95
Sweet potato	6.36	1.74	6.75	0.57	15.42	126.19	12
Other	32.04	10.48	27.37	0.94	70.83	294.77	24
Total roots and tubers	107.33	41.28	82.26	1.70	232.57	574.64	40
Banana and plantain	5.96	20.23	19.07	1.14	46.40	47.66	97

After FAO (1992).
[a]China, not included in tropical Asia, has 6.41 m ha sweet potato yielding 107.19 m t.

are for human consumption. By contrast tropical Asia is shifting away from root and tuber crops and an increasing proportion of those which continue to be grown are used in manufacturing or stockfeed.

The production of bananas, which includes sweet bananas and plantains, is well distributed throughout Africa, America and Asia: the proportion of plantains is highest in tropical Africa (about two-thirds of world production), and the proportion of sweet bananas highest in tropical America.

14.3 Role in tropical human nutrition

The primary nutritional role of non-cereal energy crops in the tropics is as carbohydrate for subsistence and local sale. Protein content is low compared with that of cereals, and although their contribution to the total protein intake of tropical peoples may in some regions be high, this is because in such areas they form a very large proportion of total food intake.

Below is given the contribution of non-cereal energy crops to total energy and protein intake in three tropical countries (Rachie, 1977):

	India	Uganda	Nigeria
% of total energy intake	1.4	43.7	33.1
% of total protein intake	1.0	18.0	29.8

The contrast between India, predominantly semi-arid and where cereals are the main source of energy (including rice in high-rainfall and irrigated areas), and Uganda and Nigeria is striking. Much of Uganda has a long enough growing season for plantains to be a staple crop, and in the higher-rainfall zones of Nigeria cassava and yams are prime sources of food energy. The influence of rainfall on the contribution of non-cereal energy crops to energy and protein intake is shown clearly in data given below from different regions of Nigeria (FAO, 1966):

	Northern Nigeria (750 mm rainfall)	Western Nigeria (1250 mm rainfall)	Eastern Nigeria (1850 mm rainfall)
% of total energy intake	17.2	53.3	68.3
% of total protein intake	9.8	41.3	65.4

An extreme instance of dependence on non-cereal crops is in the highlands of New Guinea, where sweet potatoes may comprise 80–90% of the energy intake of subsistence cultivators (Hipsley & Kirk, 1955).

In regions where cereals and root and tuber crops are both available in large quantity as alternative sources of food energy, cereals are usually the preferred item of diet. For example, in East Java, where the main carbohydrate crops grown are rice, maize and cassava, the more well-to-do sectors of the population eat rice, the somewhat less well-to-do eke out their rice with kibbled maize and consume some cassava, and the poorest sectors of the population rely heavily on cassava for their energy requirements. Cassava is a staple food of the poor in many tropical countries.

In Chapter 4 a comparison was made of the yield of food energy from cereals and non-cereal energy crops in the tropics, which indicated that on a per crop basis edible energy per hectare from non-cereals was on average substantially greater than that from cereals. However, when compared on the basis of edible energy per hectare per day the only non-cereal energy crop that appeared to approach rice and maize was sweet potato (Table 4.2).

Table 14.2, which parallels Table 4.3, compares the average protein yield from non-cereal energy crops and crop legumes in the tropics. Although percentage protein of the non-cereals is low (though bearing in mind that it is expressed on a fresh weight basis) their high yield compensates for this. Average protein production per hectare in the

Table 14.2. *Comparative protein yield of non-cereal energy crops and legume crops*

Crop	Average tropical yield (t ha^{-1})	Protein content (%)	Average protein yield (kg ha^{-1})
Non-cereal energy crops			
Cassava	9.63	1.6	154
Sweet potato	5.86	1.6	94
Yams	7.00	2.0	140
Bananas	13.00	1.1	143
Legume crops			
Soybean	1.34	38.0	509
Groundnut	0.89	25.5	227
Beans	0.60	22.0	132
Chickpea	0.66	20.0	132

tropics from cassava, sweet potato, yams and bananas is of the same order as that from beans and chickpeas, though well below the average protein yield of groundnuts and soybeans.

In Section 4.3.2, reference was made to the energetic efficiency of non-mechanised crop production, and data were presented to show that the average energy output/input ratio for hoe cultivation of rainfed cereals was normally within the range 15–20 : 1. From the limited information available (Chandra, Evenson & de Boer, 1976; Norman, 1979; Chandra, 1981) it seems that the energetic efficiency of non-cereal energy crop production is often higher than that of cereal production. Table 14.3, which gives the output/input ratio of various cropping enterprises from Fiji, illustrates this. Part of the advantage lies in the long period over which the non-cereal energy crops continue to accumulate carbohydrate without a great deal of input on the part of the cultivator. The inputs of land preparation, planting and early weeding are comparable to those of cereals, but in the latter half of the long growing period of annual root and tuber crops there is little to be done except the final harvesting.

14.4 Place in tropical cropping systems

A common feature of the non-cereal energy crops, with the exception of sweet potatoes, is their long growing period. Sweet potatoes, though perennial, are normally harvested 3–6 months after planting, but the average crop duration of yams is about 8 months and harvesting may extend up to 18 months, while cassava, a perennial, may remain in the ground for 6–24 months. (Jennings, 1970, notes that in East Africa cassava may be allowed to grow for 2–6 years as a famine reserve.) Bananas and plantains are perennials and are cultivated as such, though in shifting cultivation systems in the Zaïre rainforest, with no manure or fertiliser added, they may be harvested only once or twice (Miracle, 1967).

Table 14.3. *Energy output/input ratio of cropping enterprises in Fiji*

	Fijian farms	Indian farms
Cassava	52:1	42:1
Sweet potato	60:1	44:1
Yams	66:1	—
Rice	17:1	9:1
Maize	39:1	22:1

After Chandra (1981).

The pattern of crop production, storage and consumption in farming systems based on these crops differs from that of systems based on cereals. In cereal production, which is normally associated with tropical climates with a defined dry season or seasons, and hence defined crop maturation periods, the produce is harvested at one, sometimes two, specific times of the year and stored for later consumption. This is feasible since the moisture content of cereal grain at maturity is low and storage is not difficult. On the other hand, in farming systems based largely on tuber crops or bananas, normally associated with extended cropping seasons, the produce is available in the fresh form for the greater part of the year and is harvested as required, since storage is difficult. Even the quick-growing sweet potato may be harvested over a period of 3 months (Kassam, 1976) and its availability can be extended by staggered or sequential plantings where the rainfall regime permits. Cassava, once harvested, is extremely perishable, though simple storage methods are being developed (Lozano, Cock & Castano, 1978), and bananas must be consumed soon after harvest. Yams may be stored, though losses are frequently high (Coursey, 1967*b*); sweet potatoes are normally consumed when harvested.

The above refers, of course, to crops grown for subsistence or local sale. In high-technology plantations growing bananas for export the produce may be packed and in refrigerated storage on the day of harvest. Cassava, when grown for export as an animal feed, is first chipped and then sun-dried (Best, 1978).

With the exception of taro (*Colocasia esculenta*), which is tolerant of wet conditions and is often irrigated, and plantation bananas in climates with a defined dry season that may be irrigated by sprinklers, the main non-cereal energy crops are grown almost wholly under rainfed conditions. Their position in the cropping pattern varies greatly; perhaps the two most important factors are their fertility requirement and whether, in the spectrum of foods available to the cultivator, they are a preferred or a 'fall-back' diet item. In shifting cultivation systems in the Zaïre rainforest, Miracle (1967) recorded the following number of occurrences of individual crops grown in the first or last year of the cropping phase:

	First year	Last year
Bananas/plantains	44.1	16.7
Cassava	8.4	41.7
Yams	5.9	8.3

Here bananas and plantains are the preferred crops and cassava is a reserve food. Furthermore, cassava will grow under low-fertility conditions. There is a further reason why bananas figure prominently as the first crop: since they are robust and are planted at a wide spacing, they can be established as soon as tree-felling begins and before the scrub and felled timber is cleared up and burned. As already mentioned (Chapter 1), cassava is an ideal end-of-cropping-phase species: tolerant of low fertility, able to compete with recovering fallow vegetation, and capable of storing a food reserve for an extended period. In the yam zone of West Africa, where yams are a preferred staple food, they are usually grown in the first year of the cropping phase (Kassam, 1976).

Tuber crops are often grown on mounds or ridges to encourage root and tuber development in the loose soil and to make harvesting easy (Miracle, 1967; Irvine, 1969; Kassam, 1976). By hoeing topsoil into a mound and adding organic matter, 'islands' of higher fertility are created; less demanding, or less important, crops are grown in the intervening flat area. The sweet potato cultivators of the New Guinea highlands grow their staple crop either on mounds or in a 'gridiron' pattern (Brookfield & Hart, 1971), where drainage ditches are dug in two dimensions and the soil heaped up on the rectilinear platforms thus formed. The function of mounds in tropical cropping is discussed in detail by Denevan & Turner (1974).

As with cereals and legumes, the non-cereal energy crops are more likely to be found in crop mixtures than as sole crops, though where cassava and bananas are being farmed on a strictly commercial basis they are grown in monoculture. Furthermore, since they are associated with climatic regions of long growing season, the cropping systems are often very complex. Okigbo & Greenland (1976) and Miracle (1967) give numerous examples of African multiple cropping systems in which tuber crops and bananas are important components; Wilson *et al.* (1992) provide a detailed description of root-crop-based cropping systems in the Caribbean and of the advantages of mixed cropping; and Parsons (1970) describes the banana–coffee systems of Uganda. In Indonesia, where cassava is a major diet item of the poorer population, the crop is normally grown with upland rice and maize: the cereals are harvested during or at the end of the wet season while the cassava continues to utilise stored soil water during the dry season and to provide a year-round food reserve.

Since, except for sweet potatoes, the main non-cereal energy crops require a long growing season, they are important in climatic regimes that also support the growth of perennial tree cash crops such as coffee and cocoa, and are hence often grown in the establishment phases of such crops. In this way land that would otherwise remain unexploited by the roots of widely–spaced young trees is utilised to grow food, and in the case of the tall-growing banana or plantain the additional benefit of shade is conferred. Bananas and plantains are used extensively as shade 'trees' for coffee in tropical America and cocoa in West Africa (Burden & Coursey, 1977).

Further reading

Wilson, L.A., Rankine, L.B., Ferguson, T.U., Ahmad, N., Griffith, S. & Roberts-Nkrumah, L. (1992). Mixed root-crop systems in the Caribbean. In *Field Crop Ecosystems*, ed. C.J. Pearson, pp. 205–42. Amsterdam: Elsevier.

15

Cassava (Manihot esculenta)

15.1 Taxonomy

The genus *Manihot*, which comprises a large number of ill-defined species, is of the family Euphorbiaceae. Rogers & Appan (1970), applying taximetric methods, classified the genus into 75 species. All have 36 chromosomes and all show regular bivalent pairing, but there is evidence of polyploidy (Jennings, 1976). Magoon, Krishnan & Bai (1969) suggest that *Manihot* species are segmental allotetraploids.

Cassava, *Manihot esculenta*, is not known in the wild state. It may be crossed with a number of *Manihot* species. Rogers & Appan (1970) nominate three groups, each with two to ten members, that are related to cassava. The closest relatives appear to be *M. aesculifolia*, *M. rubricaulis* and *M. pringlei*.

Purseglove (1968) distinguished cassava cultivars by two criteria: hydrocyanic acid (HCN) content and maturity time. On the basis of HCN, cultivars may broadly be divided into: sweet cassavas, of low HCN content, in which HCN is confined to the phelloderm of the tubers; and bitter cassavas, of high HCN content, in which HCN is usually distributed throughout the tuber. However, the distinction has no taxonomic basis (Nye, 1991). Levels of HCN do not correspond with any other known morphological or ecological feature, except for the very general relationship to maturity time given below.

On the basis of maturity, cultivars may be divided into:

1. Short-season types that mature in 6–11 months and which cannot be left in the ground for longer than 9–11 months without serious deterioration; these are often sweet cassavas.
2. Long-season types that take at least 12 months to mature and have better in-ground storage capability (some types may be left in the ground for 3–4 years without serious deterioration); these tend to be bitter cassavas.

A detailed classification of cassava cultivars based on morphology has been made by Fleming & Rogers (1970): distinguishing characters include branching habit, the nature of the scars on the stem, and the number, length and width of leaf lobes.

15.2 Origin, evolution and dispersal

In view of its complex taxonomy and because, as with other root crops, the survival of plant parts for subsequent archaeological discovery is poor, the origin and evolution of cassava is somewhat speculative. According to Jennings (1976), the evidence is too tenuous to determine whether present-day cassava is descended from one or several species. Ugent, Pozorski & Pozorski (1986) favour a polyphyletic origin. Whatever its origin, it appears likely that the variability of cultivated cassava has been increased by hybridisation with several wild types.

Carter *et al.* (1992) review the domestication and spread of cassava. There are three possible areas of domestication: northeastern Brazil, Mesoamerica (Mexico, Guatemala, El Salvador, Nicaragua) and Venezuela. (The first two of these are the main centres of diversity of the genus.) De Candolle and Vavilov favoured northeastern Brazil, but Rogers (1965) is of the opinion that cassava developed in Mesoamerica. He found that intensive cultivation of cassava corresponded closely with the area of influence of ancient Mayan civilisation. Species closely related to cassava, *M. aesculifolia* and *M. pringlei*, are found wild in the area. Desiccated plant material identified as cassava and dated 2500 BP has been found in Mexico.

Renvoize (1972) suggests that sweet cassava was first domesticated in Mesoamerica and bitter cassava in Venezuela. In the American tropics, sweet types are more widely distributed than bitter types: the former appear to have been associated, as a secondary crop, with ancient civilisations, whereas the latter seem to have been domesticated later and, where they occur, tend to be the staple crop.

Selection pressure during domestication was for large storage organs, more erect and less-branched growth, and ability to strike from cuttings (Jennings, 1976). Branching occurs where inflorescences are formed; hence selection for reduced branching produces less floriferous types. This, together with the lack of propagation by seed, has led to sparse flowering and lower fertility in modern cassava cultivars.

The species spread eastward in post-Columbian times (Jennings, 1976). It was taken by the Portuguese to the west coast of Africa in the late sixteenth century, and arrived in East Africa and India in the late eighteenth century. It was also presumably taken by the Spanish to the Philippines, but whether it spread into southeast Asia by this route or from India is uncertain. However, in spite of its early appearance in the Old World, cassava remained a crop of minor importance outside tropical America until the twentieth century.

Area, yield and production data for countries producing more than 3 m t annually are given in Table 15.1. In Brazil, the biggest producer, cassava is grown for industrial alcohol as well as for human consumption and stockfeed. Productivity per hectare varies widely: the high values for Brazil and Thailand might be expected, since human pressure on land is not as heavy as in the other countries listed, while productivity is lowest where cassava is grown as an intercrop in mixed gardens, for subsistence in the wet tropics, e.g., Ghana, Zaïre and Uganda.

15.3 Crop development pattern

Cassava is a short-lived erect perennial shrub, with lobed leaves and lanceolate-obovate leaflets (Fig. 14.1). It is planted vegetatively from hardwood stem cuttings, usually about 30 cm long, and grows to 1–5 m high. Plants produce 5–20 storage roots or 'tubers'. The leaves and the

Table 15.1. *Major cassava-producing countries: area, yield and production in 1991 of countries producing more than 3 m t per annum*

Country	Area (m ha)	Yield (t ha^{-1})	Production (m t)
Brazil	1.96	12.67	24.63
Thailand	1.50	13.53	20.30
Nigeria	1.70	11.77	20.00
Zaïre	2.39	7.63	18.22
Indonesia	1.32	12.39	16.33
Tanzania	0.60	10.37	6.27
India	0.29	19.38	5.60
Paraguay	0.24	16.25	3.90
Mozambique	0.97	3.80	3.69
Ghana	0.54	6.73	3.60
Uganda	0.38	8.82	3.35
China	0.23	14.37	3.32
Vietnam	0.28	10.53	3.00

After FAO (1992).

parenchyma of the tuber may, and the tuber phelloderm does, contain HCN, within the range 10–370 mg HCN kg^{-1}, Leaf, stem and tuber HCN may deter pests and predators.

Shallow planting (to 10 cm) causes roots to form at the basal node of the cutting; roots and leaves usually form within 5–10 days of planting. Starch deposition in roots may be first observed 25 days after planting (Hunt, Wholey & Cock, 1977). The number of roots having the capacity to thicken is fixed within 3 months of planting, at about development index (DI) = 0.3 in crops of 9–12 months duration (CIAT, 1973). The onset of bulking – an increase in tuber volume – occurs in the second month after planting (Wholey & Cock, 1974). There is some evidence, mostly from small plants grown in pots in greenhouses, that short days (10–12 h) promote bulking and result in the highest tuber weight without necessarily affecting tuber number (Bolhuis, 1966; Hunt *et al.*, 1977). The anatomy and development of the tuber is reviewed in Hunt *et al.* (1977).

Although early development events are the same for many varieties, genotypes appear to have a wide maturity range depending on environment and season of planting. In the wet tropics maximum tuber dry weight and starch concentration are usually reached at the same time, about 12 months after planting (Table 15.2). In this climate time of planting has little or no effect on time to maximum yield or on yield itself. By contrast, in the wet-and-dry tropics, crops may reach acceptable tuber weight within one growing season, say 6 months, but the common developmental pattern is for crops to reach a peak biomass at the end of the first wet season, to lose leaves and (presumably) roots in the dry season, then to make further growth, mostly tuber, at relatively low leaf area index (LAI) until harvest at 14–24 months. This pattern, first described definitively by Cours (1951), probably owes more to the seasonality of environment than to ontogeny (Fig. 15.1). However, a low LAI (usually <3) when DI > 0.6 is sufficiently common to imply a developmental trend.

Growth duration may be prolonged by grafting *M. glaziovii* on to *M. esculenta* rootstock (de Bruijn & Dharmaputra, 1974). Under this system (called Mukibat after its originator, a Javanese farmer), the vigorous scion leads to greater canopy development and prolongs the period of high leaf area, so that the yield advantage per crop (not necessarily per unit time) of the Mukibat system over normal cassava increases with increasing crop duration.

Time of harvest is determined by not only the optimum for a particular genotype, planting time, environment or grafting system, as shown in Table 15.2, but also by the farmer's dietary or financial need: individual plants or tubers may be harvested progressively. In the wet tropics, harvesting is usually at 12–18 months: e.g., in Nigeria (Hahn *et al.*, 1979; Ezedinma *et al.*, 1981), Malaysia (Chew, 1974) and Venezuela (Arismendi, 1973). In the wet-and-dry tropics, crops are normally harvested at either 4–8 or 14–24 months.

Fig. 15.1. Growth and development of cassava in a wet-and-dry tropical climate. Source: IITA (1990), modified from Cours (1951).

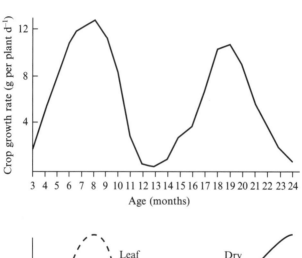

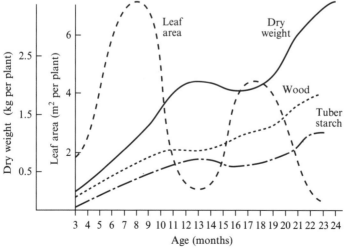

Table 15.2. *Yields of old stalk and storage roots of cassava planted in September at Nsukka, Nigeria, in relation to time of harvesting*

Yield	Age at harvesting (months after planting)										Standard error ($\pm$)
	9	9.5	10	10.5	11	11.5	12	12.5	13	13.5	
Dry weight of old stalks (t ha^{-1})	1.08	1.05	1.25	1.53	1.50	1.59	1.73	1.76	1.62	1.66	0.13[a]
Dry weight of storage roots (t ha^{-1})	2.41	2.89	3.69	4.88	5.53	5.65	7.42	7.34	6.90	6.92	0.81[a]
Dry matter content of roots (%)	31.3	31.6	32.3	32.6	39.8	41.2	39.6	42.2	44.9	41.3	2.25[a]

After Ezedinma, Ibe & Onwuchuruba (1981).
[a]Significant at 5% level of probability.

15.4 Crop/climate relations

Cassava is grown from the lowland tropics to 2300 m elevation in areas receiving more than 500 mm average annual rainfall (AAR) (Cock, 1985). The high-altitude limit of 1800 m is associated with a soil temperature of 18 °C at 09.00 hours; this appears to be the lower limit for crop establishment (Bourke, Evenson & Keating, 1984).

Cassava's wide climatic adaptation and use in a variety of cropping systems is largely related to its perennial nature and its drought tolerance or avoidance. Thus although storage root dry matter yields may be as high as 96 t ha^{-1} from the Mukibat system (de Bruijn & Dharmaputra, 1974), experimental yields 12 months after planting are usually 14–20 t ha^{-1} in the wet tropics (e.g., Enyi, 1973; Williams, 1974; Cock, 1976; Cock, Wholey & Gutierrez de las Casas, 1977; Kawano *et al.*, 1978) and 8–9 t ha^{-1} in the wet-and-dry and cool tropics (Cours, 1951; Okigbo, 1971; Boerboom, 1978), while national average yields may be as low as 4 t ha^{-1} (Table 15.1).

The higher yield in the wet tropics no doubt contributes to cassava's relative importance in hot wet tropical regions. In Africa, for example, the proportion of cassava which is grown in the hot wet tropics (where mean temperature during the growing season is above 22 °C) is twice as high as would be expected if the crop were distributed evenly according to available land area: Table 15.3 shows its predominance in wet and seasonally wet/dry areas, particularly those which have low soil fertility.

It is recommended that stem cuttings be buried horizontally in dry soil, but vertically, to avoid rotting, in wet soil. Orientation and length of the cutting affect early sprout and root growth (Hunt *et al.*, 1977), although there is a broad optimum length of about 30 cm (IDRC, 1980). Temperatures below 20 °C cause slow sprout emergence and leaf production (e.g., CIAT, 1974): the minimum, optimum and maximum for sprouting are 12–18 °C, 28–30 °C and 36–40 °C respectively (Keating & Evenson, 1979; Bourke *et al.*, 1984). Total final emergence does not appear to be affected over a broad temperature range (CIAT, 1974; Keating & Evenson, 1979). First branching occurs at about 60 days in the lowland tropics and the subsequent rate of branching is constant (44–50 days per branch) for any one genotype × location. Depending on genotype, branching may or may not be sensitive to temperature (CIAT, 1979*b*). Some cultivars do not branch (Williams, 1975; see Fig. 14.1).

As cassava is commonly grown in wet-and-dry climates, its pattern of

Table 15.3. *Cassava distribution (area, 000 ha & %) in Africa, according to climate and soil constraints*

	Soil constraint					
	Root growth (texture and depth)	Waterlogging	Low fertility	No constraint	Total area	%
Humid and semi-hot, lowland[a]	442	179	2610	1049	4280	53.6
Continental, lowland	155	58	510	424	1147	14.4
Semi-arid, lowland	96	59	469	340	964	12.1
Humid and semi-hot, highland[b]	72	149	310	484	880	11.0
Continental, highland	12	27	395	143	577	7.2
Semi-arid, highland	16	10	56	55	137	1.7
%	9.9	4.4	54.4	31.2		

Modified from Carter *et al.* (1992).
[a]'Lowland' or hot, where mean temperature during the growing season >22 °C.
[b]'Highland' mean temperature <22 °C.

growth is typically bimodal, with most active growth when reasonably-high LAIs are attained in the first and second wet seasons, perhaps 4–9 and 18–21 months after planting (Fig. 15.1).

The physiological causes of yield variation in cassava are reviewed by Veltkamp (1986). Leaf area depends, of course, on branching, leaf production and leaf longevity. Low rates of branching and leaf production cause LAI expansion to be relatively slow. Furthermore, as mentioned in Section 15.3, LAI follows a predictable pattern in the wet-and-dry tropics: it may reach 4 within 20 weeks of planting (IDRC, 1980) and possibly peak at 8 (Enyi, 1973), but thereafter it declines to as low as 1 at the end of the first season and to about 4 during the second wet season (Cours, 1951; Enyi, 1973). Thus there is, typically, hysteresis in the seasonal relationship between crop growth and leaf area (Fig. 15.2), although a feature of high-yielding varieties may be the ability to maintain high LAI late in the crop's life (Cock, 1976). Depending on cultivar,

Fig. 15.2. Seasonal relationship between crop growth rate and leaf area index (LAI) of cassava in the wet-and-dry tropics in Malagasy. Squares, year 1; circles, year 2. Source: Hunt *et al.* (1977), from data of Cours (1951).

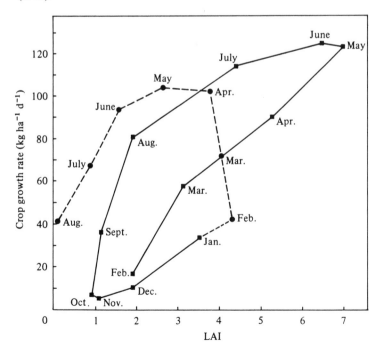

leaf longevity ranges from 80–200 days, at sites with a mean temperature of 20 °C, to 60–120 days at 24 °C and 60–80 days at 28 °C (Irikura, Cock & Kawano, 1979). Leaf longevity is reduced by drought or low temperature and under severe conditions all leaves may fall. Cassava canopies have radiation extinction coefficients of 0.72–0.88 (Veltkamp, 1986). Given that leaf area is a major limitation to yield, especially at low temperature (Irikura *et al.*, 1979), it is not surprising that experiments show positive linear relationships between crop growth or final dry weight and leaf area duration (CIAT, 1979*b*).

Crop growth rates may be relatively high (210–240 kg ha^{-1} d^{-1}; CIAT, 1973; Keating, Evenson & Fukai, 1982*c*) and are probably highest at about 29/24 °C day/night (Mahon, Lowe & Hunt, 1976). Leaves have a broad, and relatively high, temperature optimum for photosynthesis of 25–35 °C (El-Sharkaway & Cock, 1990).

Crop growth rate in the first season of growth (C,g m^{-2} d^{-1}) has been related to solar radiation (S,MJ m^{-1} d^{-1}) and LAI (L):

$$C = 0.11 \; S \times L - 0.12L^2 + 1.8S - 0.048S^2 + 0.78L - 15$$

$$(r^2 = 0.88; \text{Keating } et \; al., 1982c).$$

The photosynthetic efficiency E is 1.9–2.5 % in young plants (Veltkamp, 1986).

Dry matter partitioning is not related simply to growth rate or to ontogeny. Root/top partitioning in early growth is not particularly sensitive to temperature (Mahon *et al.*, 1976) and the crop appears to have a juvenile phase, there being a minimum of 1.8 months from planting to branching for any planting date (Keating *et al.*, 1982*b*). However, flowering and the associated change in canopy structure is accelerated by increasing temperature to about 28 °C; supra-optimal temperatures (34/28 °C, day/night) may inhibit branching (Keating *et al.*, 1982*b*). The proportion of dry matter allocated to tuber growth has been described by, for example, Williams (1972) in Malaysia. The proportion of dry matter entering the tuber may decrease with increasing LAI (CIAT, 1979*b*) while other data show that the relationship between tuber growth and LAI may be positive or show an optimum (Enyi, 1973; CIAT, 1979*b*). Such discrepancies between field observations can be interpreted as showing that dry matter distribution to the tuber is low when growth rate is high, temperature is high, days are long and LAI is high or very low.

In addition to climatic effects on dry matter distribution and storage root growth, increasing plant population (which has the desirable effect of increasing LAI) reduces both root number per plant and individual root weight in most cultivars (Williams, 1972; Godfrey-Sam-Aggrey, 1978). Wholey & Cock (1974) found that number of roots and tuber growth, not the time of onset of growth, are the main contributors to yield differences between varieties. Genotype yield rankings change with time due to differences in maturity and management; yields are commonly highest at the highest populations studied, usually about 13000 plants ha^{-1} (e.g., de Verteuil, 1971; Williams, 1972). Moreover, varying time of planting may have differential effects on root growth and top growth (Okigbo, 1971). Finally, cassava is susceptible to many diseases (Cock *et al.*, 1979; Hahn *et al.*, 1979), which may reduce tuber yield almost to zero (e.g., Nembozanga Sauti, 1981), and which contribute to yield variation not otherwise explained by climate.

The water requirements of cassava are reputed to be low, although there is no evidence to support this; rather, the crop may have a high transpiration rate (200 g m^{-2} h^{-1}; Mahon *et al.*, 1976) but avoid water deficit through leaf drop. In one study (Lal, 1981*b*), water use during 7 months of growth ranged from 240 to 1020 mm for plants at –1 MPa and at field capacity respectively; the highest rate was equivalent to 4–5 mm water per day. Significantly, growth under water deficit was associated with reduced partitioning of dry matter to the tuber relative to that to above-ground parts (Table 15.4). Water stress may or may not affect rate of tuber growth, depending on whether or not stress coincides with bulking, and growth can be greater after the stress is relieved than in

Table 15.4. *Effect of growth for 7 months at a soil water deficit of –0.01 and –1 MPa on water use efficiency (fresh, FW, or dry, DW, weight per cm water used) of cassava in Nigeria*

	Soil water deficit (MPa)			
	−0.01[a]	−1[a]	−0.01[b]	−1[b]
Tuber FW cm^{-1}	12.5	1.05	9.2	2.0
Tuber DW cm^{-1}	4.0	0.4	2.8	0.9
Shoot DW cm^{-1}	4.2	2.1	3.2	5.2

After Lal (1981).
[a] At soil bulk density of 1.3 g cm^{-3}.
[b] At soil bulk density of 1.6 g cm^{-3}.

normally watered crops (Connor, Cock & Parra, 1981). These observations are consistent with a model of cassava growth in which current photosynthate goes preferentially to above-ground parts (Cock *et al.*, 1979). They suggest that the importance of cassava in the wet-and-dry tropics lies not in physiological tolerance, but rather in drought avoidance, water capture through low plant populations, deep roots (to below 2 m; Connor *et al.*, 1981) and the ability of tubers to remain in the ground without deterioration.

15.5 Crop/soil relations

15.5.1 *Soil physical properties*

Cassava requires a soil which allows for the development of an adequate rooting volume for bulking of tubers and ease of harvest. While the optimum soil texture range is therefore light-to-medium (Hahn *et al.*, 1979), the crop can be grown successfully in clays (Harper, 1973) and drained Histosols (Williams, 1975; Wahab, Hassan & Lugo-Lopez, 1978). Clayey Oxisols are quite suitable physically if they are well aggregated. Mounding and ridging may be beneficial, though not in sandy soils (IITA, 1990). Pot studies show that increasing bulk density from 1.3 to 1.6 g cm^{-3} adversely affects plant growth and tuber yield (Lal, 1981*b*). Vine & Ahmad (1987) found in Trinidad soils that fresh tuber yield at 4.5 months in the wet season was proportional to the negative log of the time-averaged soil penetrometer resistance (r^2=0.80) provided that soil air was adequate. Yield was sharply reduced by season-mean soil air content values below 12 ml air per 100 ml soil in the 0–150 mm soil layer. Soil should not be waterlogged (Purseglove, 1968) and should be at least 50 cm deep (Table 3.11).

15.5.2 *Soil chemical properties*

Cassava is renowned as the species that will still produce a harvestable yield (5–6 t ha^{-1}) in tropical soils of low fertility where other crops will fail (Edwards, Asher & Wilson, 1977; Cock & Howeler, 1978). Edwards *et al.* (1977) concluded from solution culture studies with cassava, maize, sorghum, soybean and sunflower, that cassava was able to tolerate low calcium, nitrogen and potassium in the root environment better than any other species. These edaphic tolerances give cassava an advantage over other crops in acid infertile Oxisols, Ultisols, Alfisols and Histosols. Hahn *et al.* (1979) attributed this mainly to a well-developed root system that enables it to extract large amounts of soil nutrients, especially those located deep in the soil and unavailable to most plants. However, compared with other crops, cassava has a coarse, relatively thick and poorly branched root system with few root hairs (Howeler, Edwards & Asher, 1982). While some roots extend deeply, most are shallow: Ofori (1970) showed that actively absorbing roots were mainly in the top 10 cm, but that once they started functioning as carbohydrate sinks they were no longer active in nutrient absorption. These root characteristics are consistent with its high dependence on mycorrhizal association for nutrient uptake (see below).

While cassava can grow in infertile soils, it still removes significant quantities of nutrients (Cock & Howeler, 1978). According to CIAT (1979*b*) it is possible to obtain 30 t ha^{-1} tuber yields in the acid infertile soils of the Llanos Orientales in Colombia using improved cultivars. The considerable quantities of nutrients taken up by such a crop are shown in Table 15.5 (see also Table 17.2), although, as Howeler (1991) demonstrates, on a per unit weight basis cassava extracts much less nitrogen and phosphorus than most other crops. Continuous cassava

Table 15.5. *Approximate quantities of nutrients taken up in the tops and tubers of cassava yielding 30 t ha^{-1} tubers*

	Weight of nutrient (kg)		
Nutrient	Tops	Tubers	Total
Macronutrients			
Potassium	124	76	200
Nitrogen	126	38	164
Calcium	71	9	80
Phosphorus	21	10	31
Magnesium	22	9	31
Sulphur	—	6	6
Micronutrients			
Iron			3.6
Manganese			1.35
Zinc			1.35
Boron			0.45
Copper			0.14

After Asher, Edwards & Howeler (1980).

cultivation may lead to a reduction in soil pH (McIntosh & Effendi, 1979) through the depletion of nutrients, but the resulting yield decline may be reversed by fertilisation (Howeler, 1991).

The work of Keating, Evenson & Edwards (1982a) suggests that in infertile soils the effects of mineral nutrition are cumulative in that cuttings from plants in these soils emerge more slowly and have a lower yield potential than those from plants grown in more fertile soils. On the other hand, excessive nutrient availability, from either fertiliser or fertile soils, increases top growth at the expense of tuber growth (Sanchez, 1976).

Responses to nitrogen are common and are generally in the range 60–180 kg nitrogen ha^{-1} (e.g., IDRC, 1980; Fox, Talleyrand & Scott, 1975). In Nigeria, the optimum nitrogen rate (60–120 kg ha^{-1}) depended on time of harvest (9–15 months) and genotype (Obigbesan & Fayemi, 1976). Starch and HCN concentrations may also be affected by nitrogen level. Where soil nitrogen is adequate, or where other soil factors are limiting, tuber yield response to applied nitrogen may be negligible (e.g., Sanchez, 1973c; Haque & Walker, 1980). Negative responses to nitrogen application are sometimes obtained (Sanchez, 1973c), while very high rates, in the range 150–200 kg nitrogen ha^{-1}, may depress yield below that obtained either at low nitrogen levels (Cock & Howeler, 1978) or at nil nitrogen (Obigbesan & Fayemi, 1976). Probably because of initial low soil nitrogen status and high rates of nitrogen immobilisation, optimum nitrogen rates for cassava on Histosols may be very high: e.g., 200–270 kg nitrogen ha^{-1} (Chew, 1970a, 1974).

Available phosphorus is the main nutrient limiting cassava productivity in Oxisols and Ultisols of Latin America (CIAT, 1979b). Figure 15.3a illustrates the marked response to phosphorus that can be obtained with cassava in an acid infertile soil: yield was doubled by a rate as low as 22 kg phosphorus ha^{-1}. Cassava is quite inefficient in taking up phosphorus, requiring for near-maximum growth 20–100 times the phosphorus concentration in solution culture required by maize, soybean and cotton (Jintakanon, Edwards & Asher, 1982). It is therefore surprising that some cassava cultivars can yield well on soils highly deficient in phosphorus and other nutrients (Fig. 15.4). This is due to a high dependence on mycorrhizal association for uptake of phosphorus (Yost & Fox, 1979; Sieverding & Howeler, 1985; Table 3.6) and other nutrients such as potassium, sulphur, zinc (van der Zaag et al., 1979) and magnesium (Dodd et al., 1990a,b). The effectiveness of the association depends on the species and on soil pH (Fig 15.3b).

Fig. 15.3. (*a*) Effect of phosphorus fertiliser on storage root yield, foliage yield and harvest index of cassava grown on acid infertile soil at Carimagua, on the Llanos Orientales of Colombia. Source: Cock & Howeler (1978). (*b*) Effect of phosphorus fertiliser and inoculation with three species of vesicular-arbuscular mycorrhyzae on top dry matter yield of cassava grown in potted sterilised acid (pH 4.3) Paleadult, Colombia. Source: Howeler, Sieverding & Saif (1987).

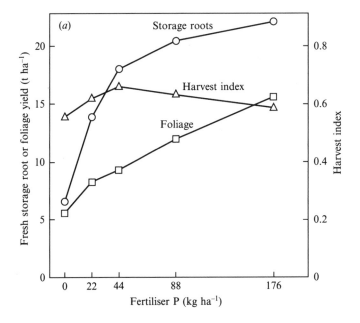

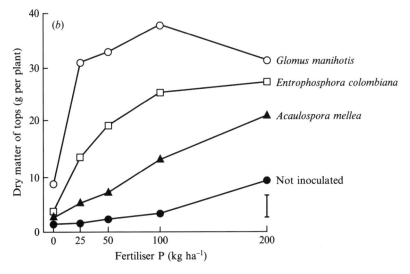

Potassium, the nutrient required in largest amounts by cassava (Table 15.5), is liable to be low in highly weathered tropical soils (Howeler, 1980). Responses to potassium are often obtained (e.g., Ezeilo, 1977; Ngongi, Howeler & MacDonald, 1977). Depending on the rate of application, potassium may improve root/top weight ratio (Obigbesan,

Fig. 15.4. Comparative phosphorus requirements of cassava, sweet potato, yams and Irish potato grown on various soils in several counties. Cassava, seven experiments, $R^2 = 0.49$; sweet potato, three experiments, $R^2 = 0.71$; yams, five experiments, $R^2 = 0.47$; Irish potato, five experiments, $R^2 = 0.81$. Source: van der Zaag (1979).

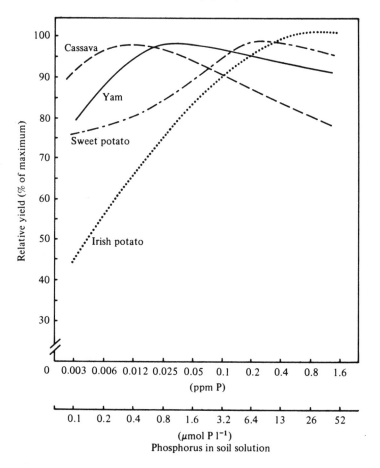

1977*a,b*), increase starch concentration and depress HCN concentration in the tubers (Obigbesan, 1974). Solution culture studies by Spear, Asher & Edwards (1978*a,b*) showed that for cassava, maize and sunflower at suboptimal potassium concentrations, potassium uptake rate per unit root weight was lowest for cassava, but that the crop was able to compensate for this by having a higher root/top ratio. Cassava was also more efficient in dry matter production per unit of potassium taken up: 1.11–1.54, 1.03 and 0.56 g dry matter (mmol potassium)$^{-1}$ for cassava, maize and sunflower respectively.

Aluminium toxicity ranks with low soil phosphorus as a major limitation to cassava yields in Oxisols and Ultisols of Latin America (CIAT, 1978*b*), though some cultivars are tolerant of high levels of exchangeable aluminium (Spain *et al.*, 1975) (see Fig. 16.2). In solution culture, cassava was far more tolerant of aluminium than maize or soybean, giving approximately 100%, 55% and 53% respectively of their maximum dry matter yield at 160 μmol aluminium 1^{-1}. It is fairly tolerant of high manganese (Table 15.6) and more tolerant of low pH *per se* than maize or tomato (Edwards *et al.*, 1977). In acid soils in Colombia, unlimed rice, maize and common bean produced virtually no yield, while unlimed cassava and cowpea (*Vigna unguiculata*) produced 54% and 60% of their maximum (limed) yields (Cock & Howeler, 1978). Cassava was found to be the most tolerant dicotyledenous crop, out of nine, to low calcium levels in solution culture (Islam, Asher & Edwards, 1987). It is not surprising, therefore, that, compared with most tropical crops, cassava requires only fairly low applications of lime: e.g., around 0.25–0.5 t ha^{-1}, compared with 2 t ha^{-1} for black beans in Oxisols of pH 4.3 on the Llanos Orientales in Colombia (Spain *et al.*, 1975). In these soils, with tolerant cultivars, it is necessary to lime only to pH 4.7, thereby supplying 0.025 cmol calcium kg^{-1} and reducing aluminium saturation to 80%, to obtain maximum cassava yields (CIAT, 1979*b*).

Liming may also increase the availability of phosphorus to cassava, but at lime rates above 0.5 t ha^{-1} micronutrient deficiencies, particularly of zinc, may be induced (Howeler, Cadavid & Calvo, 1977). On Histosols in Malaysia, low in exchangeable aluminium, Lim, Chin & Bolle-Jones (1973) found that cassava yielded relatively better than other crops at pH 3.3 and 4.0. When Chew *et al.* (1981) limed these soils, maximum tuber yields were obtained at pH 3.8.

Cassava is very susceptible to soil alkalinity and salinity. Yield declined rapidly above about pH 8.0, above 2.5% soil sodium saturation and above 0.5–0.7 dS m^{-1} electrical conductivity (Cock & Howeler, 1978),

Table 15.6. *The effect of constant solution manganese concentration on relative whole plant dry matter yield (as % of maximum) of eight crop species grown in flowing solution culture for periods of 18–31 days*

Species	Solution Mn concentration (μmol l^{-1})								Critical external concentration[a]
	1.3	1.7	3.6	10.3	42	130	394	1160	
Wheat	100	57.8	18.3	29.5	26.5	6.0	2.3	1.9	1.4
Maize	100	48.1	15.5	20.4	16.9	31.0	8.6	3.1	1.4
Cowpea	100	99.0	50.4	32.0	16.4	8.0	3.1	2.7	1.9
Common bean	96.4	100	47.2	54.8	33.2	9.3	5.3	2.9	2.2
Cassava									
cv.M Aus 7	67.2	100	81.6	64.9	88.5	7.4	4.5	3.6	4.0
cv.M Aus 10	66.9	100	73.4	90.4	78.1	16.4	5.9	3.8	8.0
cv. Nina	80.8	100	64.7	52.5	60.5	14.2	5.7	3.3	5.0
Soybean	94.4	100	99.5	87.6	51.3	20.8	10.9	5.6	8.5
Sweet potato	48.7	81.2	80.6	100	58.1	40.4	6.3	1.3	18.0
Pigeonpea	100	93.6	51.7	70.0	94.6	20.2	5.9	3.4	46.0

After Edwards & Asher (1982).
[a]Solution Mn concentration for 90% maximum whole plant yield.

whereas many crops do not show much yield reduction until sodium saturation is above 15% and conductivity above 2 dS m^{-1} (Table 11.3).

15.6 Place in cropping systems

Cassava is an important crop across a wide range of tropical environments (see Section 15.4) and is a significant component of cropping systems with a wide span of cultivation frequency, from continuous annual cropping to shifting cultivation with a long fallow phase. It is grown both as a subsistence energy crop – though often not the primary one – and as a cash crop for starch, alcohol, cattle feed, etc. (Kay, 1973), on a range of farm sizes from subsistence holdings of less than 0.2 ha to partially mechanised commercial plantations of more than 100 ha. On small farms, cassava is commonly found in intercrop and relay-crop patterns.

Because of slow early growth and low leaf area for much of its growth cycle (e.g., Fig. 15.1), cassava crop systems often result in a large proportion of bare soil. For example Aina *et al.* (1979) reported from West Africa that while sole soybean, a maize/cassava intercrop and sole pigeonpea developed 50% ground cover in 38, 45 and 50 days after planting respectively, cassava did not reach this value until 63 days. Runoff and erosion are likely to be greater when cassava is grown as a sole crop than when it is intercropped (Fig. 2.2). Indeed, the short-season crops which are commonly intercropped with cassava at the beginning of the growing season in Africa, rapidly attain high LAIs relative to cassava (Fig. 15.5). They thus both complement the growth pattern of the cassava and reduce the erosive impact of early-season rain. In Costa Rica, loss of phosphorus through soil erosion was 5 times greater than that taken up by a cassava sole crop, whereas phosphorus erosion loss when cassava was intercropped with three other crops was less than half the amount taken up by the crops (Burgos, 1980). Loss of soil and nutrients (Godon, 1985), particularly potassium (Howeler, 1991), often results from the long-term cultivation of cassava under poor soil management. Erosion may be reduced by minimum tillage or covering the soil with mulch (Table 15.7, also Ohiro & Ezumah, 1990).

The place of cassava in cropping systems is considered now according to geography.

Africa. Cassava-based cropping systems in tropical Africa are described by Ezeilo (1979), Ezumah & Okigbo (1980) and Juo & Ezumah (1992).

At least 50% of the cassava grown is intercropped (Leihner, 1982). The major associated crops in East and West African systems are summarised in Table 15.8. In a test of cassava, cocoyam and plantain in all intercrop combinations, Karikari (1981) recorded land equivalent ratios

Table 15.7. *Soil losses and cassava yields as affected by (a) method of ground preparation and (b) husbandry (fertiliser, soil cover by mulching and intercropping with a legume)*

	Soil loss (t ha^{-1})	Cassava yield (t ha^{-1})
(a) Cassava planted on 40% slope at Mondoma, Cauca; various ground preparations		
No tillage	0.8	13.5
Manually prepared planting holes	1.2	18.7
Ox-drawn ploughing	1.8	31.7
(b) Cassava planted at 80 × 80 cm spacing on 40% slope at Agua Blanca, Colombia; ploughing with oxen		
Without fertiliser	35.9	6.9
With fertiliser and maize mulch	15.1	15.9
Double-row cassava alternated with *Brachiaria humidicola*	9.8	13.1

Adapted from Howeler (1991).

Fig. 15.5. Leaf area indices of cassava and short-season intercrops in West Africa. Source: Juo & Ezumah (1992).

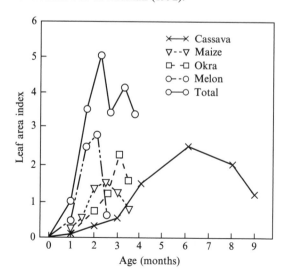

(LERs) of 2.1 for plantain/cocoyam, 2.0 for plantain/cassava, but only 1.2 for cocoyam/cassava, indicating the advantages of intercropping with species of widely different habit and canopy structure.

One characteristic feature of cassava-based cropping systems involving a fallow period is that cassava, owing to its capacity to yield tolerably well under low fertility, is usually the last crop in the cropping phase. In a survey in Zaïre, Miracle (1967) recorded cassava 5 times more frequently as the last crop than as the first in forest fallow systems, and 11 times more frequently in savanna fallow systems. Where cassava is not the farmer's primary source of energy food but is grown as a reserve, it is often left to fend for itself in the recovering fallow vegetation. It may remain in the ground for 3–4 years and, indeed, may never be harvested at all.

Another characteristic feature, though dependent to some degree on soil type and drainage, is mounding or ridging. The advantages of growing cassava on mounds or ridges include better drainage, loose soil favouring the development of the storage organ, and easier harvesting. In an example from Nigeria given by Okigbo & Greenland (1976), cassava is planted on the sides of mounds 1 m high by 3 m wide at the base; the tops of the mounds are occupied by yams and upland rice is grown in the intervening flat areas.

Finally, since cassava is adapted to wet tropical regions in which perennial tree crops flourish, it is often associated with such crops during their establishment phase. Ruthenberg (1980) gives a number of examples from Africa: e.g., in Tanzania three successive cassava crops, each of 2 years duration, planted between young coconut palms.

Table 15.8. *Associated crops in cassava-based cropping systems of tropical Africa*

East Africa	West Africa
Plantain, beans	Yams
Plantain, sweet potato	Yams, maize
Sweet potato	Cocoyam, plantain, yams
Maize, groundnut	Cocoyam, yams, maize
Maize, beans	Cocoyam, yams, maize, pigeonpea

After Ezumah & Okigbo (1980).

Asia. Descriptions of cassava-based cropping systems in Asia are given by Kumar & Hrishi (1979), Ghosh *et al.* (1987) and others for India, Effendi (1979) for Indonesia and Sinthuprama (1979) for Thailand.

Cassava is not a major crop over a very large area of semi-arid India since the growing season is too short. It is, however, of particular importance as a subsistence crop in Kerala State, where 42% of the farmers own less than 0.2 ha of land each (Kumar & Hrishi, 1979). Here it is grown in mixed-garden complexes in house compounds with coconut, mango, jackfruit, banana, taro, etc., but also as a sole crop. It is usually planted on mounds.

In Indonesia, cassava is the third most important food energy crop after rice and maize, and accounts for about 8% of the total crop production area (Effendi, 1979). Some is grown as a sole crop on plantations, and it is a frequent component of mixed-garden complexes, but the greater part is grown as a subsistence crop in intercrop patterns with upland rice and/or maize. The maize/rice/cassava intercrop (McIntosh *et al.*, 1977) has been described in Chapters 5 and 6. Intercrop cassava is planted on the flat, but as a sole crop it is generally planted on ridges. In general, mounding of cassava is uncommon in southeast Asia, though the crop is often grown on terrace banks on sloping land.

In Thailand, cassava was planted extensively in the northeast in the 1970s, for export to Europe for cattle feed (Sinthuprama, 1979). Intercropping was rarely practised while crop production was oriented to export, though nowadays intercropping with rubber or coconut, or alternative land use, e.g., horticulture, are encouraged.

There is little information on crop sequence in cassava-based cropping systems in Asia. Where it is grown as a field crop, that is, not in mixed gardens, the pattern is generally an irregular monoculture; cultivation frequency is dependent on human pressure on land. Howeler (1980) quotes an example from Thailand where, on the Sattahip soils of the southeast, cassava has been grown continuously for 25 years without fertilisation.

America. The place of cassava in the cropping systems of tropical America is discussed by Moreno & Hart (1979) for Central America, and by Porto *et al.* (1979) and Lorenzi, Normanha & Conceicao (1980) for Brazil. At least 40% of the cassava grown in tropical America is intercropped (Leihner, 1982).

In Central America, cassava is grown mainly by small farmers as a subsistence crop. In lowland wet tropical regions it may be interplanted with perennial or semi-perennial crops (e.g., plantain); in wet-and-dry climates at moderate altitudes (*c.* 700 m) it is most frequently inter-cropped with maize and common beans.

In the area southern Mexico–eastern Guatemala, farmers follow a well-organised pattern of intercropping sweet (low HCN) cassava with maize and growing bitter (high HCN) cassava as a sole crop. Intercropping experiments in Costa Rica (Moreno & Hart, 1979) are of interest since they included three food energy crops – cassava, maize, sweet potatoes – making it possible to express combined yields in energy terms (GJ ha⁻¹)(Table 15.9)

Although sole sweet potato grown late, equivalent to the second half of the cassava growing period (May–October), yielded more than when grown early, equivalent to the first half of the cassava growing period (November–April), the reverse was true when it was intercropped with cassava. This indicates greater competition between crops in the second half of the growth period of the cassava crop when its canopy cover is complete. In time-of-planting studies of the common bean/cassava inter-crop at CIAT, Colombia (Thung & Cock, 1979) and the soybean/cassava intercrop in Australia (Tsay, Fukai & Wilson, 1988), the highest LER was obtained when beans and cassava were planted at the same time. In this way the bean crop is virtually mature before complete canopy cover is attained by cassava (3 months).

Brazil produces more cassava than any other country (Table 15.1). The crop is grown over a span of latitude from the equator to 33° S, but about 75% of the crop area is within the tropics. Cassava-based farming systems range from small-scale subsistence holdings to large commercial plantations producing the crop for industrial use (Porto *et al.*, 1979). Cassava flour is a common food item throughout Brazil, and both

Table 15.9. *Energy yields (GJ ha⁻¹) per crop cycle from sole-cropped cassava, maize and sweet potato and intercrops, in Costa Rica*

Cassava sole	107
Maize (early) sole	89
Sweet potato (early) sole	92
Sweet potato (late) sole	139
Cassava/maize (early)	123
Cassava/sweet potato (early)	146
Cassava/sweet potato (late)	124

After Moreno & Hart (1979).

small- and large-scale producers grow the crop for sale to flour factories (Lorenzi *et al.*, 1980). In 1975, the Brazilian government established the National Alcohol Program to encourage the production of alcohol for fuel from crop plants such as sugarcane, cassava and sweet potatoes (Porto *et al.*, 1979).

Over half the Brazilian crop is produced in the northeastern region, in a wet-and-dry climate, largely in areas of 650–1000 mm rainfall. The most common cropping system is an intercrop of cassava with maize and common beans. In experiments in the Amazon region (2100 mm rainfall) with intercrops of cassava, beans, maize and upland rice in all combinations, three out of the five highest LERs were obtained from patterns that included cassava: cassava/maize, 2.75; cassava/maize/beans, 2.09; and cassava/beans, 1.97. This again illustrates the great advantage to be gained by growing quick-maturing seed crops while the canopy of the late-maturing root crop is still developing. The general climatological and physiological principles of intercropping with cassava are discussed by Zandstra (1979).

Further reading

Carter, S.E., Fresco, L.O., Jones, P.G. & Fairbairn, J.N. (1992). *An Atlas of Cassava in Africa: Historical, Agroecological and Demographic Aspects of Crop Distribution.* Cali, Colombia: Centro Internacional de Agricultura Tropical, 85 pp.

Cock, J.H. (1984). Cassava. In *The Physiology of Tropical Field Crops*, ed. P.R. Goldsworthy & N.M. Fisher, pp 529–50. Chichester: Wiley.

Cock, J.H. (1985). *Cassava: New Potential for a Neglected Crop.* Boulder, Colorado: Westview Press, 191 pp.

16

*Sweet potato (*Ipomoea batatas*)*

16.1 Taxonomy

The sweet potato (*Ipomoea batatas* (L.) Lam.), a perennial which is cultivated as an annual, is a member of the Convolvulaceae. There are about 500 wild species of *Ipomoea*, but the sweet potato is not known in the wild state. Yen (1976) and Austin (1978, 1988) have tackled the debated taxonomy of the genus. The revision of Austin (1978) recognised 11 species in the section *Batatas*, which includes the sweet potato. Subsequently three species have been added and one removed (Jarret, Gawel & Whittemore, 1992). The closest wild relatives of the sweet potato appear to be *I. trifida* and *I. tabascana*.

Morphological variation is treated exhaustively by Yen (1974). Economically important characteristics used in intra-specific classification include tuber shape and flesh colour. Yen recognises six basic shapes and three flesh-colour groups: white-yellow, pink-orange, and shades of purple.

16.2 Origin, evolution and dispersal

The sweet potato is of Central or tropical South American origin. Nishiyama (1971) and Martin & Jones (1972) suggest Mexico as the centre of diversity of the Batatas section of *Ipomoea*. However, the earliest archaeological record of cultivated sweet potatoes is from the Peruvian coast (4500 BP) and the crop has not been recovered from any of the ancient Mexican sites such as Tehuacán. The fact that the earliest finds are from coastal Peru is a consequence of the area's climatic aridity, favouring the preservation of crop remains: the area cannot be the centre of domestication since no wild prototypes are found there. However, the Peruvian record suggests an Andean rather than a Central American origin (Hawkes, 1989).

The presence of the sweet potato in Polynesia – whether it established itself in pre-Columbian times and if so, how – has also been a source of controversy. The historical issues are discussed by Yen (1974), who believes that the crop reached Polynesia before the eighth century. Purseglove (1965) considered that seed capsules could have been carried there by sea currents from the New World and saw no need to invoke 'Pacific regattas' to account for its probable presence in the Western Pacific before the arrival of the Spanish.

The eastward movement of the sweet potato is not debated. Columbus brought it to Europe on his return voyage and the Portuguese took it in the sixteenth century to Africa, India and east Asia to link up with introductions westward by the Spanish to Guam and the Philippines (Yen, 1976).

Data on area, yield and production for countries producing more than 1 m t of sweet potatoes are given in Table 16.1. China is responsible for more than 85% of world production. Average yields in tropical countries range from 4 to 9.5 t ha^{-1} and compare unfavourably with those from the developed temperate zone. Production in temperate and recently industrialised economies is declining (Fig. 14.2). Production per capita is highest in the Pacific and central Africa, e.g., 193 and 161 kg per capita per year in the Solomon Islands and Tonga and 150 kg per capita per year in Uganda (Horton, 1988).

16.3 Crop development pattern

Sweet potato is most commonly grown as an annual, although it is sometimes treated as a perennial, the 'tubers' (secondarily thickened

Table 16.1. *Major sweet-potato-producing countries: area, yield and production in 1991 of countries producing more than 1 m t per annum*

Country	Area (m ha)	Yield (t ha^{-1})	Production (m t)
China	6.41	16.72	107.19
Vietnam	0.32	6.48	2.10
Indonesia	0.21	9.50	1.98
Uganda	0.42	4.29	1.80
Japan	0.07	20.86	1.46
India	0.15	7.97	1.19

After FAO (1992).

roots: Artschwager, 1924) being progressively harvested without killing the parent plant. It is usually planted from stem cuttings, although it may also be propagated from tubers and, for breeding purposes, from seed. Plant habit is vine-like; herbaceous trailing or twining stems as long as 5 m are produced (Fig. 14.1). The tendency to twine is increased under shading (Martin, 1985).

Most cultivars have almost horizontal leaves, so that only a small leaf area index (LAI) of 3–4 is needed to intercept almost all radiation. Agata & Takeda (1982) and others (see Section 16.4) showed that crop growth may decrease when LAI exceeds 4. There are three-fold differences among cultivars in LAI (Bhagsari & Ashley, 1990) and presumably also in the speed with which cultivars reach canopy closure.

Root number reaches a plateau or peak within 4 weeks of planting and secondary thickening starts within 8 weeks of planting in most cultivars studied in Trinidad (Lowe & Wilson, 1974*a*). Tuber elongation is completed with 16 weeks, whereas tuber width may increase to 24 weeks or maturity (Lowe & Wilson, 1974*b*). Plants develop about ten tubers in the top 20 cm of soil, although it is common for only one to be of marketable quality (Haynes & Wholey, 1972).

As with cassava, the main harvested organ is the storage root or tuber, though leaves may be eaten in southeast Asia and New Guinea. Tubers are harvested as required; crop maturity is marked by leaf yellowing and senescence. Crops take 2–4 months from planting to maturity in the lowland tropics and first harvest is within 6 months of planting in most locations. However, at high elevations harvest· is delayed owing to lower temperature and radiation. In Papua New Guinea, Kimber (1972) divided the area of sweet potato cultivation into four according to elevation and time to first harvest, which ranges from 2–4 months to more than 10 months (Table 16.2).

Table 16.2. *Sweet potato regions in Papua New Guinea according to elevation and crop duration*

Region and elevation	Time from planting to first harvest (months)
Lowlands, to 500 m	2–4
Foothills, to 900 m	3–6
Lower montane, 900–2000 m	3–9
Intermontane valleys, 1500–2000 + m	5–10 +

After Kimber (1972).

There is an enormous number of cultivars. There may be as many as 5000 in Papua New Guinea (Bourke 1985); 88 are found in the West Indies (Gooding, 1964). Valleys in the New Guinea Highlands may each grow more than 30 varieties (Kimber, 1972), perhaps 15–20 being used by one family (French & Bridle, 1978). Some of this variation, e.g., in leaf shape and canopy structure, may be used to categorise landraces (Wholey & Haynes, 1969, cited by Lowe & Wilson, 1974*a*; Yen, 1974). Likewise, variation in some tuber characteristics, e.g., size (which is related to meristematic activity), may be developmentally important and contribute differences in yield (Lowe & Wilson, 1974*b*). Other variations, e.g., in tuber skin and flesh colour, may be due to a high mutation frequency (Hernandez, Hernandez & Miller, 1964) and have little or nothing to do with development or yield. At present, with very meagre data, it is safer to avoid identifying cultivar groups according to canopy structure or yield, but we suggest that cultivars can be broadly grouped into short- and long-season types. From the detailed study of two cultivars by Huett & O'Neill (1976), we suggest that the short-season types reach a stable harvest index at about the same time as they reach peak net assimilation rate, whereas in long-season types harvest index increases over the entire growth period (Fig. 16.1).

Variation is superimposed on these developmental patterns by daylength and growth conditions. Daylengths of ≤ 11 h stimulate flowering (McClelland, 1928) and several workers have found complex interactions in flowering response to daylength and nutrition. These are interpreted as an underlying association of flowering with high carbohydrate concentration within leaves (Kehr, Ting & Miller, 1953). Seeds are formed only when cross-compatible types are grown together: self-sterility and cross- incompatibility are common (Purseglove, 1968).

16.4 Crop/climate relations

Sweet potatoes are widely grown from 40° N to 40° S and above 2500 m at the equator (Hahn & Hozyo, 1984). They grow best where average temperatures are 24 °C ; the thermal optimum is reported to be about 24 °C compared with 25–30 °C for cassava and yams (Kay, 1973). However, differences in thermal responsiveness would be expected among the large variety of landraces; for example, in Papua New Guinea, yields are high (20–30 t per ha in 8 months) at 1600–2000 m altitude, where temperatures are typically 16–18 °C (R.M. Bourke, personal communication, 1992). In the USA there is no commercial area of

Fig. 16.1. (*a*) Harvest index, (*b*) net assimilation rate, and (*c*) leaf area index with time of two sweet potato cultivars at Alstonville, Australia (latitude 29° S). Source: Huett & O'Neill (1976).

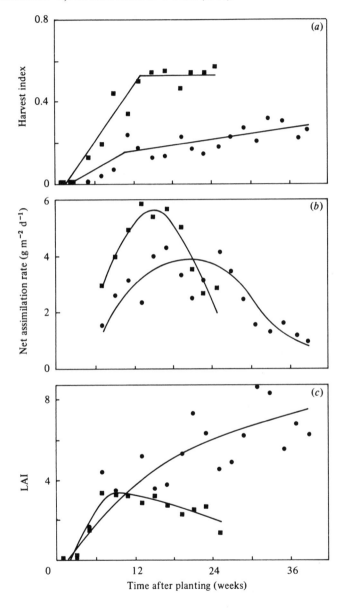

production where mean daily temperature is below 21 °C during the growing season (Steinbauer & Kushman, 1971).

In the tropics, yields decline with increasing altitude, as do the number of roots and the proportion of roots that are marketable (at least 4 cm diameter) (Ngeve, Hahn & Bouwkamp, 1992). Increasing altitude also delays maturity: Goodbody (in Bourke, 1988) found a linear relationship between time to first harvest (t, days) and altitude (A, m from 1400 to 2600 m) in Papua New Guinea ($r^2 = 0.74$):

$$t = -149 + 0.19\ A.$$

Sekioka (1964) found yields were 5–6 times higher at 25/20 °C than at 15/13 °C (day/night), and higher at soil temperatures of 30 °C than 15 °C. On the other hand, high night temperatures, by increasing carbon loss through respiration, are deleterious to growth: yield is substantially lower at 29/29 than at 29/20 °C (Yong, 1961). Seasonal plantings in northwestern Argentina suggest that flower and seed production are best when daily maxima and minima are 23–24 °C and 13–19 °C respectively (Folquer, 1974); in Puerto Rico flowering in glasshouses did not occur above 27 °C (Campbell, Hernandez & Miller, 1963).

There are a few studies of water use (e.g., Haseba & Ho, 1975). Kay (1973) suggests that the crop requires at least 500 mm during the growing season and is intolerant of water deficit during tuber initiation at 50–60 days after planting; Hahn & Hozyo (1984) suggest that at other times it has tolerance to drought. Sweet potato is intolerant of waterlogging (see Section 16.5.1), particularly during tuber initiation (Wilson, 1982; Hahn & Hozyo, 1984).

Parameters of crop growth, admittedly in a subtropical environment (latitude 29° S), are shown in Fig. 16.1. Maximum rates of dry matter accumulation were 85–170 kg ha^{-1} d^{-1}, which were reached at LAIs of 3.24 (Tsuno & Fujise, 1963; Chapman & Cowling, 1965). Substantially higher values – 260 kg ha^{-1} d^{-1} at an LAI of 6.7 and a final tuber yield of 15 t ha^{-1} – were attained when crops were trellised on wire mesh 1.2 m high (Chapman & Cowling, 1965). The time at which maximum growth rate is attained, and the duration of high growth rate, depend very much on genotype. In Trinidad (latitude 11° N) above-ground growth rate was highest at about 12 weeks after planting and whole-plant growth rate at 12–16 weeks after planting, while differences in yield between cultivars resulted largely from differences in tuber growth rates at 16–24 weeks (Lowe & Wilson, 1974a). In the subtropics a short-season cultivar may reach its peak growth rate at 14 weeks after planting whereas a long-season type may take 24 weeks (Fig. 16.1). LAI commonly increases almost linearly from 2 weeks after planting until a maximum is reached at development index (DI) = 0.4, after which it may remain constant or decline (Huett, 1975). However, in long-season cultivars under good growing conditions LAI may increase throughout the life of the crop (Fig. 16.1).

The environmental control of partitioning of dry matter to the tuber has received little attention. Relative growth rates of tubers range from 0.4 to 0.6 g g^{-1} wk^{-1} for the first half of their growth period (Austin, Aung & Graves, 1970; Huett, 1975). Growth rate and partitioning between storage root and vegetative organs are sensitive to plant structure and nutrition; translocation rates may increase during tuber growth (Hahn & Hozyo, 1984). As in most crops there is an inverse relationship between mean tuber weight and number of tubers (Lowe & Wilson, 1975). Furthermore, a large amount of fibrous root is associated with increased top growth and reduced tuber growth (Watanabe & Nakayama, 1969). Thus conditions that favour fibrous root and top growth, such as high temperature, low potassium or high nitrogen/potassium ratio (see Section 16.5.2), are detrimental to dry matter partitioning to the tuber. The vast span of yields commonly found in the tropics (1–40 t ha^{-1}; Purseglove, 1968) reflects the ability of numerous cultivars, in a wide range of garden and field situations, to respond to incompletely understood environmental factors.

16.5 Crop/soil relations

16.5.1 *Soil physical properties*

Sweet potatoes are adapted to a wide range of soil textural classes but sandy loams are considered ideal. Heavy clay soils often give low yields of poor quality (Kay, 1973) and irregular tuber shape (Martin, Leonard & Stamp, 1976). However, in spite of their high clay content (sometimes more than 60%), well-aggregated Oxisols are particularly suited to sweet potatoes (Badillo-Feliciano & Lugo-Lopez, 1977). The crop will also grow well on drained Histosols (Chew, 1970*b*).

The distribution of producers by soil types in selected sweet-potato-growing municipalities in Taiwan was found from a survey to be 6, 31, 36, 5, 2 and 11% for the soil textural classes gravel, sandy, sandy loam, loam, clay loam and clay respectively (Calkins, Huang & Hong, 1977).

The highest average yields were on sandy to clay textures (17–22.5 t h^{-1}), gravels yielding only 11 t ha^{-1}. Good yields on clays (19.5 t ha^{-1}) during the driest growing seasons were attributed to better water relations. When sweet potatoes were relay-planted into wet rice fields 20–30 days before rice harvest, sandy soils gave the highest yield of all textures (18.5 t ha^{-1}). The sandy soils were apparently better aerated and did not develop the subsurface hardpan that hinders root development in finer-textured soil (Calkins *et al.*, 1977).

Poor soil aeration damaging to sweet potatoes before harvest can occur through excessive rainfall, even (when the subsurface horizon is impermeable) on coarse-textured soils (Corey, Collins & Pharr, 1982). Wet soil conditions at harvest lead to an increase in tuber rots and adversely affect yield, storage life, nutritional and baking quality (Ton & Hernandez, 1978; Akparanta, Skaggs & Sanders, 1980). Permanent subsurface water tables can also affect yield: in a lysimeter study where water tables were maintained at 45, 30 and 15 cm below the surface in a clay loam, mean tuber yields of two cultivars were 27.2, 13.4 and 3.8 t ha^{-1} respectively (Silva & Irizarry, 1981). However, there are differences between sweet potato cultivars in their tolerance to waterlogging and post-harvest loss (Corey *et al.*, 1982; Collins & Wilson, 1988). Time of flooding can have different effects on tuber initiation and development; flooding imposed on plants raised from leaf cuttings 3 weeks after planting inhibited storage root initiation, while that imposed at 7 weeks impeded storage root enlargement (AVRDC, 1991).

The high rate of canopy development in sweet potato can result in complete ground cover in less than 35 days (Badillo-Feliciano & Lugo-Lopez, 1977). A sweet potato crop in Nigeria reached 100% ground cover 52 days from planting while sole cassava and intercropped maize/cassava achieved only 20–22% and 30% cover respectively (IITA, 1980). The crop also develops a thick mat of vines and has an extensive root system. These features help to minimise soil erosion during the cropping period, though soil losses may occur during land preparation or harvest. In a comparison of several crop combinations grown in Oxisols on slopes of 40–53%, Smith & Abruña (1955) found that least erosion occurred with combinations that included sweet potatoes. Under a traditional sweet potato cropping system in Taiwan, a mean annual soil loss of 172 t ha^{-1} was reported on 22% slopes where the annual rainfall was 2500 mm (Huang, Chang & Cheng, 1958; Huang & Cheng, 1960).

16.5.2 *Soil chemical properties*

Sweet potato, like cassava, is often considered as a crop associated with poor soils. This is probably because it is well suited to sandy soils, that are often infertile, and because tuber yields are sometimes depressed in very fertile or heavily fertilised soils. Nevertheless, good yields can be obtained only under conditions of high but balanced mineral nutrition. As with most root crops, sweet potato has a high requirement for potassium relative to nitrogen. A crop yielding 30 t ha^{-1} of tops and 22 t ha^{-1} of tubers takes up 80, 29 and 185 kg ha^{-1} of nitrogen, phosphorus and potassium respectively (AVRDC, 1975).

Bourke (1977) found that sweet potato planted after forest clearing in lowland Papua New Guinea required no fertiliser whereas crops planted after grassland required 150 kg nitrogen ha^{-1}. For sustained intensive cropping in both situations, both nitrogen and potassium fertilisers were required: sole cropping of sweet potato without fertiliser (ten crops over 5 years) led to an 86% reduction in yield (Fig. 1.2). It was not possible to separate out the relative contribution to yield decline of pests, diseases and fertility. In the highlands of Papua New Guinea soil fertility decline in sweet potato areas is reversed by growing *Casuarina* spp. for 10–20 years (Newton, 1960); casuarinas are actinorhizal nitrogen fixers (Torrey, 1982). Prior cropping with annual legumes such as soybeans gave higher sweet potato yields than prior cropping with sweet potatoes in the Philippines (Acedo & Javier, 1980).

The contribution of nitrogen from fertiliser, soil and nitrogen-fixing organisms to tuber and biomass yield is still not fully understood. Some sweet potato cultivars are capable of producing high tuber yields (21–38 t ha^{-1}) in low-nitrogen soils apparently because of nitrogen fixation by organisms in the root environment (Hill *et al.*, 1990). Inoculation of roots with nitrogen-fixing *Azospirillum* may increase storage root yield by 22% (Mortley & Hill, 1990). Nitrogen fertiliser responses are variable (Talleyrand & Lugo-Lopez, 1976; Bourke, 1985). High nitrogen rates may result in yield decline, e.g., beyond 56 kg nitrogen ha^{-1} in India (Nandpuri, Dhillon & Singh, 1971) and beyond 94 kg nitrogen ha^{-1} in Puerto Rico (Landrau & Samuels, 1951). Yield reduction at high levels of applied nitrogen is often accompanied by an increase in vine growth and a consequent reduction in tuber/top ratio (Sanchez, 1973c). However, Chapman & Cowling (1965) were able to change a negative tuber yield response from applying 112 kg nitrogen ha^{-1} (5.3–3.9 t ha^{-1}) to a positive one (12.9–18.4 t ha^{-1}) by trellising the vines. Factors such as the light relations of leaves could therefore also be involved. On the other hand, trellising × nitrogen interactions have not been consistent (e.g., Gollifer, 1973).

According to Tsuno & Fujise (1965*a*, *b*), potassium is important to the development of tubers because high concentrations in leaves (above 4%) promote translocation of photosynthate from leaves to the tuber; high photosynthate concentrations are inhibitory to photosynthesis. High fertiliser nitrogen encourages vine growth, thereby reducing potassium concentration. This accords with the finding that the greatest tuber enlargement occurs when the fertiliser nitrogen : potassium ratio is low. AVRDC (1975) recommend a ratio of < 1 : 3, though the optimum will vary with soil carbon : nitrogen and nitrogen : potassium ratios (Godfrey-Sam-Aggrey, 1976). Because of the importance of potassium to sweet potato, responses to it are frequent in the tropics (e.g., Anderson, 1974*a*; Bourke, 1985), and sometimes very high. Gollifer (1972) obtained tuber yield increases of up to 86% from 112 kg potassium ha^{-1} in the Solomon Islands.

Sweet potatoes seldom respond to phosphorus fertiliser (van der Zaag, 1979). This is because the crop, like cassava and yam, is well adapted to soils of low phosphorus availability, and is capable of 75% of its maximum yield at a soil solution concentration as low as 0.10 μmol phosphorus l^{-1} (Fig. 15.4). Results obtained by Nishimoto, Fox & Parvin (1977) showed that for 75% and 95% of maximum yield, soil solution phosphorus requirements of sweet potato were only about half those of soybean: 0.32 and 3.2 μmol phosphorus l^{-1} for sweet potato and 0.81 and 6.4 μmol phosphorus l^{-1} for soybean respectively. However, in studies on sweet potato yields in Papua New Guinea on Andisols and Entisols, soils which are commonly low in phosphorus, Goodbody and Humphreys (1986) found available phosphorus a consistently significant component of multilinear regressions predictive of yield. Van der Zaag (1979) observed that the ranking of the tolerance of Irish potato, sweet potato, yam and cassava to low soil phosphorus (Fig. 15.4) appears to be inversely related to the crop growing period; in Hawaii growing periods were 120, 150, 225 and 365 days respectively. Sweet potatoes may respond to inoculation with vesicular-arbuscular mycorrhizal fungi by improving phosphorus uptake (Negeve & Roncadori, 1985).

The liming studies of Perez-Escolar (1977) and Abruña *et al.* (1979) on Ultisols and Oxisols in Puerto Rico show that sweet potato generally reached maximum yield at soil pH values of 4.7–5.3 (see also Fig. 16.2*a*). Foster (1970) in Uganda obtained responses to liming where soil pH was less than 5.2, while Chew *et al.* (1982) found that the optimum pH for tuber yield was 5.5 in Histosols in Malaysia. Tolerance to exchangeable aluminium appears fairly good, the Puerto Rico studies sometimes showing no benefit from liming soils with aluminium saturation values as high as 34–45%. Figure 16.2(*b*) shows that the tolerance of sweet potato to aluminium is intermediate between that of cassava and yam, while Table 15.6 indicates that it is also quite tolerant of high manganese. Thus it may be considered as one of the more acid-soil-tolerant crops. Sweet potato is less tolerant of soil salinity than a number of other crops dealt with in this book (Table 11.3).

Fig. 16.2. Effect of (*a*) soil pH and (*b*) aluminium saturation on relative yield (% of maximum) of cassava, sweet potato and yam (*D. alata*) in two Ultisols and an Oxisol in Puerto Rico. In (*a*): for cassava, $Y = -1314 + 153x - 14x^2$, $r = 0.72$; for sweet potato, $Y = -938 + 387x - 36x^2$, $r = 0.81$; for yam, $Y = -234 + 57x$, $r = 0.92$. In (*b*): for cassava, $Y = 93.7 + 0.27x - 0.006x^2$, $r = 0.75$; for sweet potato, $Y = 93.5 + 0.45x - 0.021x^2$, $r = 0.89$; for yam, $Y = 96.3 - 2.53x + 0.018x^2$, $r = 0.97$. *Note*: soil pH values for unlimed soils ranged from 4.0 to 4.5. Adapted from Abruña-Rodriguez *et al.* (1982).

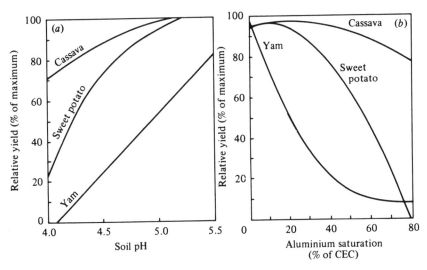

16.6 Place in cropping systems

Although the sweet potato is a common subsidiary component of cropping systems throughout tropical Asia, Africa and America, it is less
frequently recorded as the main source-of-energy food on any one farm.
It assumes its greatest importance in some cropping systems in southeast Asia and Oceania: the outstanding examples are from the highlands
of Papua New Guinea and West Irian, Indonesia. The place of the
sweet potato in the cropping systems of southeast Asia, Oceania and
tropical America is treated with great thoroughness by Yen (1974).

Southeast Asia. Sweet potatoes figure in the intensive irrigated cropping
systems of Taiwan based on wet rice and sugarcane (Kung, 1969;
Ruthenberg, 1980), normally as a winter crop after wet rice, to be followed in the next summer by another rice crop, groundnuts or soybeans.
They may be planted directly into rice stubble or after conventional
tillage, or relay-planted 20–30 days before rice harvest (Calkins, 1978).

Sweet potatoes are also grown in sequence with wet rice by the Bontoc
people of Luzon, in the Philippines (Yen, 1974). In this region of sharp
relief, the crops are grown on steep irrigated terraces. The winter sweet
potato phase must represent one of the most unusual forms of field
geometry in the world: the crop is grown on one continuous raised bed
laid out in a spiral pattern like a coiled snake. The same group of people
also crop the steep slopes in a shifting cultivation system with sweet
potatoes and upland rice. This is an instance of the frequent use of the
sweet potato as a crop for steep gradients; the dense spreading leaf
canopy provides an excellent defence against erosion (see Section 16.5.1).

Another example of upland rice/sweet potato shifting cultivation systems in the Philippines is the classic archetype of complex cropping —
that of the Hanunoo people, who grow up to 87 herbaceous and arbo-
real species in a cycle of 1–3 years cropping and 10–15 years fallow
(Conklin, 1957). Sweet potatoes form a dominant component of the
crop mixture after the rice crop is harvested.

Oceania. In Papua New Guinea, the sweet potato is grown from sea level
to altitudes greater than 2500 m, but it is more important in highland
locations. A survey of villages indicated that sweet potato contributed
25–94 % to the villagers' energy intake, and more than three-quarters of
the energy in about 60 % of the villages (Bourke, 1985). Not surprisingly,
its contribution to villagers' protein intake was lower (20–70%).

Table 16.3 summarises the characteristics of 15 cropping systems in Papua New Guinea in which sweet potato is the dominant or co-dominant crop; they represent part of a sample of 44 locations investigated by Brookfield & Hart (1971). Cultivation frequency varies widely, even on the same farm. In the more intensive systems a remarkable range of careful cultural practices is followed, to the length of 'managing' the fallow by planting nitrogen-fixing trees (*Casuarina* spp.) and selective weeding. The raising of pigs is a characteristic feature of the highland systems; Brookfield & Brown (1963) estimate that perhaps 60% of the sweet potatoes grown are used as pigfeed.

Subsistence sweet potato cultivation is also associated in Oceania with tree cash crops: for example, coffee in the New Guinea highlands and coconuts in New Caledonia and the Marquesas Islands (Yen, 1974). In Puebo, New Caledonia, the main food crops are sweet potatoes, taro, yams and cassava. As with the Bontoc people of the Philippines referred

Table 16.3. *Characteristics of 15 sweet-potato-dominant cropping systems in Papua New Guinea and West Irian, Indonesia*

	Percentage of total systems surveyed
Altitude	
0–500 m	10
500–1500 m	54
over 1500 m	36
Main associated food sources	
Taro (*Colocasia*)	93
Yams	73
Bananas	93
Wild plants	80
Hunting and freshwater fishing	60
Cultivation methods	
Mounds, ridges or squares[a]	53
Large mounds	13
Use of compost	40
Erosion control on slopes	60
Drainage furrows	53
Irrigation	20
Fallow cover control[b]	53

After Brookfield & Hart (1971).
[a] Formation of square mounds by digging furrows in a gridiron pattern.
[b] Planting of trees (e.g., *Casuarina*) at end of cropping break, and/or control of fallow vegetation by selective weeding.

to earlier, the Puebo people operate two distinct subsystems of differing cultivation frequency within the same holding: some areas are continuously cropped while adjacent hill land is under shifting cultivation. Sweet potatoes feature in both subsystems. In the intensively cropped area, large rectangular beds of mulch 1 m high are formed between coconut trees and intercropped with sweet potatoes, cassava and vegetables.

Tropical South America. Examples of South American sweet potato cropping systems given by Yen (1974) again illustrate the wide altitudinal range over which the crop is grown, from lowland tropical forests to Andean locations up to 2500 m. In the mountain valleys of Peru, sweet potatoes are often a component of intensive irrigated and rainfed cropping systems and are associated with maize, beans, squash and cassava. The crop is grown on mounds, which, as Yen notes 'are a mark of subsistence sweet potato planting throughout the world'. In contrast, the sweet potato also figures in the complex crop mixtures grown under shifting cultivation in the lowland rainforests of the upper Amazon in Brazil; other crops grown include cassava, plantains, maize and groundnuts.

The general impression gained from a review of tropical cropping systems in which the sweet potato is important is that of a crop of great versatility, not only with respect to the range of environments to which it is adapted but also with respect to cropping methods. It can be integrated as a cash crop into highly intensive wet rice systems or into the complex crop mixes of subsistence shifting cultivators in forest clearings; it may be grown as a carefully tended vegetable in house compound gardens or be roughly planted on hill slopes that are too steep to be used for any other crop; and it is equally useful to pigs and humans. Finally, as indicated in Table 4.2, the sweet potato probably produces, on average, more edible energy per hectare per day than other non-cereal crops.

Further reading

Hahn, S.K. & Hozyo, Y. (1984). Sweet potato. In *The Physiology of Tropical Field Crops*, ed. P.R. Goldsworthy & N.M. Fisher, pp 551–8. Chichester: Wiley.

17

*Yams (*Dioscorea *species)*

17.1 Taxonomy

The genus *Dioscorea* includes some 600 species, of which 50–60 are cultivated or gathered for food or pharmaceutical purposes. However, there are only about 12 species of economic significance as food plants (Coursey, 1976a). Of these, *D. rotundata*, grown in Africa, and *D. alata*, grown largely in Asia, are by far the most important, together making up about 90% of world production of food yams (Alexander & Coursey, 1969). In this chapter attention is largely confined to these two species.

D. alata, the Greater Yam, is an Asian species; *D. rotundata*, the White Guinea Yam, is African. Both African and Asian species have x =10 chromosomes; American species, e.g., *D. trifida*, have x = 9. There is a high degree of polyploidy in the genus: 2n= 3x up to 16x. Polyploidy also occurs within species, particularly those from the Old World: *D. alata* is found with 2n = 30 up to 80, though *D. rotundata* has 2n = 40 only (Coursey, 1976a).

17.2 Origin, evolution and dispersal

The family Dioscoreaceae is probably one of the oldest groups of angiosperms, and appears to have arisen in southeast Asia (Burkill, 1960). The formation of the Atlantic Ocean at the end of the Cretaceous period separated Old and New World species, which subsequently followed a divergent evolutionary path. Desiccation of the Middle East in the Miocene period separated African and Asian species, but their later evolutionary divergence was slight (Coursey, 1976a).

The origin and evolution of Old World yams is discussed by Alexander & Coursey (1969) and Coursey (1976a,b). The patterns of domestication in Africa and Asia show some similarities. In both instances the evolution of cultigens appears to have been a slow and diffuse process from tropical 'non-centres' (Harlan, 1971) spread over a wide area. It is probable that tropical rather than equatorial latitudes were important as zones of domestication: adaptation to the dry season of wet-and-dry climates would have led to the development of races with larger, more dormant tubers than those adapted to wet tropical climates. The evolutionary changes during domestication would appear to be largely negative: loss of toxicity, spininess and sexual fertility. Modern developments through breeding and selection have been negligible.

D. alata and *D. rotundata* are not known in the wild state. *D. alata* is likely to have developed, perhaps through hybridisation, from wild species 'in the north-central parts of the southeast Asian peninsula' (Coursey, 1976a). The date of domestication is unknown, but it was certainly before 3500 BP, since cultivated Asian yams spread across the Pacific in the Polynesian migrations of that period. Although *D. alata* reached the east coast of Africa *c.* 1500 BP from Malaysia, it does not seem to have played any part in the evolution of cultivated African yams.

It is likely that *D. rotundata* arose by hybridisation of a savanna-zone *Dioscorea* species with *D. cayenensis*, a forest-zone species, along the West African forest–savanna ecotone (Coursey, 1976b). *D. cayenensis* has been domesticated as the Yellow Guinea Yam, and is the second most important species grown in Africa after *D. rotundata*. Intermediates between the species cause them to be often termed the *Dioscorea cayenensis–rotundata* complex. Nowadays this complex is separated into groups (e.g., 12 'cultivar' groups, by Hamon & Toure, 1990) according to morphology and enzyme patterns. Akoroda & Chheda (1983) list the main morphological differences between the extreme commonly recognised types.

Asiatic and, later, African yams were used extensively as ships' victuals in the post-Columbian period: their vitamin C content made them valuable as antiscorbutics. In this way, *D. alata* reached West Africa from Asia. During the slave trade both *D. alata* and *D. rotundata* were taken from West Africa to the Caribbean. There has been virtually no movement of American food yams to the Old World, nor of African species to Asia, until modern transfers of experimental material (Coursey, 1976a).

Production statistics for yams are given in Table 17.1. Nigeria provides two-thirds, and West Africa broadly 85% of world production. Other tropical regions of some importance – but not individually exceeding 500 000 t annum – are Caribbean nations, particularly

Jamaica and Haiti, Ethiopia and Papua New Guinea. Productivity exceeds 10 t per ha^{-1} per annum in some West African countries (Table 17.1), Jamaica, Dominica and Barbados (FAO, 1992).

17.3 Crop development pattern

Yams are twining herbs grown from small tubers (seed yams), tuber pieces (setts), stem cuttings and from seed. The shoots do not have tendrils and climb by twining (to 3 m, Fig. 14.1); they bear mostly dioecious flowers which form from cells in the hypocotyl region, i.e., at the stem base (Passam, 1977). Plants usually reach maturity in 6–11 months: *D. rotundata* needs 6–7 months, and 8 months for 'good' yields in West Africa (Kassam, 1976), while *D. rotunda* cultivars take 8–10 months from planting to harvest in Jamaica (Wilson *et al.*, 1992). In the wet tropics in the Pacific one large tuber piece (most probably *D. alata*) may be allowed to grow as a perennial for 2–5 years and in rare cases up to 20 years (Hiyane & Hadley, 1977).

The peculiar developmental characteristics of yams are a relatively long period from planting to shoot emergence (and, as a consequence, asynchronous emergence) and slow leaf area development (Onwueme, 1981): 50–60% of the crop growth cycle elapses before a leaf area index (LAI) of 3 is reached, compared with about 30% in *Ipomoea* (Wilson, 1977). Main shoot growth stops through dieback – senescence of the main shoot apex – which may or may not lead to a further wave of secondary branching and lateral shoot growth (Okezie, Okonkwo & Nweke, 1981). Tuber formation begins about 10 weeks after planting and continues until or throughout the dieback of the shoot. Short days (less than 10–11 h) increase tuber formation (Njoku, 1963), although

Table 17.1. *Major yam-producing countries: area, yield and production of yams in 1991, of countries producing more than 500 000 t per annum*

Country	Area (m ha)	Yield (t ha^{-1})	Production (m t)
Nigeria	1.50	10.67	16.00
Ivory Coast	0.27	9.51	2.56
Benin	0.10	11.82	1.21
Ghana	0.23	4.40	1.00

After FAO (1992).

daylength and daylength–temperature interactions have not been fully studied. Traditionally the crop is managed as an annual and tubers are harvested at the end of the wet season or early dry season, although in some areas it is customary to leave tubers in the ground in the belief that there is little deterioration after maturity. After-ripening dormancy lasts 20–120 days (Passam,1977), during which time there is appreciable tuber weight loss through respiration (0.15–0.4% d^{-1}: Coursey, 1961; Passam, Read & Rickard, 1978).

The different growth durations of *D. alata*, *D. rotundata* and *D. cayanensis* allow for flexibility in planting time in response to a variable onset of seasonal rains in the wet-and-dry tropics, and different development patterns allow for a year-round spread in harvest time. This is illustrated by production patterns of the three species in Jamaica (Fig. 17.1), where the rainy season is April to October: *alata* yams are commonly planted in April–June, *rotundata* in the drier season, November–March, and *cayenensis* are planted throughout the year with a peak harvest in March.

17.4 Crop/climate relations

As already mentioned, the most important area for yam production is the wet-and-dry tropics of West Africa, particularly Nigeria (Table 17.1). Yams are most intensively grown in the lowland, high-rainfall wet-and-dry tropics in areas with at least 1150 mm of rain during the growing season and a dry season of 2–4 months. Here yields may reach 60–70 t ha^{-1} (in Ghana, Gurnah, 1974; in Puerto Rico, Martin, 1972). In the wet tropics, as in the Pacific (3000–8000 mm average annual rainfall (AAR)) yields are 10–50 t ha^{-1} (Gollifer, 1971; van der Zaag & Fox, 1981). In the Caribbean, where AAR ranges from above 3500 to less than 800 mm, yam yields range from 18 to 2.5 t ha^{-1} (Wilson *et al.*, 1992). On dry margins of the 'yam zone' where AAR may be as little as 400 mm, and on soils of low water-holding capacity, yields decline to less than 5 t ha^{-1}. Tubers contain about 25% dry matter (e.g., Sobulo, 1972*b*).

High yields of yam depend on good planting material and husbandry, particularly weed control to establish a reasonable crop leaf area, and on a favourable environment, particularly adequate water and near-optimum temperatures (25–30 °C) during the period of maximum potential growth at 14–20 weeks after planting. Early growth is slow: *D. rotundata* may produce only one leaf in 4 weeks after germination

(Okezie *et al.*, 1981). Hence increasing the weight of seed yams or setts reduces the time to sprouting and increases both vegetative growth and eventual tuber yield (Miege, 1957; Barker, 1964). Likewise, early planting, up to 3 months before the start of the wet season, may increase yield (Waitt, 1963; Kay, 1973), presumably by providing a longer period of near-optimum conditions for vegetative growth.

Fig. 17.1. Seasonal harvest production of *alata*, *rotundata*, *cayenensis* and *trifida* yams in Jamaica, reflecting crop development patterns and variable times of planting. Source: Wilson *et al.* (1992).

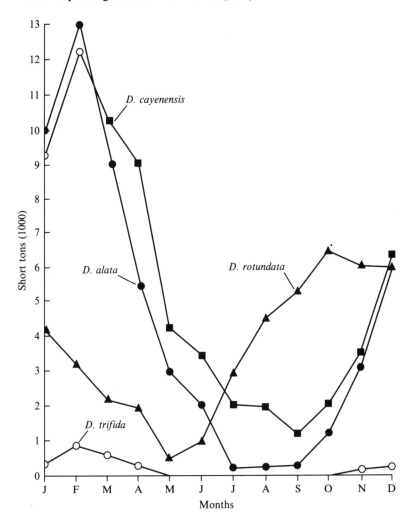

In *D. rotundata* root growth is greatest in the first 12 weeks after planting pre-sprouted setts (Unamma, Akobundu & Fayemi, 1981). During this time the crop's carbon economy is probably still dependent on the planting material (Nwoke & Okonkwo, 1978). Leaf area and weight increase most rapidly 8–14 weeks after planting, and in at least one study at Ibadan, Nigeria, final tuber yield was highly correlated with leaf and shoot weight ($r = 0.99$ and 0.89 respectively: Unamma *et al.*, 1981). Similarly, tuber yield has been correlated, though rather poorly, with leaf area duration ($r = 0.45$–0.55: Chapman, 1965), and low yields are usually ascribed to poor rainfall during the period of maximum leaf area index (LAI) and early tuber growth (Fig. 17.2). In contrast, inverse relationships have been established between total yield and the proportion of dry weight allocated to tuber growth (Okoli, 1980), which suggests that, particularly under unfavourable water regimes, there may be supra-optimal leaf and stem growth in long-season landraces.

The general importance of leaf growth, or presumably leaf area, in determining yam yield may be attributed partly to low net assimilation rates. Young plants sprout from tuber pieces within 2–3 weeks and the net assimilation rate and relative growth rate decline as the weight of the parent tuber diminishes, to a relatively constant net assimilation rate of about 30 g m^{-2} wk^{-1} at weeks 10–20 (Njoku, Nwoke & Okonkwo, 1984).Values of 12,16 and 40 g m^{-2} wk^{-1} have been calculated for crops in Nigeria and the Caribbean (Chapman, 1965; Okezie *et al.*, 1981). Low net assimilation rate could, in theory, be compensated for by high leaf area through high plant population. However, maximum LAI is usually 3.5–4 and only very rarely up to 10 (Chapman, 1965; Ferguson & Haynes, 1970). Sole-crop yams are planted at 7000–10 000 plants ha^{-1}, despite Gurnah's (1974) finding that yield increased linearly to the highest population tested: 35 000 plants ha^{-1}. In these experiments in Ghana (latitude 7° N), mean tuber weight declined with increasing plant population but the number of tubers per plant was the same at all populations, increasing only with increasing weight of sett. Despite the possibility of increasing leaf area and tuber yield through higher plant population, yams are traditionally grown at low populations or as intercrops, and effective LAI can be maximised only through staking or trellising.

Flowering and seed formation have received relatively little attention because seeds are not the favoured planting material. Yams flower profusely, although flower abortion is heavy; e.g., 38–86% in three cultivars of *D. rotundata* in Nigeria (Sadik & Okereke, 1975). Seed longevity presumably exceeds that of vegetative planting material and seed germinability is comparable with that of most small-seeded species (≥ 80%: Sadik & Okereke, 1975), whereas the germinability of setts is highly variable (e.g., Sobulo, 1972*a*). It seems that yams could become an important crop outside the 'yam zone' only with further research on seed production and nursery and field planting methods, and by the substitution of high plant populations (with weed control) for present

Fig. 17.2. Leaf area and tuber development of *Dioscorea alata* in two seasons in Trinidad, latitude 10° N. *Note*: the lower yield in 1962–3 was attributed to lower rainfall during the period of high leaf area index. Source: Chapman (1965).

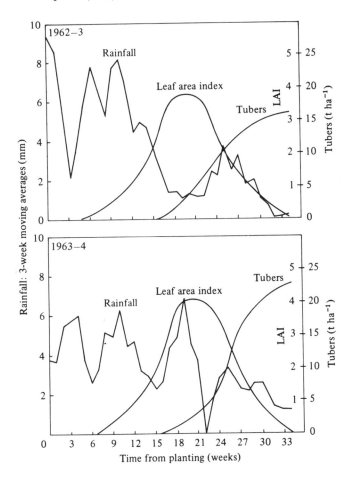

staking techniques. Slow crop growth and inefficient husbandry appear to be more of a constraint on production than crop/climate relations.

17.5 Crop/soil relations

17.5.1 *Soil physical properties*

Three important soil properties for yam growing are friability, depth and drainage. A friable soil is particularly important, because while other root crops such as cassava initially penetrate the soil with relatively thin roots which later expand, the yam tuber penetrates the soil as it expands (Onwueme, 1978). Thus deep sandy loams (Kay, 1973) and deep loams (Kassam, 1976) are preferred soils. The presence of a layer of coarse material within the tuber zone has been shown at Ibadan, Nigeria, to distort tubers and prevent further downward growth (Mansfield, 1979). *D. esculenta*, however, can be grown in gravelly soil (Irvine, 1969).

In West Africa seedbed preparation may involve the construction of mounds, ridges or trenches (Onwueme, 1978), while in Micronesia species of different tuber size are matched with the depth of soil: species with small tubers are grown in shallow soils (Sproat, 1968). Mounding improves drainage, which is important because yams are generally intolerant of waterlogging (Quin, 1985). Mounding, commonly practised in the rainforest and Guinea Savanna zone (Igwilo, 1989), also mitigates poor soil physical conditions and hence increases the response to fertilisers (Kang & Wilson, 1981), but if the subsoil is acid and infertile, there may be no benefit (Maduakor, Lal & Opara-Nadi, 1984). Planting on the flat in West Africa is generally confined to river flood plains where the soil is deep and has good physical properties.

17.5.2 *Soil chemical properties*

Yams require soils of high fertility and in West Africa they are therefore traditionally grown as the first crop after clearing. In contrast to cassava and sweet potato, yams are not recommended for marginal soils, although *D. alata* is more tolerant of poor soils than most other edible yam species (Irvine, 1969; Kay, 1973). While nutrient uptake per hectare varies with yield, as affected by season, species and cultivar (e.g., Obigbesan & Agboola, 1978), the quantity of nutrients removed when tubers are harvested is similar to that removed by cassava and sweet potato (Table 17.2).

In a review of world literature, Ferguson & Haynes (1970) concluded that responses by yams to nitrogen fertiliser were relatively low. In some instances, low levels of potassium gave small increases in yield, but responses to phosphorus were uncommon. In West Africa generally, the order of frequency and degree of response is also nitrogen > potassium >> phosphorus (Irving, 1956; Koli, 1973; Kpeglo, Obigbesan & Wilson, 1981). Compared with cereals, Nye (1954) found that yams were less sensitive to early nitrogen shortage and suggested this may be due to reserves of nitrogen in the seed tuber.

Two of the difficulties of interpreting early studies on nutrient response by yams are, first, that fertiliser rates were often quite low and, second, that the species used were not identified, even though yam species may differ markedly in fertiliser response; e.g., Igwilo (1989) and Fig. 17.3. In a 3-year study in an extensive yam-growing area around Tamale in the Guinea Savanna of northern Ghana, where farmers' yields had declined, Koli (1973) found that the main responses of *D. rotundata* were to nitrogen (22% at 67 kg nitrogen ha^{-1}) and phosphorus (6% at 15 kg phosphorus ha^{-1}); there was no response to potassium. However, Obigbesan, Agboola & Fayemi (1977) at Ibadan, Nigeria, obtained yield responses to potassium at 25–100 kg potassium ha^{-1} in *D. rotundata*, *D. alata* and *D. cayenensis*.

Table 17.2. *Approximate quantities of nutrients in 30 t ha^{-1} of tubers of yams, cassava and sweet potato*

Crop	Nutrients (kg ha^{-1})					Source of data
	N	P	K	Ca	Mg	
Yams						
D. alata	107	14	135	2	7	Calculated for 30 t ha^{-1}
D. cayenensis	95	13	125	3	9	tubers from yam crops
D. rotundata						yielding 25–36 t ha^{-1};
cv. Efuru	134	16	152	3	9	Obigbesan & Agboola,
cv. Aro	116	15	128	3	9	1978
Cassava	38	10	76	9	9	Asher *et al.*, 1980[a]
	120	40	187	77	40	Sanchez, 1976
	117	27	297	11	73	Obigbesan, 1977[b]
Sweet potato	131	15	160	—	—	Sanchez, 1976

[a]See Table 15.5.

Response to fertiliser is affected by the time of application in relation to the onset of rains and the development pattern of the particular yam. In Trinidad, Chapman (1965) obtained a tuber yield response (of 30%) only when nitrogen fertiliser application was delayed until 3 months after planting. This delay reduced leaching of nitrate by heavy rain and also coincided with a rapid increase in LAI and secondary shoot development. The response can also depend on whether the yams are intercropped or staked and, as mentioned above, on the species. In the rainforest zone of Southeast Nigeria, Odurukwe (1986) obtained a response to NPK fertiliser (of 29%) only when *D. rotundata* was intercropped with maize.

Studies by Kpeglo *et al.* (1981) show that while high rates of nitrogen fertiliser promote sprouting of tubers during storage, phosphorus and potassium fertilisers suppress it. *D. rotundata*, *D. alata* and *D. esculenta* have all been shown to utilise phosphorus efficiently at low concentrations of soil solution phosphorus (van der Zaag *et al.*, 1980; see also Fig. 15.4). Phosphorus uptake, although relatively low (Table 17.2), is assisted by mycorrhizal infection (van der Zaag *et al.*, 1980). Yield responses to applied phosphorus have, however, been obtained in Trinidad (Brown, 1931), Ghana (Koli, 1973), Nigeria (Ekpete, 1978) and Hawaii (van der Zaag *et al.*, 1980). On the other hand, responses to

Fig. 17.3. Response of *D. alata* and *D. esculenta* to (*a*) nitrogen and (*b*) potassium fertiliser. Adapted from Ferguson & Haynes (1970).

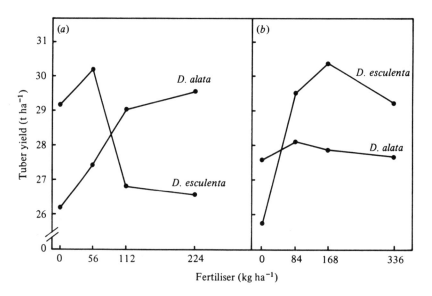

phosphorus in Nigeria have sometimes been negligible (Kayode, 1985) or negative (Ekpete, 1978).

Where aluminium saturation is high, yams are sensitive to low soil pH (Fig. 16.2). Fig. 16.2(*b*) shows that only 40% of maximum yield was obtained with 30% aluminium saturation and negligible yield with an aluminium saturation of 50%. Manganese toxicity did not appear to be involved (Abruña-Rodriguez *et al.*, 1982). In the same study, yam yield was reduced to zero at an aluminium : exchangeable bases ratio of 1.8, that of sweet potato at a ratio of 2.6, while cassava still gave about 80% of maximum yield at a ratio of 3.0. Abruña-Rodriguez *et al.* (1982) concluded that for maximum yam yield soils should be limed to about pH 5.5 to leave essentially no exchangeable aluminium. Although some yam species and cultivars may be shown to be more tolerant of high exchangeable aluminium and low fertility, it would seem that without liming and fertilising only low yields can be expected on the large areas of acid soils in the tropics.

17.6 Place in cropping systems

The area of West Africa in which yams are an important crop is clearly enough defined to have become known as the 'yam zone'. It is approximately a rectangle bounded by latitudes 4° N and 10° N and longitudes 5° W and 10° E (see Coursey, 1976*b*).

Traditionally, yams in West Africa are planted in mounds 0.6–1.3 m high and 0.9–1.3 m apart and are staked as the young shoots appear with poles up to 2 m long. Tuber yields are positively related to the size of mound (Kang & Wilson, 1981). Where yams are planted after maize, the dried cereal stalks may be bent over after harvest about 1 m above the ground and the vines trained along them. In plots cleared from forest, the crop is grown around trees left after clearance and the vines trained up strings attached to branches (Kowal & Kassam, 1978). Other vine support practices are listed by Coursey (1967). In areas where farms are partially mechanised, yams are planted on high broad ridges 1–1.2 m apart formed by tractor-drawn implements (Onwueme, 1978).

In Nigeria, which grows about two-thirds of the world tonnage of yams (Table 17.1), about 60% of the yam crop area is in mixed cropping (Okigbo & Greenland, 1976). Where the cropping system includes a fallow rest period, yams are nearly always one of the first crops planted after the fallow is terminated, since their fertility demand is relatively high.

Uzozie (1971) made a detailed survey of cropping systems in the three
eastern states of Nigeria. He recognised nine basic cropping patterns,
yams figuring in seven of them. These are given below, the dominant
crop first:

1. Sole yam
2. Yam/maize
3. Cassava/yam
4. Yam/cassava/maize
5. Cocoyam/plantain/cassava/yam
6. Cassava/yam/cocoyam/maize
7. Yam/cassava/cocoyam/maize/pigeonpea

Fig. 17.4. Yam-based cropping systems at Ibadan, Nigeria. Source:
Okigbo & Greenland (1976).

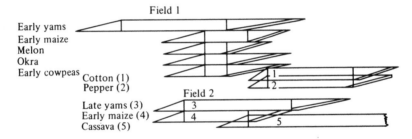

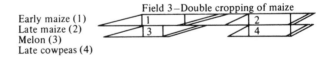

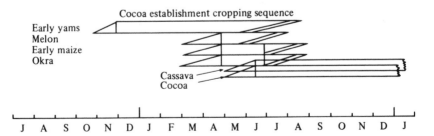

The spatial patterns of these crop combinations are complex. Uzozie recognised three main types of mixed crop plot:

1. *Boundary type.* Cassava is planted round the edges of the plot and also demarcates interior subplots. Within these subplots, yams, maize, melon, squash, etc., are intercropped on mounds and beds.

2. *Subplot type.* The plot is divided into subplots by ridges. Within each subplot, only one or two crops are grown, on mounds, ridges or beds. Crops include yams, cassava, cocoyam and maize.

3. *Interplanted type.* There is no subdivision within the plot. Many crops are grown together on mounds and short ridges in an irregular pattern. Yams are planted on the tops of mounds, maize on the sides and cassava in the intervening flat areas.

The temporal patterns of planting and harvesting within the growing

Table 17.3. *Characteristics of 13 yam-dominant cropping systems in Papua New Guinea and West Irian, Indonesia*

	Percentage of total systems surveyed
Altitude (m)	
0–500	82
500–1500	12
Over 1500	6
Main associated food sources	
Taro *(Colocasia)*	100
Sweet potato	85
Bananas	92
Wild plants	100
Hunting and freshwater fishing	69
Sea and reef fishing	54
Cultivation methods	
Clearing without fire	31
Clearing with fire	69
Mounds, ridges or squares	85
Large mounds	23
Deep holing[a]	54
Use of compost	62
Erosion control on slopes	23
Drainage furrows	15
Irrigation	31
Fallow cover control	15

After Brookfield & Hart (1971).
[a]Holes > 50 cm deep dug for yam planting.

season are also complex. Okigbo & Greenland (1976) give a number of diagrams to illustrate the range: three examples from Ibadan are shown in Fig. 17.4. Yams are the first crop, or in the group of first crops, planted in the seasonal cycle; they are rarely relay-planted into established crops. Figure 17.4 (bottom) illustrates the use of yams as an intercrop in the establishment phase of a perennial crop, in this instance cocoa.

Papua New Guinea is one of the few countries outside West Africa where yams are of significance as a food crop. Table 17.3 summarises the characteristics of a group of 13 cropping systems where yams are the dominant or co-dominant crop. Most of the yam-based systems, in contrast to those based on sweet potatoes (Table 16.3), are confined to the lowlands. Although in general cultivation is somewhat less intensive than in the sweet potato systems – in respect of erosion control, drainage and fallow cover control – tillage is equally thorough. In addition to mounding and ridging, deep holes are prepared for yam-planting, in some instances to encourage the growth of very large tubers for ritual purposes (e.g., Lea, 1966). Because yams are less tolerant of low fertility than sweet potatoes, the Wain tribesmen normally grow them as the first crop in a cycle on less-frequently cultivated areas away from the village, whereas sweet potatoes are grown in the depleted areas near the villages. The yam areas may also be cropped to sweet potatoes, but only as a second crop following yams (Jackson, 1965).

Another isolated centre of yam production is Ethiopia. A number of *Dioscorea* species are grown in complex intensive cropping systems with cereals and other non-cereal carbohydrate crops, in particular the indigenous ensat (*Ensete ventricosa*, Musaceae), a staple energy food in southern and southeastern Ethiopia at altitudes of 1500–3000 m; yams, though, are not grown above 2000 m. Westphal (1975) describes these ensat-based cropping systems in detail; yams are listed as a frequent associate crop in six of the ten systems described.

Further reading

Miege, J. & Lyonga, S.N. (1982). *Yams*. Oxford: Oxford University Press.
Wilson, L.A., Rankine, L.B., Ferguson, T.U., Ahmad, B.N., Griffith, S. & Roberts-Nkrumah, L. (1992). Mixed root-crop systems in the Caribbean. In *Field Crop Ecosystems*, ed. C.J. Pearson, pp. 205–42. Amsterdam: Elsevier.

18

*Bananas (*Musa *species)*

18.1 Taxonomy

The sweet banana and cooking banana or plantain belong to the genus *Musa* in the family Musaceae. The genus comprises about 40 species, all perennial. Wild *Musa* species are all diploid, with x = 10 or 11, though there are a few doubtful species with x= 7 or 9 (Stover & Simmonds, 1987). The wild species are grouped into four sections:

Section	2n = 2x	No. of spp.	Location
Eumusa	*22*	13–15	Southeast Asia
Rhodochlamys	22	5–7	Southeast Asia
Australimusa	20	5–7	Papua New Guinea
Callimusa	20	6–10	Southeast Asia

The great majority of cultivated bananas are derived from two species of the Eumusa group, *M. acuminata* and *M. balbisiana*, and are triploid (2n = 3x = 33). Although in earlier literature the sweet bananas were designated *M. sapientum* L. and plantains *M. paradisiaca* L., the taxonomy of edible bananas is so complex that, following Stover & Simmonds (1987), specific names have been abandoned and all are now designated by their genome complement.

There are three major groups: AAA, AAB and ABB (A, *acuminata* genome; B, *balbisiana* genome). Including somatic mutants, there are perhaps 500 recognisable clones in existence, including also AA, AB, AABB, AAAB and ABBB. AAAA clones have only been produced by breeding. The important commercial types of sweet banana are all AAA. The types of bananas grown and consumed as local subsistence food are derived from all three of the major groups (Burden & Coursey, 1977):

1. AAA. Mainly types with low starch and high sugar content when ripe and which, if cooked, are cooked green. This group also includes red bananas for cooking.
2. AAB. The true plantains, which are generally starchy even when ripe and are usually eaten cooked.
3. ABB. Starchy cooking and eating bananas that are found in all banana-growing areas under a wide variety of local names.

18.2 Origin, evolution and dispersal

All wild *Musa* spp. are native to south and southeast Asia and the southwest Pacific, from the eastern coast of India to Papua New Guinea (Simmonds, 1972, 1976; Stover & Simmonds, 1987). No date can be assigned to their domestication. Edibility first evolved in wild *M. acuminata*, and edible diploids are still widely but thinly spread through southeast Asia and Melanesia. There is no evidence of edibility evolving within *M. balbisiana*. It is therefore believed that hybrids evolved by the outward migration of edible diploid *M. acuminata* into the area of *M. balbisiana*, followed by hybridisation and polyploidy. Stover & Simmonds (1987) consider that the *balbisiana* genome gave a degree of resistance to drought in the wet-and-dry climates north of the primary wet tropical centre of origin.

Edible bananas were probably taken to the east coast of Africa from Indonesia rather than from India; the likely date is approximately 1500 BP. Around the same period they also moved eastward into the Pacific. There is no good evidence of bananas in the New World before Columbus. Bananas were found in West Africa by the Portuguese in the fifteenth century and were taken by them to the Canary Islands. The first known introduction to the New World was from the Canary Islands to Haiti in 1516.

The commercial banana industry developed in the West Indies and Central America in the nineteenth century, based largely on the AAA types Gros Michel and various mutant members of the Cavendish group. With the spread of Panama disease (*Fusarium oxysporum* f. *cubense*) and, later, Sigotoka leaf spot (*Mycosphaerella musicola*), programmes to breed disease-resistant types developed. Modern breeding is reviewed by Menendez & Shepherd (1975) and Stover & Simmonds (1987). The genetic basis is narrow; the only available female parent is Highgate, a semi-dwarf mutant of Gros Michel, since the Cavendish group, resistant to Panama disease, is completely seed-sterile (Stover & Simmonds, 1987).

Countries producing more than 1.5 m t of bananas plus plantains are listed in Table 18.1. Major producers are Uganda (which lists all its production as 'plantains') and India and Brazil (where production is recorded as 'bananas'). About 70% of the reported world production of plantain is from Africa.

18.3 Crop development pattern

Bananas are tall, rhizomatous perennials. The short underground rhizome grows horizontally but slowly; aerial shoots (suckers or followers) arise from lateral buds on the rhizome (Fig. 14.1). Within the AAA group, the aerial shoots of many landraces and Gros Michel grow 4–9 m tall and the Cavendish subgroup 2.5–4 m. Dwarf Cavendish (introduced from China in 1926; Purseglove, 1972) is widely cultivated because its aerial shoots grow only to about 2 m.

The aerial shoots are pseudostems built of overlapping leaf bases rolled tightly around each other. The pseudostem commonly has 11 unexpanded leaves within it (Summerville, 1944) and carries about 10

Table 18.1. *Production of bananas and plantains: countries in 1991 producing more than 1.5 m t per annum*

Country	Production (m t)
Uganda	8.31
India	6.40
Brazil	5.63
Colombia	4.34
Ecuador	3.95
Philippines	3.54
Rwanda	3.03
Indonesia	2.40
Zaïre	2.22
China	2.10
Mexico	1.87
Venezuela	1.68
Costa Rica	1.64
Thailand	1.62
Burundi	1.58
Tanzania	1.50

After FAO (1992).

expanded leaves of which more than half may be 'nonfunctional' owing to wind damage, senescence and disease (Heenan, 1973). When the pseudostem is 5.5–10 months old and has produced 30–40 leaves (but up to 50 leaves in subtropical climates) the meristem becomes reproductive. Each shoot is determinate; once the apex becomes reproductive no further leaves are initiated and leaves remaining within the pseudostem emerge at a slightly reduced rate (e.g., Anno & Lambert, 1976). The number of leaves produced before flowering is not predetermined and can vary widely (Robinson, 1981); high temperature reduces the number so that fewer leaves are recorded on bananas in tropical than in subtropical climates (Olsson, Cary & Turner, 1984). The floral phase (Stover & Simmonds, 1987), from floral initiation to inflorescence (bunch) emergence from the throat of the pseudostem, is associated with inflorescence differentiation as the true stem elongates and the inflorescence is moving upward within the pseudostem. Emergence of the bunch from the pseudostem is the start of Simmonds's 'fruiting phase'. The fruit develop parthenocarpically from ovaries of the female flowers.

Inflorescence growth is sigmoidal during the floral phase. The most rapid growth takes place 4–6 weeks before emergence, when the female flower primordia are being differentiated and the final three or four leaves are emerging (Turner, 1972*a*). Inflorescence growth after emergence involves cell division and, later, cell expansion of the ovaries (Ram, Ram & Steward, 1962) in a uni- or bi-sigmoidal growth pattern (Wardlaw, Leonard & Barnell, 1939; Turner, 1972*a*; Ganry, 1973).

The timing of development is difficult to generalise on because of perennation and the range of environments in which bananas are grown. Cultivars take 6–8 months from planting to bunch emergence in Malaysia and 9.5–13 months from sucker emergence to bunch emergence in New Guinea (Heenan, 1973). Stover & Simmonds (1987) cite data from Martinique of 6–7 months to bunch emergence at up to 140 m altitude, 9–10 months at 365 m and 11–13 months at 400–640 m. Irrigation reduces the time to bunch emergence and the reduction is cumulative: i.e., more pronounced in the second and third pseudostem growth cycles (ratoons) than in the primary cycle or plant crop (Moreau, 1965). Most of the period to bunch emergence is associated with vegetative growth: at latitude 29° S in Australia cv. Williams normally takes 11 months to reach floral initiation and only 2.5 months from floral initiation to bunch emergence (Turner, 1972*a*). The period from bunch emergence to harvest is about 100–120 days in the tropics and as long as 220 days in subtropical Africa and Australia. If pruned to three pseudostems at various stages of development, a banana plant will produce a bunch every 6 months in the lowland wet tropics.

18.4 Crop/climate relations

Bananas are grown in the wet, wet-and-dry and cool tropics. The main determinants of their distribution are rainfall in excess of 1250 mm per year (Stover & Simmonds, 1987; although Purseglove, 1972, cites 2000–2500 mm) and mean minimum temperature above 15.5 °C. They are mostly grown in the lowland wet tropics, as seen from the distribution of cooking bananas within 7° of the equator in Central and West Africa (Flinn & Hoyoux, 1976). However, they are sometimes very important in the wet-and-dry tropics, occupying more than 50% of the cropped land in Western Tanzania (Malima, 1976) and being important cash crops under unimodal and bimodal wet-and-dry water regimes in Gabon and Ivory Coast respectively (Aubert, 1971). In Papua New Guinea, triploid bananas are grown to about 2200 m and diploid genotypes to 1800 m (R.M. Bourke, personal communication, 1992).

Bananas are planted as rhizome pieces or aerial shoots at densities usually below 2000 ha^{-1}. Growth of the primary crop is thus limited by radiation interception; yields respond linearly to increasing population to above 2500 plants ha^{-1} (Stover & Simmonds, 1987). However, by the time the bunch emerges from the main pseudostem the crop may have attained a reasonable leaf area because the second pseudostem (which later becomes the ratoon or second cycle crop) has a leaf area comparable to that of the main pseudostem. Crop LAI at bunch emergence is 4–5 (Turner, 1972*b*). Canopy extinction coefficients (k) range from 0.46 to 0.75 (Turner, 1990).

Leaves make an appreciable proportion of their total growth within the pseudostem (Barker, 1969). They emerge at rates of one per 6–15 days in the tropics (Barker, 1969; Anno & Lambert, 1976). Rate of leaf emergence is positively correlated with temperature: emergence ceases below 10 °C (Turner, 1970) and production increases by one leaf per month per 3.3–3.7 °C rise in minimum or mean temperature from 10 to 20 °C or 13.5 to 25 °C respectively (Turner, 1971; Robinson, 1981). In the tropics the average life of a leaf from emergence to fall is 100–110 days, increasing to 150 days for the last leaves formed (Stover, 1974). Leaf spot (*Mycosphaerella musicola*), low temperature or water deficit accelerate leaf senescence (Turner, 1971; Stover, 1974; Turner & Barkus, 1980).

Dry matter partitioning and rhizome and root growth have received little attention. The rhizome grows slowly in the first month, but thereafter it constitutes 20–25% of whole-plant dry weight until rhizome growth ceases at about the time the second pseudostem begins rapid growth (Turner, 1972*b*). Root extension growth is very seasonal, at least in subtropical environments, from zero to over 20 mm d^{-1} for primary roots (Robinson & Bower, 1988). Root growth ceases and senescence is rapid after bunch emergence (Champion & Olivier, 1961). Thus, at the end of three crop cycles, rhizomes and roots in lysimeters weighed only 1.5 and 2.5 kg compared with 20 kg for whole plants (Turner & Barkus, 1980).

Bananas are often, but not inherently, shallow-rooted (Olsson, Carey & Turner, 1984, see also Section 18.5.1). Shallow roots make the crop susceptible to lodging and water deficit. Lodging is a problem, particularly in the Caribbean and West Africa, where losses due to high winds are regular and unavoidable; Stover & Simmonds (1987) estimated losses in Cameroun due to lodging at 20% per annum. Good soil water conditions reduce susceptibility to wind damage by increasing pseudostem thickness so that the proportion of broken pseudostems is reduced, as well as directly increasing fruit yield and the proportion of fruit which is harvestable (Holder & Gumbs, 1983*a, b*). Conversely, drought or over-wet soil increases wind damage, the first by, presumably, reducing the strength of the pseudostem and the second by increasing the amount of uprooting. Wind-caused ripping and segmentation of leaf laminae, which is universal, does not have much effect on photosynthetic rate unless the leaves are killed (Taylor & Sexton, 1972). Yield losses due to lodging and loss of leaf area have caused Stover (1982) and others to propose that yields will be highest from short-statured varieties, planted at high populations to achieve the same LAI as traditional, relatively tall varieties.

Shmueli (1953) found that 65% of the total water taken up was from the top 30 cm of soil and only 5% from deeper than 60 cm. Bunch weight often responds to increasing amount or frequency of irrigation (e.g., Arscott, Bhangoo & Karon, 1965; Ghavami, 1974), presumably because evaporation (and thus photosynthesis) is sensitive to available soil water. Evaporation declines sharply when soil water is reduced from two-thirds to half of field capacity (Shmueli, 1953; Arscott *et al.*, 1965). Evaporation has been approximated by:

$$E = 0.9E_p \times W_a,$$

where E and E_p are crop and pan evaporation and W_a is available soil water as a fraction of field capacity (Turner, 1972b). Banana crop evaporation increases by 0.7 mm per °C rise in temperature (Arscott et al., 1965).

There is a general correlation between inflorescence (bunch) weight and size of the vegetative plant, whether assessed by the number of leaves (Stover & Simmonds, 1987), the number of 'functional' leaves (Heenan, 1973) or the size of pseudostem (Fernandez-Caldas & Garcia, 1972). In bananas it is not clear to what extent this correlation depends on environmental effects, on current photosynthate production or on redistribution of dry matter and minerals to the fruit. It is reasonable to speculate that redistribution is relatively important: substantial quantities of material (e.g., 39% of dry weight) can be lost from the pseudostem after bunch harvest, presumably to other growing points (Turner & Barkus, 1973). Moreover, individual fruit weight and bunch yield are all greater in ratoon than in primary crops (Robinson, 1981; Turner & Barkus, 1982), no doubt owing to retranslocation as well as to higher leaf area during ratoon cycles.

Turner & Barkus (1982) showed that individual fruit weight varied between crop cycles relatively independently of other yield components. However, compensation between yield components occurs in the field. Champion (1967) found near-parallel negative linear correlations between individual fruit weight and number of hands per inflorescence in bananas from six African countries. These observations are reconcilable if we consider that fruit number is determined shortly after floral initiation whereas individual fruit weight is determined by environmental effects on photosynthesis and redistribution after bunch emergence.

Environment affects timing of emergence and rate of individual fruit growth. Late planting — that is, half-way through rather than at the start of the wet season — may not delay the attainment of first flowering but it spreads the onset of flower emergence for the crop population. This causes a compounding delay in flowering in subsequent cycles (Moreau, 1965). On the other hand, drought accelerates maturation and reduces fruit filling, thereby reducing fruit quality (Eastwood & Jeater, 1949). Fruit growth increases with temperature, being positively related to day-degrees above 14.5 °C (over the range 18–29 °C: Ganry & Meyer, 1975) or to mean daily temperature from 13 to 22 °C (Turner & Barkus, 1982). Ganry & Meyer (1975) found an optimum at 28–30 °C, which is perhaps the basis for agreement, at least among reviewers, that the optimum temperature for banana fruit yield is about 27 °C (Purseglove, 1972).

18.5 Crop/soil relations

18.5.1 *Soil physical properties*

Stover & Simmonds (1987) concluded that the only factor common to the wide range of soils on which bananas are grown is good drainage. Good drainage is of course dependent on a favourable soil structure and pore size distribution, and is essential to soil aeration. Texturally the soils vary greatly from very coarse to fine volcanic materials (Andisols), through sands and loams within various orders to clays within orders that include Inceptisols, Oxisols and Vertisols. Some soils are very high in organic matter (Twyford, 1967). Alluvial soils used for bananas in Central America vary from stony sands to heavy clays, but all possess adequate pore space. However, the unaggregated compact clays and fine impermeable sands of Trinidad and eastern Venezuela are unsuitable (Tai, 1977). Deep well-drained loams and light clay loams, shown to give consistently high yields in Central America (Lahav & Turner, 1983), are probably ideal with respect to both physical and chemical properties.

The distribution of the several hundred roots that arise from the banana rhizome is affected by conditions in the soil profile, but they may penetrate to 1.5 m (Lahav & Turner, 1983) and spread laterally for about 2–5 m. In Israel, most roots of irrigated bananas were found between 2 and 40 cm depth; the greatest number were between 20 and 40 cm and only a few vertical roots directly beneath the plants reached 60–80 cm (Shmueli, 1953). In acid soils (Ultisols, Alfisols, Entisols and Inceptisols) in Puerto Rico, Irizarry, Vincente-Chandler & Silva (1981) found that all roots of plantains were above 30–60 cm soil depth. In plantains (*Musa* AAB), successive ratoons form progressively nearer the soil surface, with the consequence that while shallow-planted parent crops may give good yields, the ratoon crops yield around 30% less because of uprooting by wind (Obiefuna, 1983).

Water tables also affect depth of rooting (Avilan, Meneses & Sucre, 1982). Lysimeter studies with plantains in a clay loam showed that lowering the water table from 12 to 36 cm increased fruit yield from 5.6 to 37.8 t ha^{-1} and increased roots in the 15–30 cm layer from 2200 to 4900 m m^{-3} (Irizarry, Silva & Vincente-Chandler, 1980). Field studies confirm the adverse effect of waterlogging on banana growth and yield (Table 18.2). Harvested fruit yield was 43% greater on the drained site than on the waterlogged site.

Cultivation on slopes ensures good surface drainage but may result in erosion. Clean-cultivated bananas growing in a clay loam on a 16° slope (28%) in Taiwan suffered a mean annual soil loss of 92 t ha^{-1} over 4 years (JCRR, 1977).

18.5.2 *Soil chemical properties*

The uptake of nutrients by bananas is affected by site factors (Twyford & Walmsley, 1974a), including soil water (Ssali, 1977) and temperature (Turner & Lahav, 1985). Hedge & Srinivas (1989) found when bananas were grown at various soil water deficits (by irrigating when matric potentials reached –25, –45, –65, and –85 kPa) in a well-drained sandy clay loam (Haplustalf) uptake of nitrogen, phosphorus, calcium and magnesium decreased with increasing water deficit but potassium

Table 18.2. *Growth and yield parameters of banana cv. 'Robusta' grown in drained sites and sites prone to waterlogging in a Udic Chromustert in St Lucia, Caribbean*

Parameter	Waterlogged (Mean and SE)	Drained (Mean and SE)
Time from planting to bunch emergence of 50% of the sample (days)	196.8 (9.8)	180.8 (6.8)
Time from planting to harvesting of 50% of the sample (days)	296.9 (6.7)	273.9 (6.9)
Final psuedostem height (cm)	215.5 (16.3)	245.9 (9.5)
Final pseudostem girth (cm)	46.6 (2.5)	52.9 (2.6)
Number of leaves at bunch emergence	9.28 (0.53)	14.1 (0.83)
Number of leaves at harvesting	6.74 (0.58)	6.75 (0.31)
Hands per bunch	7.36 (0.19)	7.79 (0.15)
Fingers per bunch	115.5 (8.1)	132.3 (6.1)
Bunch weight (kg)	18.4 (1.27)	21.8 (1.30)
Harvest (%)	76.3 (5.2)	92.5 (6.8)
Broken pseudostems (%)	18.8 (3.2)	6.9 (5.5)
Actual yield (t ha^{-1})	23.7	33.9

After Holder & Gumbs (1983b).

Waterlogging was induced by over-irrigation; the drained sites were not irrigated and experienced 6 months of severe soil water deficit. The average water table depth during the trial fluctuated between 10 and 30 cm (mean 18.8 cm) for the site prone to waterlogging and between 40 and 101 cm (mean 65.8 cm) for the drained site.

uptake was highest at –65 kPa. When yields are heavy, uptake and removal of nutrients in harvested fruit is high (Table 18.3). Only exceptional soils will sustain high yields without fertilisation (Stover & Simmonds, 1987).

A good nutrient supply at an early age is required to develop an effective leaf area index (LAI), which in turn promotes growth and bunch weight. Murray (1961), for instance, found a close relationship between bunch weight and the area of the third youngest leaf at 6 months. However, a good nutrient supply is also required to accelerate the development of secondary pseudostems (followers) and thereby increase the rate of bunch production per hectare (Table 18.4).

Some minerals required for growth immediately after planting are contained in the rhizome (Twyford & Walmsley, 1974*b*), but uptake from the soil is the major source of nutrients in a plant crop. Followers in an established crop are far less dependent on the soil as a nutrient source. The ^{32}P studies of Walmsley & Twyford (1968*a*) reveal that, for a period after harvest of the parent, a follower can obtain nutrients stored in the rhizome and pseudostem of the parent and can also obtain nutrients currently being taken up by the roots of the parent plant. Turner & Barkus (1973) estimated that the loss of nitrogen, phosphorus, potassium, calcium, manganese and copper from the pseudostem after bunch harvest was equivalent to as much as 40% of the needs of a young follower over a 10-week period. Significant quantities of potas-

Table 18.3. *Average amount of nutrients in Cavendish bananas*[a]

Element	Nutrients removed at harvest (kg ha^{-1})	Nutrients remaining in plants (kg ha^{-1})	Total (kg ha^{-1})	Proportion removed at harvest (%)
N	189	199	388	49
P	29	23	52	56
K	778	660	1438	54
Ca	101	126	227	45
Mg	49	76	125	39
S	23	50	73	32
Cl	75	450	525	14
Na	1.6	9	10.6	15

After Lahav & Turner (1983).
[a] Based on a population of 2000 mother plants ha^{-1}, yielding 50 t ha^{-1} fresh fruit with an average bunch weight of 25 kg (roots not included).

sium and magnesium may be leached out of the leaves of growing plants and become available for uptake by followers (Bhan, Wallace & Lunt, 1958). Later, during decomposition of the large quantities of pseudostems and leaves (up to about 200 t ha^{-1} fresh weight per year), nutrients are released for uptake by followers. It may take about 2 years from first planting for this recycling process to become established, but the consequences are that less fertiliser is required for ratoon crops than for the plant crop. For example, Twyford & Walmsley (1974*b*) recommended on infertile soils on islands in the eastern Caribbean an application of 6.25 t ha^{-1} of 9 : 4 : 29 (nitrogen : phosphorus : potassium) fertiliser for a plant crop but only 0.65 t ha^{-1} yr^{-1} for ratoon crops. The quantity of nutrients cycled varies not only with environment but also with differences in uptake between clones (e.g., Marchal & Mallessard, 1979). Nutrient losses by leaching on the Ivory Coast over 8 years amounted to 60–85% of the quantity applied in fertilisers (Godefroy, Roose & Muller, 1975).

Bananas and plantains growing around villages receive nutrients in various forms of organic waste. Its fertiliser value can be quite significant

Table 18.4. *The effect of nitrogen, phosphorus and potassium on bunch weight, yield and rate of bunch production of bananas*

Nutrient	Location	Fertiliser rate (kg ha^{-1}yr^{-1})	Bunch weight (kg)	Yield (t ha^{-1})	Bunches (ha^{-1})
N[a]	India	0	9.5	15.2	1600 (565 days)[b]
		96	12.2	19.5	1598 (506 days)[b]
		192	16.0	25.6	1600 (493 days)[b]
		384	11.5	31.2	2713 (500 days)[b]
P[c]	Windward	0	8.8	16.2	1840
	Islands	70	12.3	25.2	2049
		140	11.7	22.6	1932
K[d]	Panama	0	23.3	70.4	2741
		450	28.8	88.6	2797
		900	31.2	100.8	2932

[a] N data adapted from Chappopadhyay *et al.* (1980). Yield and bunches ha^{-1} data are per crop.
[b] Time from planting to harvest.
[c] P data adapted from Twyford (1965). Yield and bunches ha^{-1} data are per year.
[d] K data adapted from Rodriguez-Gomez (1980). Yield and bunches ha^{-1} data are per 18 months.

in increasing yields. Table 18.5 shows that the full benefit was not evident until the ratoon crop because of the time taken for the material to decompose and mineralise.

Nitrogen is required in large quantities by the banana (Table 18.3) and together with potassium is the most commonly required fertiliser nutrient (Martin-Prevel, 1980a, b), even in the very fertile soils of Central America (Butler, 1960). It is particularly important in promoting dry matter increase (Martin-Prevel & Montagut, 1966) but high levels may result in tall, weak psuedostems which break during high winds (Holder & Gumbs, 1983b). Lahav & Turner (1983) showed that nitrogen uptake was correlated with total dry weight (DW) production over a range of varieties, soils and environments:

$$\text{nitrogen uptake} = 6.76 \, DW + 24.7; \, r^2 = 0.79.$$

The range of tissue nitrogen concentration is therefore small: 1–2% in most tissues during the vegetative stage and about 3% in the leaves (Lahav & Turner, 1983). Thus the capacity of the banana to exploit high levels of available soil nitrogen and store it in the tissue for use during later growth is negligible in contrast to that of cereals (e.g., maize: Terman, 1974). The implication is that fertiliser nitrogen should be given in split applications rather than as a single large dressing. Split

Table 18.5. *Effect of fertiliser and organic material on plant yield and fruit number per bunch in Nigeria*[a,b]

Treatment	Yield (t ha^{-1})		Fruit no. per bunch	
	Plat crop	Ratoon crop	Plant crop	Ratoon crop
No fertiliser	13.5 a	14.8 a	25.9 ab	26.4 a
Fertiliser[c]	14.0 a	16.2 a	28.0 c	28.3 a
Mulch[d]	14.9 a	20.3 bc	27.5 bc	32.9 bc
Mulch + fertiliser	15.0 a	19.3 b	27.8 bc	31.6 b
Mulch + ash + household refuse[e]	15.6 a	22.0 c	29.0 c	35.3 c

After IITA (1980).
[a] Crop spacing 2 m × 3 m.
[b] Values with different letters significantly different at *P*<0.05.
[c] 1 kg of 15–15–15 per stool spread over 10 application per year.
[d] Mainly grass.
[e] Yam and plantain peel.

applications are also appropriate where pseudostems are at different stages of development.

Relative to nitrogen and potassium, the phosphorus requirement of bananas is low (Table 18.3), probably because they are able to remobilise phosphorus within the plant (Martin-Prevel, 1978). However, responses to phosphorus are not uncommon in low-phosphorus soils. In the tropics the rate of phosphorus uptake is highest during the 2–3 months after planting; after bunching the rate drops to about 20% of that in the vegetative phase (Walmsley & Twyford, 1968b).

Very large quantities of potassium are taken up by the banana plant (Table 18.3). With continued banana cropping, potassium deficiency has become more common on fertile coastal soils of Central America (Rodriguez-Gomez, 1980), though soils in the Canary Islands still supply high levels of potassium after many years of cropping (Fernandez-Caldas & Garcia, 1970). In Martinique volcanic ash soils, the frequency with which potassium is applied is inversely related to the capacity of the soils to adsorb it (Fontaine *et al.*, 1989). Deep roots may facilitate uptake of potassium (Warner & Fox, 1977). Low potassium availability reduces total dry matter production, leaf size and longevity, delays flower initiation, reduces fruit number per bunch and, in particular, reduces fruit size (Turner & Barkus, 1980, 1982). Turner & Barkus (1980) showed in lysimeters with cv. Williams that while low potassium supply halved total dry matter yield, bunch yield was reduced by 80%. In contrast to nitrogen, potassium concentration in the whole plant on a dry matter basis changes with time, typically decreasing from the sucker to the harvest stage (Martin-Prevel, 1967). High levels of potassium fertiliser can depress magnesium and calcium uptake (Messing, 1974; Lahav & Turner, 1983).

Bananas are grown over a wide range of soil pH, to values as low as 3.4 (Godefroy *et al.*, 1978). In Jamaica vigorous and productive plants grow in the pH range 4.5–8.0 (Tai, 1977) and in Fiji calcareous soils (pH 8–8.5) yield well when supplied with organic matter (Twyford, 1967). Occasional effects of soil pH on yield have been recorded (see Champion *et al.*, 1958). Aluminium and manganese toxicity do not appear to be problems of any significance (Plucknett, 1978; Perez-Escolar & Lugo-Lopez, 1979). Liming may be beneficial in very acid soils where micronutrient availability is reduced (e.g., Bhangoo & Karon, 1962). Low soil pH and low exchangeable calcium are often associated with the incidence of Panama disease (Alvarez *et al.*, 1981). Bananas tolerate acid sulphate soils (Moormann, 1963).

Bananas have moderate tolerance of salinity, but salt problems have been reported from a number of tropical countries. Dunlap & McGregor (1932) found that when soluble salt levels were below 0.5 g l^{-1} growth was satisfactory, but that when they rose above 1.0 g l^{-1} plants were stunted or killed; between these values plants and fruit were visibly affected.

18.6 Place in cropping systems

The cropping systems under which bananas are grown may be broadly divided into two groups: commercial plantation production of sweet bananas on a medium (10–100 ha) or large (>100 ha) scale, usually for export to nations with advanced economies, and small-scale production of a range of types for subsistence or local marketing. Since this book is focussed largely on food crops produced and consumed in the tropics as staple items of diet, medium- and large-scale production systems will not be considered.

Stover & Simmonds (1987) estimate that about half the world's production of bananas is eaten cooked. Plantains (AAB) are a major diet component in Central and West Africa. Both AAA and ABB types are important in East Africa. In south and southeast Asia bananas are not commonly cooked; production is largely of sweet AAA types. Plantains and other cooking bananas are of importance in Colombia and Venezuela, and in the West Indies AAA bananas are eaten raw and all three types are eaten cooked (Burden & Coursey, 1977).

Bananas are a ubiquitous crop in the better-watered areas of the tropics. The reasons for their popularity are summarised by Ruthenberg (1980), who points out that they combine many of the advantages of both annual and perennial crops:

1. Land clearance and cultivation inputs are low and the latter are well spread through the year.
2. Fruit is produced within 1 year of planting.
3. Except in markedly seasonal climates, fruit is harvestable for most months of the year.
4. Gross return per unit area, whether in money or food energy, is high.
5. The crop has many end-uses: fresh and cooked food for humans, feed for livestock, beer, roofing material, etc.
6. It returns substantial quantities of organic matter to the soil and protects slopes against serious erosion.

7. It responds well to fertiliser and animal manure.

On small holdings, bananas may or may not be spatially integrated into the annual cropping system. When first planted, they are normally intercropped with annuals, though it is also common to find them planted as a row or as an isolated clump within or adjacent to the annual crop area. They are a characteristic component of mixed-garden communities in house compounds. Intercropping plantains with cocoyam or with maize and cassava in Nigeria did not have any significant effect on plantain yield (Devos & Wilson, 1979); the intercropping of plantains with cassava or with cocoyam is also referred to in Section 15.6. Bananas or plantains may be established as an intercrop with shade-demanding perennial crops such as cocoa in West Africa and coffee in tropical America; short-lived diploids are used to provide temporary shade for cocoa in New Guinea.

Since it is not possible within the compass of this section to attempt to categorise, let alone to describe, all the cropping systems of which bananas are a significant component, the section will conclude instead with an account of one important and clearly defined type, the banana–coffee system of East Africa, which illustrates the features of intercropping with both annual and perennial crops and which has been described in detail elsewhere (Parsons, 1970; Ruthenberg, 1980).

In southeastern Uganda and in Tanzania bordering on Lake Victoria, in two-peak sub-equatorial rainfall regimes, coffee (mainly robusta) is the major cash crop and bananas are the major food crop. Subsidiary food crops include maize, beans, groundnuts and cassava. Fig. 18.1 illustrates the spatial relations of crops in a holding in Bukoba, Tanzania. Around the domicile and kraal are dense vigorous bananas (with vegetables), vigorous because they receive most of the animal manure. Moving outwards the bananas become less dense and other associated crops – maize, coffee, beans – are grown, coffee becoming dominant towards the periphery of the holding (Ruthenberg, 1980). Crop sequence within the annual components of such systems is very ill-defined (Parsons, 1970). In general, the nearer the domicile and kraal the higher the cultivation frequency, since it is the inner areas that receive most of the kraal manure and domestic refuse.

There are variants of this system. In the Buganda region of Uganda, annual crops include yams, cocoyam, beans and other vegetables grown between the bananas. The application of animal manure is uncommon, but the ground is heavily mulched with banana leaves and split pseudostems cut down after the fruit has been harvested (Mukasa &

Thomas, 1970). In somewhat drier regions of Uganda, the main crops are bananas, finger millet (*Eleusine coracana*) and cotton (Parsons, 1970). These crops are grown separately and finger millet and cotton are grown in rotation in open fields. Livestock are more important than in the banana–coffee system but little animal manure is applied to crops.

This brief description of some patterns of banana cropping makes an appropriate end to our account of the ecology of tropical food crops, since it illustrates the diversity and complexity of the environments and farming systems in which such crops are grown and emphasises the differences between cropping enterprises in the developed temperate world and the developing tropical world. The conventional temperate-zone picture of large sole-crop fields of a mechanised farming unit growing a narrow specialised range of crops for sale has little relevance to the tropics, where the typical scene is a small family farm growing a wide

Fig. 18.1. Spatial arrangement of crops in a banana–coffee holding in Bukoba, Tanzania. 1, Hut; 2, boma; 3, banana; 4, coffee; 5, maize; 6, beans; 7, vegetables. Source: Ruthenberg (1980).

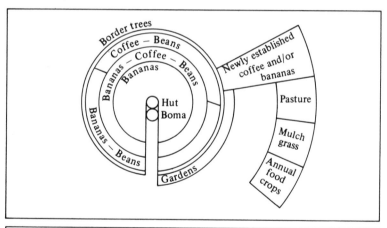

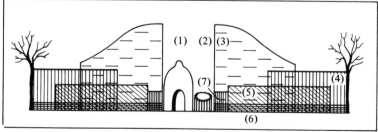

range of subsistence and cash crops with hand labour, perhaps assisted by draft animals, in a variety of mixed cropping patterns.

The inputs for production – land preparation, irrigation, drainage, fertilisation, protection against weeds, pests, diseases and soil erosion – represent modifications to the crop environment. Since, owing to limitations of capital, income and technology, input levels for food crop production in the tropics are generally much lower than in temperate developed regions, the direct impact of the natural environment on productivity is correspondingly greater. We have attempted to provide an insight into the environmental relations of tropical food crops, but our approach has been descriptive rather than prescriptive; it has not been our purpose to review knowledge on how crops can or should be grown. However, there can be no argument that if the peoples of the tropical world are not to slip further behind in standards of human welfare, reliable prescriptive answers based on a quantitative understanding of crop/environment relations are still urgently needed.

Further reading

Stover, R.H. & Simmonds, N.W. (1987). *Bananas*. Harlow, UK: Longman.

References

Abrol, I.P., Bhumbla, D.R. & Meelu, O.P. (1985). Influence of salinity and alkalinity on properties and management of ricelands. In *Soil Physics and Rice*, pp.183–198. Los Baños, Philippines: International Rice Research Institute.

Abruña, F. (1980). Responses of soybeans to liming on acid tropical soils. In *World Soybean Research Conference II Proceedings*, ed. F.T. Corbin, pp. 35–46. Boulder, Colorado: Westview Press.

Abruña, F., Pearson, R.W. & Perez-Escolar, R. (1975). Lime responses of corn and beans grown on typical Ultisols and Oxisols of Puerto Rico. In *Soil Management in Tropical America*, ed. E. Bornemisza & A. Alvarado, pp. 261–81. Raleigh, North Carolina: North Carolina State University.

Abruña, F., Vincente-Chandler, J., Rodriguez, J., Badillo, J. & Silva. S. (1979). Crop response to soil acidity factors in Ultisols and Oxisols in Puerto Rico. V. Sweet potato. *University of Puerto Rico Journal of Agriculture*, 63, 250–67.

Abruña-Rodriguez, F., Vincente-Chandler, J., Rivera, E. & Rodriguez, J. (1982). Effect of soil acidity factors on yields and foliar composition of tropical root crops. *Soil Science Society of America Journal*, 46, 1004–7.

Acedo, A.L. & Javier, R.R. (1980). Residual nitrogen from legumes and its effects on the succeeding crop of sweet potato. *Annals of Tropical Research*, 2, 72–9.

Acuna, E.J. & Sanchez, P.C. (1969). Response of the groundnut to application of nitrogen, phosphorus, and potassium on the light sandy savanna soils of the state of Monegas. *Fertilité*, 35, 3–9.

Adams, F. & Pearson, R.W. (1970). Differential response of cotton and peanuts to subsoil acidity. *Agronomy Journal*, 62, 9–12.

Adams, J.E. (1967). Effect of mulches and bed configuration. *Agronomy Journal*, 59, 595–9.

Adams, M.W. (1973). Plant architecture and physiological efficiency in the field bean. In *Potentials of Field Beans and other Food Legumes in Latin America*, pp. 266–78. Cali, Colombia : Centro Internacional de Agricultura Tropical.

Adams, M.W. & Pipolly, J.J. (1980). Biological structure, classification and distribution of economic legumes. In *Advances in Legume Science*, ed. R.J. Summerfield & A.H. Bunting, pp. 1–16. Kew: Royal Botanic Gardens.

Adeoye, K.B. & Mohamed-Saleem, M.A. (1990). Comparison of effects of some tillage methods on soil physical properties and yield of maize and stylo in a degraded ferruginous tropical soil. *Soil and Tillage Research*, 18, 63–72.

Adepetu, J.A. & Corey, R.B. (1977). Changes in N and P availability and P fractions in Iwo soil from Nigeria under intensive cultivation. *Plant and Soil*, **46**, 309–16.

Adiningsih, J.S., Sudjadi, M. & Setyorini, D. (1988). Overcoming soil fertility constraints in acid upland soils for food crop based farming systems in Indonesia. *Indonesian Agricultural Research and Development Journal*, **10**, 49–58.

Ageeb, O.A.A. & Ayoub, A.T. (1976). Effect of sowing date and soil type on plant survival and grain yield of chickpeas (*Cicer arietinum* L.). *Journal of Agricultural Science*, **88**, 521–7.

Agrawal, R.P. & Sharma, S.K. (1984). Effect of triple superphosphate and polyvinyl alcohol on crust strength, soil physical properties and seedling emergence of pearl millet. *Tropical Agriculture (Trinidad)*, **61**, 269–72.

Aguilar, M.I., Fisher, R.A. & Kohashi, S.J. (1977). Effects of plant density and thinning on high yielding dry beans, *Phaseolus vulgaris* L. *Experimental Agriculture*, **13**, 325–35.

Ahmad, F. & Tan, K.H. (1986). Effect of lime and organic matter on soybean seedlings grown in aluminium-toxic soil. *Soil Science Society of America Journal*, **50**, 656–61.

Ahn, P.M. (1970). *West African Soils*. London: Oxford University Press, 332 pp.

Ahn, P.M. (1979). Microaggregation in tropical soils: its measurement and effects on the maintenance of soil productivity. In *Soil Physical Properties and Crop Production in the Tropics*, ed. R. Lal & D.J. Greenland, pp. 75–86. Chichester: Wiley.

Aina, P.O., Lal, R. & Taylor, G.S. (1979). Effects of vegetal cover on soil erosion on an Alfisol. In *Soil Physical Properties and Crop Production in the Tropics*, ed. R. Lal & D.J. Greenland, pp. 50–8. Chichester: Wiley.

Akoroda, M.O. & Chheda, H.R. (1983). Agro-botanical and species relationships of Guinea yams. *Tropical Agriculture (Trinidad)*, **60**, 242–5.

Akparanta, S.E., Skaggs, R.W. & Sanders, D.C. (1980). Drainage requirements for sweet potato at harvest. *Journal of the American Society for Horticultural Science*, **105**, 447–51.

Alagawadi, A.R. & Gaur, A.C. (1988). Associative effect of *Rhizobium* and phosphate-solubilizing bacteria on the yield and nutrient uptake of chickpea. *Plant and Soil*, **105**, 241–6.

Alagarswamy, G., Maiti, R.K. & Bidinger, F.R. (1977). *Physiology Program. International Pearl Millet Workshop, ICRISAT.* Hyderabad, India: International Crops Research Institute for the Semi-Arid Tropics.

Alegre, J.C., Cassel, D.K. & Bandy, D.E. (1986*a*). Effects of land clearing and subsequent management on soil physical properties. *Soil Science Society of America Journal*, **50**, 1379–84.

Alegre, J.C., Cassel, D.K. & Bandy, D.E. (1986*b*). Reclamation of an Ultisol damaged by mechanical land clearing. *Soil Science Society of America Journal*, **50**, 1026–31.

Alegre, J.C., Cassel, D.K. & Makarim, M.K. (1987). Strategies for reclamation of degraded lands. In *Tropical Land Clearing for Sustainable Agriculture*, International Board for Soil Research and Management Inaugural Workshop, Bangkok, Thailand, pp. 77–91.

Alegre, J.C., Cassel, D.K. & Bandy, D.E. (1988). Effect of land clearing method on chemical properties of an Ultisol in the Amazon. *Soil Science Society of America Journal*, **52**, 1283–8.

Alegre, J.C., Cassel, D.K. & Bandy, D.E. (1990). Effect of land-clearing method and soil management on crop production in the Amazon, *Field Crops Research*, **24**, 131–41.

Alexander, J. & Coursey, D.G. (1969). The origins of yam cultivation. In *The Domestication and Exploitation of Plants and Animals*, ed. P.J. Ucko & G.W. Dimbleby, pp. 405–25. London: Duckworth.

Alexander, K.G. & Miller, M.H. (1991). The effect of soil aggregate size on early growth and shoot–root ratio of maize (*Zea mays* L.). *Plant and Soil*, **138**, 189–91.

Allan, W. (1965). *The African Husbandman*. Edinburgh: Oliver & Boyd, 505 pp.

Allen, O.N. & Allen, C.K. (1981). *The Leguminosae*. Madison, Wisconsin: University of Wisconsin Press, 812 pp.

Allison, F.E. (1973). *Soil Organic Matter and its Role in Crop Production*. Amsterdam: Elsevier, 637 pp.

Altieri, M.A., Francis, C.A., Schoonhoven, A. van & Doll, J.D. (1978). A review of insect prevalence in maize (*Zea mays* L.) and bean (*Phaseolus vulgaris* L.) polycultural systems. *Field Crops Research*, **1**, 33–49.

Alva, A.K., Asher, C.J. & Edwards, D.G. (1986*a*). The role of calcium in alleviating aluminium toxicity. *Australian Journal of Agricultural Research*, **37**, 375–82.

Alva, A.K., Edwards, D.G., Asher, C.J. & Blamey, F.P.C. (1986*b*). Effects of phosphorus/aluminium molar ratio and calcium concentration on plant responses to aluminium toxicity. *Soil Science Society of America Journal*, **50**, 133–7.

Alva, A.K., Edwards, D.G., Asher, C.J. & Blamey, F.P.C. (1986*c*). An evaluation of aluminium indices to predict aluminium toxicity to plants grown in nutrient solutions. *Communications in Soil Science and Plant Analysis*, **17**, 1271–80.

Alvarado, A.D. (1972). Determination of the optimum sowing date for chickpea in the Lagunera district. *Agricultura Tecnica en Mexico*, **3**, 197–200.

Alvarado, A. & Buol, S.W. (1985). Field estimation of phosphate retention by Andepts. *Soil Science Society of America Proceedings*, **49**, 911–14.

Alvarez, C.E., Garcia, V., Robles, J. & Diaz, A. (1981). Influence of soil characteristics on the incidence of Panama disease. *Fruits*, **36**, 71–81.

Ambak, K. & Tadano, T. (1991). Effect of micronutrient application on the growth and occurrence of sterility in barley and rice in a Malaysian deep peat soil. *Soil Science and Plant Nutrition*, **37**, 715–24.

Ambak, K., Bakar, Z.A. & Tadano, T. (1991). Effect of liming and micronutrient application on the growth and occurrence of sterility in maize and tomato plants in a Malaysian deep peat soil. *Soil Science and Plant Nutrition*, **37**, 689–98.

Anden-Lacsina, T. & Barker, R. (1978). The adoption of modern varieties. In *Interpretive Analysis of Selected Papers from 'Changes in Rice Farming in Selected Areas of Asia'*, pp. 13–33. Los Baños, Philippines: International Rice Research Institute.

Anderson, G.D. (1974*a*). Potassium responses of various crops in East Africa. In *Potassium in Tropical Crops and Soils*, pp. 413–37. Berne, Switzerland: International Potash Institute.

Anderson, G.D. (1974*b*). Bean responses to fertilizers on Mt Kilimanjaro in relation to soil and climatic conditions. *East African Agricultural and Forestry Journal*, **39**, 272–88.

Anderson, W.K. (1979). Sorghum. In *Australian Field Crops*, vol. 2, ed. J.V. Lovett & A. Lazenby, pp. 37–69. London: Angus & Robertson.

Andrade, A.M. & Sedijama, T. (1977). Efeitos do espacamento e da densidade de plantio sobre a variedade de soja UFV-I no Triangulo Mineiro. *Revista Ceres (Brazil)*, **24**, 412–9.

Andrew, C.S. (1977). Nutritional restraints on legume symbiosis. In *Exploiting the Legume–Rhizobium Symbiosis in Tropical Agriculture*, ed. J.M. Vincent, A.S. Whitney & J. Bose, pp. 253–74. Honolulu, Hawaii: University of Hawaii Press.

Andriesse, J.P. (1974). Tropical lowland peats in South-East Asia. *Royal Tropical Institute Communication*, No. 63, 63 pp.

Angus, J.F., Cunningham, R.B., Moncur, M.W. & Mackenzie, D.H. (1980). Phasic development in field crops. I. Thermal response in the seedling phase. *Field Crops Research*, 3, 365–78.

Anno, A. & Lambert, C. (1976). Caractéristiques de croissance et les phases de développement chez le bananier plantain (var. Corne). B. *Fruits*, 31, 678–83.

App, A., Santiago, T., Daez, C., Menguito, C., Ventura, W., Tirol, A., Po, J., Watanabe, I., Datta, S.K. de & Roger, P. (1984). Estimation of the nitrogen balance for irrigated rice and the contribution of phototrophic nitrogen fixers. *Field Crops Research*, 9, 17–27.

Arismendi, L.G. (1973). Planting dates and harvesting dates for cassava (*Manihot esculenta* Crantz.) in the Jusepin savanna. *Boletin Informativo, Instituto de Investigaciones Agropecuarias, Universidad de Oriente, Agronomia*, 6 pp.

Armstrong, W. (1971). Radial oxygen losses from intact rice roots as affected by distance from the apex, respiration and waterlogging. *Physiologia Plantarum*, 25, 192–7.

Arndt, W. (1965). The impedance of soil seals and the forces of emerging seedlings. *Australian Journal of Soil Research*, 3, 55–69.

Arndt, W. & McIntyre, G.A. (1963). Initial and residual effects of superphosphate and rock phosphate for sorghum and peanuts on a lateritic red earth. *Australian Journal of Agricultural Research*, 14, 785–95.

Arora, Y., Mulongoy, K. & Juo, A.S.R. (1986). Nitrification and mineralisation potentials in a limed Ultisol in the humid tropics. *Plant and Soil*, 92, 153–7.

Arscott, T.G., Bhangoo, M.S. & Karon, M.L. (1965). Irrigation investigations of the Giant Cavendish banana: I. *Tropical Agriculture (Trinidad)*, 42, 139–44.

Artschwager, E. (1924). On the anatomy of the sweet potato root with notes on the internal breakdown. *Journal of Agricultural Research*, 27, 157–66.

ASA (1976). *Multiple Cropping*. Madison, Wisconsin: American Society of Agronomy, 378 pp.

Asher, C.J. & Edwards, D.G. (1978). Relevance of dilute solution culture studies to problems of low fertility tropical soils. In *Mineral Nutrition of Legumes in Tropical and Subtropical Soils*, ed. C.S. Andrew & E.J. Kamprath, pp. 131–52. Melbourne: Commonwealth Scientific and Industrial Research Organization.

Asher, C.J., Edwards, D.G. & Howeler, R.H. (1980). *Nutritional Disorders of Cassava*. St Lucia, Queensland: University of Queensland Department of Agriculture, 48 pp.

Ashraf, M. & McNeilly, T. (1987). Salinity effects on five cultivars/lines of pearl millet (*Pennisetum americanum* [L] Leeke). *Plant and Soil*, 103, 13–19.

Aubert, B. (1971). Action du climat sur le comportement du bananier en zones tropicales et subtropicales. *Fruits*, 26, 175–88.

Austin, D.F. (1978). The *Ipomoea batatas* complex. I. Taxonomy. *Bulletin of the Torrey Botanical Club*, 105, 114–29.

Austin, D.F. (1988). The taxonomy, evolution and genetic diversity of sweet potatoes and related wild species. In *Exploration, Maintenance and Utilization of Sweet Potato Genetic Resources*, ed. P. Gregory, pp. 27–60. Lima, Peru: International Potato Centre.

Austin, M.E., Aung, L.H. & Graves, B. (1970). Some observations on the growth and development of sweet potato (*Ipomoea batatas*). *Journal of Horticultural Science*, **45**, 257–64.

Avilan, L.R., Meneses, L.R. & Sucre, R.E. (1982). Distribucion radical del banano bajo diferentes sistemas de manejo de suelos. *Fruits*, **37**, 103–10.

AVRDC (1975). *Annual Report for 1974*. Shanhua, Taiwan: Asian Vegetable Research and Development Center, 142 pp.

AVRDC (1986) *Soybean in Tropical and Subtropical Cropping Systems*. Shanhua, Taiwan: Asian Vegetable Research and Development Center, pp. 217–18.

AVRDC (1991). *1990 Progress Report*. Shanhua, Taiwan: Asian Vegetable Research and Development Center, 313 pp.

Awadhwal, N.K. & Smith, G.D. (1990). Performance of low-draft tillage implements on a hard setting Alfisol of the SAT in India. *Soil Use and Management*, **6**, 28–31.

Awai, J. (1981). Inoculation of soya bean (*Glycine max* (L.) Merr.) in Trinidad. *Tropical Agriculture (Trinidad)*, **58**, 313–18.

Awan, A.B. (1964). Effects of lime on the availability of P on Zamorano soils. *Soil Science Society of America Proceedings*, **28**, 672–3.

Ayanaba, A. (1977). Towards better use of inoculants in the humid tropics. In *Biological Nitrogen Fixation in Farming Systems of the Tropics*, ed. A. Ayanaba & P.J. Dart, pp. 189–204. New York: Wiley.

Ayanaba, A., Asanuma, S. & Munns, D.N. (1983). An agar plate method for rapid screening of *Rhizobium* for tolerance to acid-aluminium stress. *Soil Science Society of America Journal*, **47**, 256–8.

Ayyangar, G.N.R., Vijiaraghavan, C. & Pillai, V.C. (1933). Studies on *Pennisetum typhoideum* (Rich.) – the pearl millet. I. Anthesis. *Indian Journal of Agricultural Science*, **3**, 688–94.

Babalola, O. & Lal, R. (1977a). Subsoil gravel horizon and maize root growth. I. Gravel concentration and bulk density effects. *Plant and Soil*, **46**, 337–46.

Babalola, O. & Lal, R. (1977b). Subsoil gravel horizons and maize root growth. II. Effects of gravel size, inter-gravel texture and natural gravel horizon. *Plant and Soil*, **46**, ·347–57.

Babalola, O. & Oputa, C. (1981). Effects of planting patterns and population on water relations of maize. *Experimental Agriculture*, **17**, 97–104.

Bache, B. (1979). Base saturation. In *The Encyclopaedia of Soil Science*, ed. R.W. Fairbridge & C.W. Finkl, pp. 38–42. Stroudsberg, Pennsylvania : Dowden, Hutchinson & Ross.

Badillo-Feliciano, J. & Lugo-Lopez, M.A. (1977). Sweet potato production in Oxisols under a high level of technology. *University of Puerto Rico Agricultural Experiment Station Bulletin*, No. 256.

Bagnall & King (1991a) Response of peanut (*Arachis hypogaea*) to temperature, photoperiod and irradiance. I. *Field Crops Research*, **26**, 263–77.

Bagnall & King (1991b) Response of peanut (*Arachis hypogaea*) to temperature, photoperiod and irradiance. II. *Field Crops Research*, **26**, 279–93.

Balasubramanian, V. & Sinha, S.K. (1976). Nodulation and nitrogen fixation in chickpea (*Cicer arietinum* L.) under salt stress. *Journal of Agricultural Science*, **87**, 465–6.

Baligar, V.C., Nash, V.E., Whisler, F.D. & Myhre, D.L. (1981). Sorghum and soybean growth as influenced by synthetic pans. *Communications in Soil Science and Plant Analysis*, **12**, 97–107.

Baligar, V.C., Santos, H.L.D., Pitta, G.V.E., Filho, E.C., Vasconcellos, C.A. & Filho, A.F. De C.B. (1989). Aluminium effects on growth, yield and nutrient use efficiency ratios in sorghum genotypes. *Plant and Soil*, **116**, 257–64.

Bandy, D.E. & Sanchez, P.A. (1986). Post-clearing soil management alternatives for sustained production in the Amazon. In *Land Clearing and Development in the Tropics*, ed. R. Lal, P.A. Sanchez & R.W. Cummings, pp. 347–61. A.A. Rotterdam: Balkema.

Barker, E.F.I. (1964). Plant population and crop yield. *Nature (London)*, **204**, 856–7.

Barker, W.G. (1969). Growth and development of the banana plant: gross leaf emergence. *Annals of Botany*, **33**, 523–35.

Barry, D.A.J. & Miller, M.H. (1989). Phosphorus nutritional requirements of maize seedlings for maximum yield. *Agronomy Journal*, **81**, 95–9.

Bathke, G.R. & Blake, G.R. (1984). Effects of soybeans on soil properties related to soil erodibility. *Soil Science Society of America Journal*, **48**, 1398–401.

Bawazir, A.A.A. & Idle, D.B. (1989). Drought resistance and root morphology in sorghum. *Plant and Soil*, **119**, 217–21.

Bawden, R. & Ison, R.L. (1992). The purpose of field-crop ecosystems: social and economic aspects. In *Field Crop Ecosystems*, ed. C.J. Pearson, pp. 11–36. Amsterdam: Elsevier.

Bazan, R. (1975). Nitrogen fertilization and management of grain legumes in Central America. In *Soil Management in Tropical America*, ed. E. Bornemisza & A. Alvarado, pp. 228–45. Raleigh, North Carolina: North Carolina State University.

Beecher, H.G. (1991). Effect of saline water on rice yield and soil properties in the Murrumbidgee Valley. *Australian Journal of Experimental Agriculture*, **31**, 819–23.

Beer, J.F. de (1963). *Influences of Temperature on Arachis hypogaea L. with Special Reference to its Pollen Viability*. Wageningen: Centrum Landbouwpublikatias en Landbouwdocumentatie, 81 pp.

Begg, J.E. (1965). The growth and development of a crop of bulrush millet (*Pennisetum typhoides S. & H.*). *Journal of Agricultural Science*, **65**, 341–9.

Begg, J.E. & Burton, G.W. (1971). Comparative study of five genotypes of pearl millet under a range of photoperiods and temperatures. *Crop Science*, **11**, 803–5.

Begg, J.E., Bierhuizen, J.F., Lemon, E.R., Misra, D.K., Slatyer, R.O. & Stern, W.R. (1964). Diurnal energy and water exchanges in bulrush millet in an area of high solar radiation. *Agricultural Meteorology*, **1**, 294–312.

Beinroth, F.H. (1975). Relationship between US Soil Taxonomy, the Brazilian Soil Classification System and FAO/UNESCO Soil Units. In *Soil Management in Tropical America*, ed. E. Bornemisza & A. Alvarado, pp. 92–108. Raleigh, North Carolina: North Carolina State University.

Bell, L.C. & Gillman, G.P. (1978). Surface charge characteristics and soil solution composition of highly weathered soils. In *Mineral Nutrition of Legumes in Tropical and Subtropical Soils*, ed. C.S. Andrew & E.J. Kamprath, pp. 37–58. Melbourne: Commonwealth Scientific and Industrial Research Organization.

Bell, M.J., Middleton, K.J. & Thompson, J.P. (1989). Effects of vesicular-arbuscular mycorrhizae on growth and phosphorus and zinc nutrition of peanut (*Arachis hypogaea L.*) in an Oxisol from subtropical Australia. *Plant and Soil*, **117**, 49–57.

Berglund-Brücher, O. & Brücher, H. (1976). The South American wild bean (*Phaseolus aborigineus Burk.*) as an ancestor of the common bean. *Economic Botany*, **30**, 257–72.

Bernstein, L. (1974). Salt tolerance of plants. *US Department of Agriculture, Information Bulletin*, No. 283.

Best, R. (1978). Processing cassava for animal feed. In *Proceedings of a Cassava Harvesting and Processing Workshop, Cali, Colombia, April 1978*, ed. E.J. Weber, J.G. Cock & A. Chouinard, pp. 12–20. Ottawa: International Development Research Centre.

Bethlenfalvay, G.J., Ulrich, J.M. & Brown, M.S. (1985). Plant response to mycorrhizal fungi: host, endophyte, soil effects. *Soil Science Society of America Proceedings*, **49**, 1164–8.

Beyrouty, C.A. , Wells, B.R., Norman, R.J., Marvel, J.N. & Pillow, J.A. (1988). Root dynamics of a rice cultivar grown at two locations. *Agronomy Journal*, **80**, 1001–4.

Bezuneh, T. (1975). Status of chickpea production and research in Ethiopia. In *International Workshop on Grain Legumes*, pp. 95–101. Hyderabad, India: International Crops Research Institute for the Semi-Arid Tropics.

Bhagsari, A.S., & Ashley, D.A. (1990). Relationship of photosynthesis and harvest index to sweet potato yield. *Journal of the American Society for Horticultural Science*, **115**, 288–93.

Bhan, K.C., Wallace, A. & Lunt, O.R. (1958). Some mineral losses from leaves by leaching. *Journal of the American Society for Horticultural Science*, **73**, 289–93.

Bhangoo, M.S. & Karon, M.L. (1962). Investigations on the Giant Cavendish banana: II. *Tropical Agriculture (Trinidad)*, **39**, 203–10.

Bhatnagar, M.P. & Kumar, K. (1960). Anthesis studies in Rajasthan bajra (*Pennisetum typhoideum*). *Indian Journal of Agricultural Science*, **30**, 185–95.

Bhivare, V.N. & Chavan, P.D. (1987). Effect of salinity on translocation of assimilates in French bean. *Plant and Soil*, **102**, 295–7.

Biderbost, E.B.J., Rodriquez. A.A., Deromedis, R. & Lasso, R. (1974). Floral development and a breeding technique for chickpea (*Cicer arietinum* L.). *Rivista Industrial y Agricola de Tucuman*, **51**, 1–9.

Bidin, A.A. & Barber, S.A. (1985). Phosphate in Malaysian Ultisols and Oxisols as evaluated by a mechanistic model. *Soil Science*, **139**, 500–4.

Bidinger, F.R., Mahalakshmi, V., Talukdar, B.S. & Alagarswamy, G. (1981). *Improvement of Drought Resistance in Pearl Millet*. ICRISAT Conference Paper, No. 44. Hyderabad, India: International Crops Research Institute for the Semi-Arid Tropics.

Birch, H.F. (1964). Mineralization of plant nitrogen following alternate wet and dry conditions. *Plant and Soil*, **20**, 43–9.

Black, A.S. & Waring, S.A. (1979). Adsorption of nitrate, chloride and sulphate by some highly weathered soils from south-east Queensland. *Australian Journal of Soil Research*, **27**, 333–51.

Black, J.N. (1971). Energy relations in crop production: a preliminary survey. *Annals of Applied Biology*, **67**, 272–8.

Blacklow, W.M. (1972). Influence of temperature on germination and elongation of the radicle and shoot of corn (*Zea mays* L.). *Crop Science*, **12**, 647–50.

Bleeker, P. (1983). *Soils of Papua New Guinea*. Canberra : CSIRO/Australian National University Press, 352 pp.

Bloodworth, M.E., Burleson, C.A. & Cowley, W.R. (1958). Root distribution of some irrigated crops using undisrupted soil cores. *Agronomy Journal*, **50**, 317–20.

Blum, A. & Ritchie, J.T. (1984). Effect of soil surface water content on sorghum root distribution in the soil. *Field Crops Research*, **8**, 169–76.

Blumenthal, M.J., Quach, V.P. & Searle, P.G.E. (1988). Effect of soybean population density on soybean yield, nitrogen accumulation and residual nitrogen. *Australian Journal of Experimental Agriculture*, **28**, 99–106.

Boddey, R.M. & Dobereiner, J. (1988). Nitrogen fixation associated with grasses and cereals: recent results and perspectives for future research. *Plant and Soil*, **108**, 53–65.

Boddey, R.M., Urquiaga, S., Reis, V. & Dobereiner, J. (1991). Biological nitrogen fixation in association with sugar cane. *Plant and Soil*, **137**, 111–17.

Boerboom, B.W.J. (1978). A model of dry matter distribution in cassava (*Manihot esculenta* Crantz.). *Netherlands Journal of Agricultural Science*, **26**, 267–77.

Bohlool, B.B., Ladha, J.K., Garrity, D.P. & George, T. (1992). Biological nitrogen fixation for sustainable agriculture: a perspective. *Plant and Soil*, **141**, 1–11.

Bolhuis, G.C. (1966). Influence of length of the illumination period on root formation in cassava (*Manihot utilissima* Pohl.). *Netherlands Journal of Agricultural Science*, **14**, 251–4.

Bolhuis, G.C. & Groot, W. de (1959). Observations on the effect of varying temperatures on the flowering and fruit set in three cultivars of groundnut. *Netherlands Journal of Agricultural Science*, **7**, 317–26.

Bonnett, O.T. (1940). Development of inflorescences of sweet corn. *Journal of Agricultural Research*, **60**, 25–37.

Bonnett, O.T. (1966). *Inflorescences of Maize, Wheat, Rye, Barley and Oats: Their Initiation and Development*. Agriculture Experiment Station Bulletin, No. 721. Urbana-Champaign, Illinois: University of Illinois College of Agriculture.

Bonsu, M. (1991). Effect of liming on maize production and erosion on an acid soil in SW Ghana. *Tropical Agriculture (Trinidad)*, **68**, 271–3.

Boonjawat, J., Chaisiri, P., Limpananont, J., Soontaros, S., Pongsawasdi, P., Chaopongpang, S., Pornpattkul, S., Wongwaitayakul, B. & Sangduan, L. (1991). Biology of nitrogen fixing Rhizobacteria. *Plant and Soil*, **137**, 119–25.

Boote, K.J., Jones, J.W., Smerage, G.H., Barfields, C.S. & Berger, R.D. (1980). Photosynthesis of peanut canopies as affected by leafspot and artificial defoliation. *Agronomy Journal*, **72**, 247–52.

Boote, K.J., Stansell, J.R., Schubert, A.M. & Stone, J.F. (1982). Irrigation, water use, and water relations. In *Peanut Science and Technology*, ed. H.E. Pattee & C.T. Young, pp. 164–205. Yoakum, Texas: American Peanut Research and Education Society.

Borgonovi, R.A., Shaffert, R.E. & Pitta, G.V.E. (1987). Breeding aluminium-tolerant sorghums. In *Sorghum for Acid Soils*, ed. L.M. Gourley & J.G. Salinas, pp. 271–92. Cali, Colombia: Centro Internacional de Agricultura Tropical.

Boserup, E. (1965). *The Conditions of Agricultural Growth*. London: Allen & Unwin, 124 pp.

Bouldin, D.R. (1979). The influence of subsoil acidity on crop yield potential. *Cornell International Agriculture Bulletin*, No. 34, 17 pp.

Bouman, S.A.M. & Driessen, P.M. (1985). Physical properties of peat soils affecting rice-based cropping systems. In *Soil Physics and Rice*, pp 70–83. Los Baños, Philippines: International Rice Research Institute.

Bourke, R.M. (1977). Sweet potato (*Ipomoea batatas*) fertilizer trials on the Gazelle Peninsula of New Britain: 1954–1976. *Papua New Guinea Agricultural Journal*, **28**, 73–95.

Bourke, R.M. (1985). Sweet potato (*Ipomoea batatas*) production and research in Papua New Guinea. *Papua New Guinea Journal of Agriculture, Forestry and Fisheries*, **33**, 89–108.

Bourke, R.M., Evenson, J.P. & Keating, B.A. (1984). Relationship between altitudinal limit of cassava and soil temperature in Papua New Guinea. *Tropical Agriculture (Trinidad)*, **61**, 315–16.

Bowden, J.W., Posner, A.M. & Quirk, J.P. (1981). Adsorption and charging phenomena in variable charge soils. In *Soils with Variable Charge*, ed. B.K.G. Theng, pp. 147–66. Lower Hutt, New Zealand: New Zealand Society of Soil Science.

Bowen, C.R. & Rodgers, D.M. (1987). Evaluation of a greenhouse screening technique for iron-deficiency chlorosis of sorghum. *Crop Science*, **27**, 1024–9.

Bowen, W.T., Quintana, J.O., Periera, J., Bouldin, D.R., Reid, W.S. & Lathwell, D.J. (1988). Screening green manures as nitrogen sources for succeeding non-legume crops. *Plant and Soil*, **111**, 75–80.

Boyer, J.S. (1970). Leaf enlargement and metabolic rates of corn, soybean, and sunflower at various leaf water potentials. *Plant Physiology*, **46**, 233–5.

Boyer, J.S. (1976). Water deficits and photosynthesis. In *Water Deficits and Plant Growth*, vol. 4, ed. T.T. Kozlowski, pp. 154–90. New York: Academic Press.

Brady, N.C. (1974). *The Nature and Properties of Soils*, 8th edn. New York: Macmillan, 639 pp.

Bramley, R.G.V. & White, R.E. (1990). The variability of nitrifying activity in field soils. *Plant and Soil*, **126**, 203–8.

Brammer, H. (1977). Incorporation of physical determinants in cropping pattern design. In *Cropping Systems Research and Development for the Asian Rice Farmer*, pp. 83–95. Los Baños, Philippines: International Rice Research Institute.

Braun, H. (1974). Shifting cultivation in Africa (evaluation of questionnaires). In *Shifting Cultivation and Soil Conservation in Africa, No. 24*, pp. 21–36. Rome: Swedish International Development Authority/Food and Agriculture Organization.

Breemen, N. van (1980). Acidity of wetland soils, including Histosols, as a constraint to food production. In *Priorities for Alleviating Soil-Related Constraints to Food Production in the Tropics*, pp. 189–202. Los Baños, Philippines: International Rice Research Institute.

Brenes, E. & Pearson, R.W. (1973). Root response of three Gramineae species to soil acidity in an Oxisol and an Ultisol. *Soil Science*, **116**, 295–302.

Bressani, R. & Elias, L.G. (1980). Nutritional value of legume crops for humans and animals. In *Advances in Legume Science*, ed. R.J. Summerfield & A.H. Bunting, pp. 135–55. Kew: Royal Botanic Gardens.

Bressani, R., Flores, M. & Elias, L.G. (1973). Acceptability and value of food legumes in the human diet. In *Potentials of Field Beans and Other Food Legumes in Latin America*, pp. 17–48. Cali, Colombia: Centro Internacional de Agricultura Tropical.

Bromfield, E.S.P. & Ayanaba, A. (1980). The efficacy of soybean inoculation on acid soil in tropical Africa. *Plant and Soil*, **54**, 95–106.

Brookfield, H.C. & Brown. P. (1963). *Struggle for Land: Agriculture and Group Territories among the Chimbu of the New Guinea Highlands*. Melbourne: Oxford University Press, 193 pp.

Brookfield, H.C. & Hart, D. (1971). *Melanesia: A Geographical Interpretation of an Island World*. London: Methuen. 464 pp.

Brooking, I.R. (1976). Male sterility in *Sorghum bicolor* (L.) Moench induced by low night temperature: I. *Australian Journal of Plant Physiology*, **3**, 589–96.

Brouk, B. (1977). *Plants Consumed by Man*. London: Academic Press, 479 pp.

Brown, A.D. (1981). Groundwater transport of nitrogen in rice fields in northern Thailand. In *Nitrogen Cycling in South-east Asian Wet Monsoonal Ecosystems*, ed. R. Wetselaar, J.R. Simpson & T. Rosswell, pp. 165–70. Canberra: Australian Academy of Science.

Brown, W.L. & Goodman, M.M. (1977). Races of corn. In *Corn and Corn Improvement*, ed. G.F. Sprague, pp. 49–88. Madison, Wisconsin: American Society of Agronomy.

Bruce, R.C., Warrell, L.A., Edwards, D.G. & Bell, L.C. (1988). Effects of aluminium and calcium in the soil solution of acid soils on root elongation of *Glycine max* cv. Forrest. *Australian Journal of Agricultural Research*, **38**, 319–38.

Bruijn, G.G. de & Dharmaputra, T.S. (1974). The Mukibat system, a high-yielding method of cassava production in Indonesia. *Netherlands Journal of Agricultural Science*, **22**, 89–100.

Brun, L.J., Kanemasu, E.T. & Powers, W.L. (1972). Evapotranspiration from soybean and sorghum fields. *Agronomy Journal*, **64**, 145–8.

Brunken, J.N. (1977). A systematic study of *Pennisetum* sect. *Pennisetum* (Gramineae). *American Journal of Botany*, **64**, 151–75.

Brunken, J.N., de Wet, J.M.H. & Harlan, J.R. (1977). The morphology and domestication of pearl millet. *Economic Botany*, **31**, 163–74.

Bryant, P.M. & Humphries, L.R. (1976). Photoperiod and temperature effects on the flowering of *Stylosanthes guyanensis*. *Australian Journal of Experimental Agriculture and Animal Husbandry*, **16**, 506–13.

Budyko, M.I. (1968). Solar radiation and the use of it by plants. In *Agroclimatological Methods*, pp. 39–54. Paris: United Nations Educational, Scientific and Cultural Organization.

Buerkert, A., Cassman, K.G., de la Piedra, R. & Munns, D.N. (1990). Soil acidity and liming effects on stand, nodulation, and yield of common bean. *Agronomy Journal*, **82**, 749–54.

Bui, E.N., Mermut A.R. & Santos, M.C.C. (1989). Microscopic and ultramicroscopic porosity of an Oxisol as determined by image analysis and water retention. *Soil Science Society of America Journal*, **53**, 661–5.

Bunting, A.H. & Anderson, B. (1960). Growth and nutrient uptake of Natal Common groundnuts in Tanganyika. *Journal of Agricultural Science*, **55**, 35–46.

Bunting, A.H. & Curtis, D.K. (1968). Local adaptation of sorghum varieties in northern Nigeria. In *Agroclimatological Methods*, pp. 101–6. Paris: United Nations Educational, Scientific and Cultural Organization.

Bunting, A.H. & Elston, J. (1980). Ecophysiology of growth and adaptation in the groundnut: an essay on structure, partition and adaptation. In *Advances in Legume Science*, ed. R.J. Summerfield & A.H. Bunting, pp. 495–500. Kew: Royal Botanic Gardens.

Buol, S.W., Hale, F.D. & McCracken, R.J. (1980). *Soil Genesis and Classification*, 2nd edn. Ames, Iowa: Iowa State University Press, 404 pp.

Burden, O.J. & Coursey, D.G. (1977). Bananas as a food crop. In *Food Crops of the Lowland Tropics*, ed. C.L.A. Leakey & J.B. Wills, pp. 97–100. Oxford: Oxford University Press.

Buresh, R.J. & Datta, S.K. de (1991). Nitrogen dynamics and management in rice–legume cropping systems. *Advances in Agronomy*, **45**, 2–59.

Buresh, R.J., Woodhead, T., Shepherd, K.D., Flordelis, E. & Cabangon, R.C. (1989).

Nitrate accumulation and loss in a mungbean/lowland rice cropping system. *Soil Science Society of America Journal*, **53**, 477–82.

Burgos, C.F. (1980). Soil-related intercropping practices in cassava production. In *Cassava Cultural Practices*, ed. E.J. Weber, M.J.C. Toro & M. Graham, pp. 75–81. Ottawa, Ontario: International Development Research Centre.

Burkill, I.H. (1960). The organography and the evolution of the Dioscoreaceae, the family of the yams. *Journal of the Linnean Society (Botany)*, **56**, 319–412.

Burnside, O.C. (1977). Control of weeds in non-cultivated, narrow row sorghum. *Agronomy Journal*, **69**, 851–4.

Burris, J.S., Wahab, A.H. & Edge, O. (1971). Effects of seed size on seedling performance in soybeans: I. *Crop Science*, **11**, 492–6.

Burris, J.S., Edge, O.T. & Wahab, A.H. (1973). Effects of seed size on seedling performance in soybeans: II. *Crop Science*, **13**, 207–10.

Butler, A.F. (1960). Fertilizer experiments with the Gros Michel banana. *Tropical Agriculture (Trinidad)*, **37**, 31–50.

Buyanovsky, G.A. & Wagner, G.H. (1986). Post-harvest residue input to cropland. *Plant and Soil*, **93**, 57–65.

Caddel, J.L. & Weikel, D.E. (1972). Photoperiodism in sorghum. *Agronomy Journal*, **64**, 473–6.

Calkins, P.H. (1978). Why farmers plant what they do: a study of vegetable production technology in Taiwan. *Asian Vegetable Research and Development Center Technical Bulletin*, No. 8.

Calkins, P.H., Huang, S.Y. & Hong, J.F. (1977). Farmers' viewpoint on sweet potato production in Taiwan. *Asian Vegetable Research and Development Center Technical Bulletin*, No. 4.

Campbell, G.M., Hernandez, T.P. & Miller, J.C. (1963). The effect of temperature, photoperiod and other related treatments on flowering in *Ipomoea batatas*. *Proceedings of the American Society for Horticultural Science*, **83**, 618–22.

Carlson, J.B. (1973). Morphology. In *Soybeans: Improvement, Production, and Uses*, ed. B.E. Caldwell, pp. 17–96. Madison, Wisconsin: American Society of Agronomy.

Carlson, J.B. & Lersten, N.R. (1987). Reproductive morphology. *Agronomy*, **16**, 95–134.

Carr, D.J. & Skene, K.G.M. (1961). Diauxic growth curves of seeds with special reference to French beans (*Phaseolus vulgaris* L.). *Australian Journal of Biological Sciences*, **14**, 1–12.

Carter, S.E., Fresco, L.O., Jones, P.G. & Fairbairn, J.N. (1992). *An Atlas of Cassava in Africa. Historical, agroecological and demographic aspects of crop distribution.* Cali: Centro Internacional de Agricultura Tropical, 85 pp.

Cassman, K.G. & Munns, D.N. (1980). Nitrogen mineralisation as affected by soil moisture, temperature and depth. *Soil Science Society of America Journal*, **44**, 1233–7.

Cassman, K.G., Munns, D.N. & Beck, D.P. (1981*a*). Growth of *Rhizobium* strains at low concentrations of phosphate. *Soil Science Society of America Journal*, **45**, 520–3.

Cassman, K.G., Whitney, A.S. & Fox, R.L. (1981*b*). Phosphorus requirements of soybean and cowpea as affected by mode of N nutrition. *Agronomy Journal*, **73**, 17–22.

Catsky, J. & Ticha, I. (1980). Ontogenetic changes in the internal limitation to bean-leaf photosynthesis: VI. *Photosynthetica*, **14**, 489–96.

Chabrolin, R. (1977). Rice in West Africa. In *Food Crops of the Lowland Tropics*, ed. C.L.A. Leakey & J.B. Wills, pp. 7–25. Oxford: Oxford University Press.

Chalk, P.M. (1991). The contribution of associative and symbiotic nitrogen fixation to the nitrogen nutrition of non-legumes. *Plant and Soil*, **132**, 29–39.

Champion, J. (1967). *Notes et documents sur les bananiers et leur culture.* Paris: Institut Francais de Recherches Fruitières Outre-mer, 214 pp.

Champion, J., Dugain, F., Maignien, R. & Dommergues, Y. (1958). Le sols de bananeraies et leur amelioration en Guinée. *Fruits*, **13**, 415–52.

Chandra, R. & Pareek, R.P. (1985). Role of host genotype in effectiveness and competitiveness of chickpea (*Cicer arietinum* L) rhizobium. *Tropical Agriculture (Trinidad)*, **62**, 90–4.

Chandra, S. (1981). Energetics and subsistence affluence in traditional agriculture. *Australian National University Development Studies Centre, Occasional Paper*, No. 24, 38 pp.

Chandra, S., Evenson, J.P. & Boer, A.J. de (1976). Incorporating energetic measures in an analysis of crop production practices in Sigotoka Valley, Fiji. *Agricultural Systems*, **1**, 301–11.

Chang, J.-H. (1981). Corn yield in relation to photoperiod, night temperature, and solar radiation. *Agricultural Meteorology*, **24**, 253–62.

Chang, S.C. (1971). Chemistry of paddy soils. *ASPAC Food and Fertilizer Technology Centre Extension Bulletin*, No. 7, 26 pp.

Chang, T.T. (1976). Rice. *Oryza sativa* and *Oryza glaberrima* (Gramineae–Oryzae). In *Evolution of Crop Plants*, ed. N.W. Simmonds, pp. 98–104. London & New York: Longman.

Chang, T.T. & Oka, H.I. (1976). Genetic information on the climatic adaptability of rice cultivars. In *Climate and Rice*, pp. 87–111. Los Baños, Philippines: International Rice Research Institute.

Chapman, A.L. & Petersen, M.L. (1962). The seedling establishment of rice under water in relation to temperature and dissolved oxygen. *Crop Science*, **2**, 391–5.

Chapman, E.A. (1965). Some investigations into factors limiting yields of White Lisbon yams (*Dioscorea alata*) under Trinidad conditions. *Tropical Agriculture (Trinidad)*, **42**, 145–51.

Chapman, S.C., Ludlow, M.M., Blamey, F.P.C. & Fischer, K.S. (1993a). Effect of drought during early reproductive development on the growth of cultivars of groundnut (*Arachis hypogaea* L.). I. Utilization of radiation and water during drought. *Field Crops Research*, **32**, 193–210.

Chapman, S.C., Ludlow, M.M., Blamey, F.P.C. & Fischer, K.S. (1993b). Effect of drought during early reproductive development on the growth of cultivars of groundnut (*Arachis hypogaea* L.). II. Biomass production, pod development and yield. *Field Crops Research*, **32**, 211–25.

Chapman, S.C., Ludlow, M.M., Blamey, F.P.C. & Fischer, K.S. (1993c). Effect of drought during early reproductive development on the dynamics of yield development of cultivars of groundnut (*Arachis hypogaea* L.). III. *Field Crops Research*, **32**, 227–42.

Chapman, T. & Cowling, D.J. (1965). A preliminary investigation into the effect of leaf distribution on the yields of sweet potato (*Ipomoea batatas*). *Tropical Agriculture (Trinidad)*, **42**, 199–203.

Chappopadhyay, P.K., Halder, N.C., Maiti, S.C. & Bose, T.K. (1980). Effect of nitrogen nutrition on growth, yield and quality of Giant Governor banana. In *Banana Production Technology*, ed. C.R. Muthukrishnan & J.B. Abdul Khader, pp. 109–12. Coimbatore, India: Tamil Nadu Agricultural University.

Charles-Edwards, D.A. (1982). *Physiological Determinants of Crop Growth*. Sydney: Academic Press.

Charoenchamratcheep, C., Smith, C.J., Satawathananont, S. & Patrick, W.H. (1987). Reduction and oxidation of acid sulphate soils of Thailand. *Soil Science Society of America Proceedings*, **51**, 630–4.

Charreau, C. (1974). Systems of cropping in the dry tropical zone of west Africa, with special reference to Senegal. In *International Workshop on Farming Systems*, pp. 443–68. Hyderabad, India: International Crops Research Institute for the Semi-Arid Tropics.

Chatterjee, B.N. & Sen, H. (1977). Yield performance and moisture extraction pattern of winter crops under rainfed condition in the Gangetic Plains of West Bengal. *Journal of Soil and Water Conservation in India*, **27**, 101–6.

Chaudhary, T.N., Bhatnagar, V.K. & Prihar, S.S. (1975). Corn yield and nutrient uptake as affected by water-table depth and soil submergence. *Agronomy Journal*, **67**, 745–9.

Chesney, H.A.D. (1975). Fertilizer studies with groundnuts on the brown sands of Guyana. II. Effect of nitrogen, phosphorus, potassium, and gypsum and timing of phosphorus application. *Agronomy Journal*, **67**, 10–13.

Chew, W.Y. (1970a). Varieties and NPK fertilizers for tapioca (*Manihot utilissima* Pohl.) on peat. *Malaysian Agricultural Journal*, **47**, 483–91.

Chew, W.Y. (1970b). Effects of length of growing season and NPK fertilizers on the yield of five varieties of sweet potatoes (*Ipomoea batatas* Lam.) on peat. *Malaysian Agricultural Journal*, **47**, 453–64.

Chew, W.Y. (1974). Yields of some varieties of tapioca (*Manihot utilissima* Pohl.) grown on Malaysian peat as affected by different planting methods, plant densities, fertilizers and growth periods. *Malaysian Agricultural Journal*, **49**, 393–402.

Chew, W.Y., Joseph, K.T., Ramli, K. & Majid, A.B.A. (1981). Influence of liming and soil pH on cassava *(Manihot esculenta)* in tropical oligotrophic peat. *Experimental Agriculture*, **17**, 171–8.

Chew, W.Y., Joseph, K.T., Ramli, K. & Majid, A.B.A. (1982). Liming needs of sweet potato in Malaysian acid peat. *Experimental Agriculture*, **18**, 65–71.

Chong, K., Wynne, J.C., Elkan, G.H. & Schneeweis, T.J. (1987). Effects of soil acidity and aluminium content on *Rhizobium* inoculation, growth and nitrogen fixation of peanuts and other grain legumes. *Tropical Agriculture (Trinidad)*, **64**, 97–104.

Choudhari, S.D., Udaykumar, M. & Sastry, K.S.K. (1985). Physiology of bunch peanuts (*Arachis hypogaea* L.). *Journal of Agricultural Science*, **104**, 309–15.

Chowdhury, S.L. (1974). Cropping systems in the semi-arid tropics of India. In *International Workshop on Farming Systems*, pp. 373–84. Hyderabad, India: International Crops Research Institute for the Semi-Arid Tropics.

Chowdhury, S.L., Ram, S. & Giri, G. (1975). Effect of P, N and inoculum on root, nodulation and yield of gram. *Indian Journal of Agronomy*, **20**, 290–1.

Christianson, C.B., Bationo, A. & Baethgen, W.E. (1990a). The effect of soil tillage and fertilizer use on pearl millet yields in Niger. *Plant and Soil*, **123**, 51–8.

Christianson, C.B., Bationo, A., Henao, J. & Vlek, P.L.G. (1990b). Fate and efficiency of N fertilizers applied to pearl millet in Niger. *Plant and Soil*, **125**, 221–31.

Chundawat, G.S., Sharma, R.G. & Shekawat, G.S. (1976). Effect of nitrogen, phosphorus and bacterial fertilization on growth and yield of gram grown in Rajasthan. *Indian Journal of Agronomy*, **21**, 127–30.

CIAT (1973). *Annual Report, 1972.* Cali, Colombia: Centro Internacional de Agricultura Tropical, 192 pp.

CIAT (1974). *Annual Report, 1973.* Cali, Colombia: Centro Internacional de Agricultura Tropical, 254 pp.

CIAT (1978*a*). *Annual Report, 1977, Bean Program.* Cali, Colombia: Centro Internacional de Agricultura Tropical, 85 pp.

CIAT (1978*b*). *Annual Report, 1977, Cassava Program.* Cali, Colombia: Centro Internacional de Agricultura Tropical, 68 pp.

CIAT (1979*a*). *Annual Report, 1978. Bean Program.* Cali, Colombia: Centro Internacional de Agricultura Tropical, 75 pp.

CIAT (1979*b*). *Annual Report, 1978, Cassava Program.* Cali, Colombia: Centro Internacional de Agricultura Tropical, 100 pp.

CIAT (1980*a*). *Annual Report, 1979, Bean Program.* Cali, Colombia: Centro Internacional de Agricultura Tropical, 111 pp.

CIAT (1980*b*). *Annual Report, 1979, Cassava Program.* Cali, Colombia: Centro Internacional de Agricultura Tropical, 93 pp.

CIAT (1980*c*). *Annual Report. 1979, Tropical Pastures Program.* Cali, Colombia: Centro Internacional de Agricultura Tropical, 156 pp.

CIAT (1981). *Potentials for Field Beans in Eastern Africa. Regional Workshop, Lilongwe, Malawi, 1980.* Cali, Colombia: Centro Internacional de Agricultura Tropical.

CIAT (1987). *Annual Report, 1987, Bean Program.* Cali, Colombia: Centro Internacional de Agricultura Tropical.

CIMMYT (1975). *CIMMYT Report on Maize Improvement.* El Batan, Mexico: Centro Internacional de Mejoramiento de Maiz y Trigo.

Clark, C. & Haswell, M. (1970). *The Economics of Subsistence Agriculture*, 4th edn. London: MacMillan, 245 pp.

Clarkson, N.M. & Russell, J.S. (1979). Effect of temperature on the development of two annual medics. *Australian Journal of Agricultural Research*, **30**, 909–16.

Clayton, W.D. & Renvoize, S.A. (1986). *Genera Graminum. Grasses of the World.* London: Her Majesty's Stationery Office.

Clegg, C.G. (1947). Notes on chickpea and its cultivation in the Lake Province, Tanganyika. *East African Agricultural Journal*, **13**, 27–8.

Cline, G.R. & Kaul, K. (1990). Inhibitory effects of acidified soil on the soybean/*Bradyrhizobium* symbiosis. *Plant and Soil*, **127**, 243–9.

Coaldrake, P.D. (1985). Leaf area accumulation of pearl millet as affected by nitrogen supply. *Field Crops Research*, **11**, 185–92.

Coaldrake, P.D. & Pearson, C.J. (1985*a*). Development and dry weight accumulation of pearl millet as affected by nitrogen supply. *Field Crops Research*, **11**, 174–84.

Coaldrake, P.D. & Pearson, C.J. (1985*b*). Panicle differentiation and spikelet number related to size of panicle in *Pennisetum americanum*. *Journal of Experimental Botany*, **36**, 833–40.

Coaldrake, P.D., Pearson, C.J. & Saffigna, P.G. (1987). Grain yield of *Pennisetum americanum* adjusts to nitrogen supply by changing rates of grain filling and root uptake of nitrogen. *Journal of Experimental Botany*, **38**, 558–66.

Cochemé, J. & Franquin, P. (1967). An agroclimatological survey of a semi-arid area in Africa south of the Sahara. *World Meteorological Organization Technical Note*, No. 86, 146 pp.

Cochrane, T.T. (1975). Land use classification in the lowlands of Bolivia. In *Soil Management in Tropical America*, pp. 109–25. Raleigh, North Carolina: North Carolina State University.

Cock, J.G. (1976). Characteristics of high yielding cassava varieties. *Experimental Agriculture*, **12**, 135–43.

Cock, J.H. (1984). Cassava. In *The Physiology of Tropical Field Crops*, ed. P.R. Goldsworthy & N.M. Fisher, pp. 529–550. Chichester: Wiley.

Cock, J.H. (1985). *Cassava : New Potential for a Neglected Crop*. Boulder, Colorado : Westview Press, 191 pp.

Cock, J.H. & Howeler, R.H. (1978). The ability of cassava to grow on poor soils. In *Crop Tolerance to Suboptimal Land Conditions*, ed. G.A. Jung, pp. 145–54. Madison, Wisconsin: American Society of Agronomy.

Cock, J.H., Wholey, D. & Gutierrez de las Casas, O. (1977). Effects of spacing on cassava (*Manihot esculenta*). *Experimental Agriculture*, **13**, 289–99.

Cock, J.H., Franklin, D., Sandoval, E. & Juri, P. (1979). The ideal cassava plant for maximum yield. *Crop Science*, **19**, 271–9.

Collier, W.L. (1979). *Social and Economic Aspects of Tidal Swamp Land Development in Indonesia*. Occasional Paper, No. 15, Development Studies Centre. Canberra: Australian National University Press, 42 pp.

Collins, W.W. & Wilson, L.G. (1988). Reactions of sweet potatoes to flooding. *Hortscience*, **23**, 1079.

Conklin, H.C. (1957). *Hanunoo agriculture*. FAO Forestry Development Paper, No. 12. Rome: Food and Agriculture Organization.

Connor, D.J., Cock, J.H. & Parra, G.E. (1981). Response of cassava to water shortage. I. Growth and yield. *Field Crops Research*, **4**, 181–200.

Conrad, J.P. (1938). Distribution of sugars, root enclosed, in the soil following corn and sorghums and their effects on the succeeding wheat crop. *Journal of the American Society of Agronomy*, **30**, 475–83.

Conway, G.R. (1987). The properties of agroecosystems. *Agricultural Systems*, **24**, 95–118.

Cookston, R.K., O'Toole, J., Lee, R., Ozbun, J.L. & Wallace, D.H. (1974). Photosynthetic depression in beans after exposure to cold for one night. *Crop Science*, **14**, 457–64.

Cooper, J.P. (1975). Control of photosynthetic production in terrestrial systems. In *Photosynthesis and Productivity in Different Environments*, ed. J.P. Cooper, pp. 593–621. Cambridge: Cambridge University Press.

Cooper, P.J.M. (1979). The association between altitude, environmental variables, maize growth and yield in Kenya. *Journal of Agricultural Science*, **93**, 635–49.

Corbin, E.J., Brockwell, J. & Gault, R.R. (1977). Nodulation studies on chickpea (*Cicer arietinum*). *Australian Journal of Experimental Agriculture and Animal Husbandry*, **17**, 126–34.

Corey, K.A., Collins, W.W. & Pharr, D.M. (1982). Effect of duration of soil reduction on ethanol concentration and storage loss of sweet potato roots. *Journal of the American Society for Horticultural Science*, **107**, 195–8.

Cours, G. (1951). Le Manioc à Madagascar. *Mémoires de l'Institut Scientifique de Madagascar, Series B,3*, 203–416.

Coursey, D.G. (1961). The magnitude and origins of storage losses in Nigerian yams. *Journal of the Science of Food and Agriculture*, **12**, 574–80.

Coursey, D.G. (1967a). Yam storage: a review of yam storage practices and of information on storage losses. *Journal of Stored Products Research*, **2**, 229–44.

Coursey, D.G. (1967b). *Yams*. London: Longman, 230 pp.

Coursey, D.G. (1976a). Yams. *Dioscorea* spp. (Dioscoreaceae). In *Evolution of Crop Plants*, ed. N.W. Simmonds, pp. 70–4. London: Longman.

Coursey, D.G. (1976b). The origins and domestication of yams in Africa. In *Origins of African Plant Domestication*, ed. J.R. Harlan, J.M.J. de Wet & A.B.L. Stemler, pp. 383–408. The Hague: Mouton.

Couto, W., Sanzonowicz, C. & Barcellos, A. De O. (1985). Factors affecting oxidation–reduction processes in an Oxisol with a seasonal water table. *Soil Science Society of America Proceedings*, **49**, 1245–8.

Cox, F.R. (1973a). Micronutrients. In *A Review of Soils Research in Tropical Latin America*, ed. P.A. Sanchez, pp. 182–97. Raleigh, North Carolina: University of North Carolina.

Cox, F.R. (1973b). Potassium. In *A Review of Soils Research in Tropical Latin America*, ed. P.A. Sanchez, pp. 162–78. Raleigh, North Carolina: University of North Carolina.

Cox, F.R. (1978). Effect of quantity of light on the early growth and development of the peanut. *Peanut Science*, **5**, 27–30.

Cox, F.R. (1979). Effect of temperature treatment on peanut vegetative and fruit growth. *Peanut Science*, **6**, 14–17.

Cox, F.R., Adams, F. & Tucker, B.B. (1982). Liming, fertilization and mineral nutrition. In *Peanut Science and Technology*, ed. H.E. Pattee & C.T. Young, pp. 139–63. Yoakum, Texas: American Peanut Research and Education Society.

Craswell, E.T. & Pushparajah (1989). *Management of Acid Soils in the Humid Tropics of Asia*. Australian Centre for International Agricultural Research Monograph, No. 13, 118 pp.

Craswell, E.T. & Vlek, P.L.G. (1979). Fate of fertilizer nitrogen applied to wetland rice. In *Nitrogen and Rice*, pp. 175–92. Los Baños, Philippines: International Rice Research Institute.

Craufurd, P.Q. & Bidinger, F.R. (1988). Effect of duration of the vegetative phase on crop growth, development and yield in two contrasting pearl millet hybrids. *Journal of Agricultural Science*, **110**, 71–9.

Criswell, J.G. & Humè, D.J. (1972). Variation in sensitivity to photoperiod among early maturing soybean strains. *Crop Science*, **12**, 657–60.

Cummings, R.W. (1976). *Food Crops in Low-income Countries: The State of Present and Expected Agricultural Research and Technology*. New York: The Rockefeller Foundation, 103 pp.

Curtis, D.L. (1968). The relationship between the date of heading of Nigerian sorghums and the duration of the growing season. *Journal of Applied Ecology*, **4**, 215–26.

Daage, F.C. (1987). Current programs, problems, and strategies for land clearing and development on volcanic ash soils. In *Tropical Land Clearing for Sustainable Agriculture*, pp. 195–206. Bangkok, Thailand: International Board for Soil Research and Management Inaugural Workshop.

Dadson, R.B. & Acquaah, G. (1984). *Rhizobium japonicum*, nitrogen and phosphorus effects on nodulation, symbiotic nitrogen fixation and yield of soybean (*Glycine max* L. Merrill) in the southern savanna of Ghana. *Field Crops Research*, **9**, 101–8.

Dalal, R.C. (1989). Long-term effects of no-tillage, crop residue and nitrogen application on properties of a Vertisol. *Soil Science Society of America Journal*, **53**, 1511–15.

Dalrymple, D.G. (1971). *Survey of Multiple Cropping in Less Developed Nations.* Washington, DC: Economic Research Service, US Department of Agriculture, 108 pp.

Dalrymple, D.G. (1978). *Development and Spread of High Yielding Varieties of Wheat and Rice in the Less Developed Nations*, 6th edn. Washington. DC: US Department of Agriculture, Office of International Cooperation and Development, 134 pp.

Danso, S.K.A. (1977). The ecology of *Rhizobium* and recent advances in the study of the ecology of *Rhizobium*. In *Biological Nitrogen Fixation in Farming Systems of the Tropics*, ed. A. Ayanaba & P.J. Dart, pp. 115–25. New York: Wiley.

Dart, P.J. & Krantz, B.A. (1977). Legumes in the semi-arid tropics. In *Exploiting the Legume–Rhizobium Symbiosis in Tropical Agriculture*, pp. 119–54. Honolulu, Hawaii: University Press of Hawaii.

Dart, P.J., Islam, R. & Eaglesham, A. (1975). The root nodule symbiosis of chickpea and pigeonpea. In *International Workshop on Grain Legumes*, pp. 63–83. Hyderabad, India: International Crops Research Institute for the Semi-Arid Tropics.

Dasberg, S. & Bakker, C. (1970). Characterising soil aeration under changing soil moisture conditions for bean growth. *Agronomy Journal*, **62**, 689–96.

Datta, N.P. & Scrivastva (1963). Influence of organic matter on the intensity of phosphate bonding on some acid soils. *Journal of the Indian Society of Soil Science*, **11**, 189–94.

Datta, S.K. de (1975). Upland rice around the world. In *Major Research in Upland Rice*, pp. 2–11. Los Baños, Philippines: International Rice Research Institute.

Datta, S.K. de (1981). *Principles and Practices of Rice Production.* Los Baños, Philippines: International Rice Research Institute, 618 pp.

Datta, S.K. de (1987). Nitrogen transformation processes in relation to improved cultural practices for lowland tropics. *Plant and Soil*, **100**, 47–69.

Datta, S.K. de & Feuer, R. (1975). Soils on which upland rice is grown. In *Major Research in Upland Rice*, pp. 27–39. Los Baños, Philippines: International Rice Research Institute.

Datta, S.K. de & Malabuyoc, J. (1976). Nitrogen response of lowland and upland rice in relation to tropical environmental conditions. In *Climate and Rice*, pp. 509–39. Los Baños, Philippines: International Rice Research Institute.

Datta, S.K. de & Vergara, B.S. (1975). Climates of upland rice regions. In *Major Research in Upland Rice*, pp. 14–26. Los Baños, Philippines: International Rice Research Institute.

Datta, S.K. de, Krupp, W.K., Alvarez, E.l. & Modgal, S.C. (1973). Water management practices in flooded tropical rice. In *Water Management in Philippine Irrigation Systems*, pp. 11–18. Los Baños, Philippines: International Rice Research Institute.

Datta, S.K. de, Fillery, I.R.P. & Craswell, E.T. (1983). Recent results on nitrogen fertilizer efficiency studies on wetland rice. *Outlook in Agriculture*, **12**, 125–34.

Davis, J.H.C., Woolley, J. & Moreno, R.A. (1986). Multiple cropping with legumes and starchy roots. In *Multiple Cropping Systems*, ed. C.A. Francis, pp. 133–60. New York: Macmillan.

Davis, S.D. & McCree, K.J. (1978). Photosynthetic rate and diffusion conductance as a function of age in leaves of common bean plants. *Crop Science*, **18**, 280–2.

Daynard, T.B. & Duncan, W.G. (1969). The black layer and grain maturity in corn. *Crop Science*, **9**, 473–6.

Daynard, T.B., Tanner, J.W. & Hume, D.J. (1969). Contribution of stalk soluble carbohydrates to grain yield in corn (*Zea mays* L.). *Crop Science*, **9**, 831–4.

Daynard, T.B., Tanner, J.W. & Duncan, W.G. (1971). Duration of the grain filling period and its relation to grain yield in corn (*Zea mays* L.). *Crop Science*, **11**, 45–8.

Debouck, D. (1991). Systematics and morphology. In *Common Beans: Research for Crop Improvement*, ed. A. van Schoonhoven & O. Voysest, pp. 55–118. Wallingford, UK: CAB International/Centro Internacional de Agricultura Tropical.

Deibert, E.J., Bijeriego, M. & Olson, R. A. (1979). Utilization of ^{13}N fertilizer by nodulating and non-nodulating soybean isolines. *Agronomy Journal*, **71**, 717–23.

Delgado Salinas, A., Bonet, A. & Gepts, P. (1988). The wild relative of *Phaseolus vulgaris* in Middle America. In *Genetic Resources of Phaseolus Beans: Their Maintenance, Domestication, Evolution, and Utilization*, ed. P. Gepts, pp. 163–84. Dordrecht: Kluwer.

Delouche, J.C. (1953). Influence of moisture and temperature levels on the germination of corn, soybeans and watermelons. *Proceedings of the Association of Official Seed Analysts of North America*, **43**, 117–26.

Denevan, W.M. & Turner, B.L. (1974). Forms, function and associations of raised fields in the old world tropics. *Journal of Tropical Geography*, **39**, 24–34.

Denmead, O.T. (1976). Temperate cereals. In *Vegetation and the Atmosphere*, vol. 2, ed. J.L. Monteith, pp. 1–32. London: Academic Press.

Dennison, E.B. (1961). The value of farmyard manure in maintaining fertility in north Nigeria. *Empire Journal of Experimental Agriculture*, **29**, 330–6.

Devos, P. & Wilson, G.F. (1979). Intercropping of plantains with food crops: maize, cassava and cocoyams. *Fruits*, **34**, 169–74.

D'Hoore, J.L. (1964). *Soil Map of Africa, Scale 1 to 5 000 000, Explanatory Monograph*. Lagos: CCTA, 205 pp.

Dias, A.C. & Nortcliff, S. (1985). Effects of two land clearing methods on the physical properties of an Oxisol in the Brazilian Amazon. *Tropical Agriculture (Trinidad)*, **62**, 207–12.

Diepen, C.A. van, Wolf, J., Keulen, H. van & Rappoldt, C. (1989). WOFOST: a simulation model of crop production. *Soil Use and Management*, **5**, 16–24.

Diepen, C.A. van, Keulen, H. van, Wolf, J. & Berkhout, J.A.A. (1991). Land evaluation: from intuition to quantification. *Advances in Soil Science* **15**, 139–204.

Dingkuhn, M., Datta S.K. de, Javellana, C., Pamplona, R. & Schnier, H.F. (1992*a*). Effect of late-season nitrogen fertilisation on photosynthesis and yield of transplanted and direct-seeded tropical flooded rice. I. Growth dynamics. *Field Crops Research*, **28**, 223–34.

Dingkuhn, M., Datta S.K. de, Javellana, C., Pamplona, R. & Schnier, H.F. (1992*b*). Effect of late-season nitrogen fertilisation on photosynthesis and yield of transplanted and direct-seeded tropical flooded rice. II. A canopy stratification study. *Field Crops Research*, **28**, 235–49.

Dodd, J.C., Arias, I., Koomen, I. & Hayman, D.S. (1990*a*). The management of populations of vesicular-arbuscular mycorrhizal fungi in acid-infertile soils of a savanna ecosystem. I. The effect of pre-cropping and inoculation with VAM-fungi on plant growth and nutrition in the field. *Plant and Soil*, **122**, 229–40.

Dodd, J.C., Arias, I., Koomen, I. & Hayman, D.S. (1990*b*). The management of populations of vesicular-arbuscular mycorrhizal fungi in acid-infertile soils of a savanna ecosystem. II. The effects of pre-crops on the spore populations of native and introduced VAM-fungi. *Plant and Soil*, **122**, 241–7.

Doebley, J. (1990). Molecular evidence and the evolution of maize. *Economic Botany*, **44**, 6–27.

Doebley, J., Goodman, M.M. & Stuber, C.W. (1987). Patterns of isozyme variation between maize and Mexican annual teosinte. *Economic Botany*, **41**, 234–44.

Doggett, H. (1976). Sorghum: *Sorghum bicolor* (Gramineae, Andropogoneae). In *Evolution of Crop Plants*, ed. N.W. Simmonds, pp. 112–17. London: Longman.

Doggett, H. (1988). *Sorghum*, 2nd edn. Harlow, UK: Longman.

Donald, R.G., Kay, B.D. & Miller, M.H. (1987). The effect of aggregate size on early shoot growth and root growth of maize. *Plant and Soil*, **103**, 251–9.

Donovan, P.A. (1963). Groundnut investigation at Matopos Research Station. *Rhodesian Agricultural Journal*, **60**, 121–2.

Downes, R.W. (1968). The effect of temperature on tillering of grain sorghum seedlings. *Australian Journal of Agricultural Research*, **19**, 59–64.

Downes, R.W. (1970). Effect of light intensity and leaf temperature on photosynthesis and transpiration in wheat and sorghum. *Australian Journal of Biological Sciences*, **23**, 775–82.

Downey, L.A. (1971). Water requirements of maize. *Journal of the Australian Institute of Agricultural Science*, **37**, 32–41.

Driessen, P.M. (1978). Peat soils. In *Soils and Rice*, pp. 736–79. Los Baños, Philippines: International Rice Research Institute.

Driessen, P.M. & Moormann, F.R. (1985). Soils on which rice-based cropping systems are practiced. In *Soil Physics and Rice*, pp. 46–86. Los Baños, Philippines: International Rice Research Institute.

Dugas, W.A., Myer, W.S., Barrs, H.D. & Fleetwood, R.J. (1990). Effects of soil type on soybean crop water use in weighing lysimeters. II. Root growth, soil water extraction and water table contributions. *Irrigation Science*, **11**, 77–81.

Duke, J.A. (1978). The quest for tolerant germplasm. In *Crop Tolerance to Suboptimal Land Conditions*, ed. G.A. Jung, pp. 1–61. American Society of Agronomy Special Publication, No. 32.

Duke, J.A. (1981). *Handbook of Legumes of World Economic Importance*. New York: Plenum Press, 344 pp.

Duncan, R.R., Bockholt, R.J. & Miller, F.R. (1981). Descriptive comparison of senescent and non-senescent sorghum genotypes. *Agronomy Journal*, **73**, 849–53.

Duncan, W.G. (1975). Maize. In *Crop Physiology: Some Case Histories*, ed. L.T. Evans, pp. 25–50. Cambridge: Cambridge University Press.

Duncan, W.G., Shaver, D.L. & Williams, W.A. (1973). Insolation and temperature effects on maize growth and yield. *Crop Science*, **13**, 187–91.

Duncan, W.G., McCloud, D.E., McGraw, R.L. & Boote, K.J. (1978). Physiological aspects of peanut yield improvement. *Crop Science*, **18**, 1015–20.

Dunlap, V.C. & McGregor, J.D. (1932). The relationship between soil alkalinity and banana production in St Catherine District, Jamaica. *United Fruit Company Bulletin*, No. 45.

Dunphy, E.J., Hanway, J.J. & Green, D.E. (1979). Soybean yields in relation to days between specific developmental stages. *Agronomy Journal*, **71**, 917–20.

Eaglesham, A.R.J., Ayanaba, A., Rao, V.R. & Eskew, D.L. (1981). Improving the nitrogen nutrition of maize by intercropping with cowpea. *Soil Biology and Biochemistry*, **13**, 169–71.

Earley, E.B. & Cartter, J.L. (1945). Effect of temperature of the root environment on growth of soybean plants. *Journal of the American Society of Agronomy*, **37**, 727–35.

Eastin, J.A. (1969). Leaf position and leaf function in corn: carbon-14 labelled photosynthate distribution in corn in relation to leaf function. In *Proceedings of the 24th Annual Corn and Sorghum Research Conference*, pp. 181–9. American Seed Trade Association.

Eastin, J.D., Hultquist, J.H. & Sullivan, C.V. (1973). Physiologic maturity in grain sorghum. *Crop Science*, **13**, 175–8.

Eastwood, H.W. & Jeater, J.W. (1949). Supplementary watering of bananas to overcome a major hazard to efficient production. *Agricultural Gazette of New South Wales*, **60**, 89–92.

Eck, H.V., Wilson, G.C. & Martinez, T. (1975). Nitrate reductase activity of grain, dry matter and nitrogen. *Crop Science*, **15**, 557–61.

Edwards, C.J. & Hartwig, E.E. (1972). Effect of seed size upon rate of germination in soybeans. *Agronomy Journal*, **63**, 429–30.

Edwards, D.G. & Asher, C.J. (1982). Tolerance of crop and pasture species to manganese toxicity. In *Plant Nutrition, 1982. Proceedings of the IX International Plant Nutrition Colloquium*, Vol. I, ed. A. Scaife, pp. 145–50. Slough: Commonwealth Agricultural Bureaux.

Edwards, D.G., Asher, C.J. & Wilson, G.L. (1977). Mineral nutrition of cassava and adaptation at low fertility conditions. In *Fourth Symposium of the International Society of Tropical Root Crops*, ed. J.H. Cock, R. MacIntyre & M. Graham, pp. 124–30. Ottawa, Ontario: International Development Research Centre.

Edwards, D.G., Sharifuddin, H.A.H., Yusoff, M.N.M., Grundon, N.J., Shamshuddin, J. & Norhayati, M. (1991). The management of soil acidity for sustainable crop production. In *Plant–Soil Interactions at low pH*, ed. R.J. Wright *et al.*, pp. 383–96. Dordrecht: Kluwer.

Effendi, S. (1979). Cassava intercropping patterns and management practices in Indonesia. In *Intercropping with Cassava*, ed. E. Weber, B. Nestel & M. Campbell, pp. 35–6. Ottawa, Ontario: International Development Research Centre.

Egharevba, P.N. (1978). A review of millet work at the Institute of Agricultural Research, Samaru. *Samaru Miscellaneous Paper*, No. 77, 17 pp.

Egli, D.B. (1981). Species differences in seed growth characteristics. *Field Crops Research*, **4**, 1–12.

Egli, D.B. & Wardlaw, I.F. (1980). Temperature response of seed growth characteristics in soybean. *Agronomy Journal*, **72**, 560–4.

Eira, P.A. da, Passahna, G.G., Britto, D.P.P. & Carbajal, A.R. (1974). Phosphorus and potassium fertilizers for black beans and their residual effects. *Pesquisa Agropecuaria Brasileira, Agronomia*, **9**, 121–4.

Ekern, P.C. (1965). Evapotranspiration of pineapple in Hawaii. *Plant Physiology*, **40**, 736–9.

Ekpete, D.M. (1978). Fertilizer requirements of yam *(Dioscorea rotundata)* in Nsukka acid sandy soils. *East African Agricultural and Forestry Journal*, **43**, 378–85.

El-din, S.M.S.B. & Moawad, H. (1988). Enhancement of nitrogen fixation in lentil, faba bean, and soybean by dual inoculation with Rhizobia and mycorrhizae. *Plant and Soil*, **108**, 117–24.

El-Sharkawy, M. A. & Cock, J.H. (1990). Photosynthesis of cassava (*Manihot esculenta*). *Experimental Agriculture*, **26**, 325–40.

El-Sharkawy, M.A. & Hesketh, J.D. (1965). Photosynthesis among species in relation to characteristics of leaf anatomy and carbon dioxide diffusion resistances. *Crop Science*, **5**, 517–21 .

Elsheikh, E.A.E. & Wood, M. (1990). Salt effects on survival and multiplication of chickpea and soybean rhizobia. *Soil Biology and Biochemistry*, **22**, 343–7.

El-Swaify, S.A., Singh, S. & Pathak, P. (1987). Physical and conservation constraints and management components for SAT Alfisols. In *Alfisols in the Semi-Arid Tropics*, pp. 15–30. Patancheru, India: International Crops Research Institute for the Semi-Arid Tropics.

Enyi, B.A.C. (1973). Growth rate of three cassava varieties (*Manihot esculenta* Crantz.) under varying population densities. *Journal of Agricultural Science*, **81**, 15–28.

EPA (1990). Policy options for stabilizing global climate. Draft report to Congress Office of Policy Analysis. Washington, DC: Environmental Protection Agency, 45 pp.

Escalada, R.G. & Plucknett, D.L. (1975). Ratoon cropping of sorghum: I. *Agronomy Journal*, **67**, 473–8.

Esechie, H.A. (1985). Relationship of stalk morphology and chemical composition to lodging resistance in maize (*Zea mays* L.) in a rainforest zone. *Journal of Agricultural Science*, **104**, 429–33.

Etasse, C. (1977). Sorghum and pearl millet. In *Food Crops of the Lowland Tropics*, ed. C.L.A. Leakey & J.B. Wills, pp. 27–39. Oxford: Oxford University Press.

Evans, A.M. (1976). Beans. *Phaseolus* spp. (Leguminosae–Papilionatae). In *Evolution of Crop Plants*, ed. N.W. Simmonds, pp. 168–72. London: Longman.

Evans, L.T. & Datta, S.K. de (1979). The relation between irradiance and grain yield of irrigated rice in the tropics, as influenced by cultivar, nitrogen fertilizer application and month of planting. *Field Crops Research*, **2**, 1–17.

Ezedinma, F.O.C., Ibe, D.G. & Onwuchuruba, A.I. (1981). Performance of cassava in relation to time of planting and harvesting. In *Tropical Root Crops: Research Strategies for the* 1980s, ed. E.R. Terry, K.A. Oduro & F. Caveness, pp. 111–15. Ottawa, Ontario: International Development Research Centre.

Ezeilo, W.N.O. (1977). The effect of fertilizers and other inputs on yield and nutritive value of cassava and other tropical root crops. In *Fertilizer Use and Production of Carbohydrates and Lipids*, pp. 193–208. Berne, Switzerland: International Potash Institute.

Ezeilo, W.N.O. (1979). Intercropping with cassava in Africa. In *Intercropping with Cassava*, ed. E. Weber, B. Nestel & M. Campbell, pp. 49–56. Ottawa, Ontario: International Development Research Centre.

Ezumah, H.C. & Okigbo, B.N. (1980). Cassava planting systems in Africa. In *Cassava Cultural Practices*, ed. E.J. Weber, J.C. Toro & M. Graham, pp. 44–9. Ottawa, Ontario: International Development Research Centre.

Fagaria, N.K., Wright, R.J. & Baligar, V.C. (1988). Rice cultivar evaluation for phosphorus use efficiency. *Plant and Soil*, **111**, 105–9.

Fagaria, N.K. (1989). Effects of phosphorus on growth, yield and nutrient accumulation in the common bean. *Tropical Agriculture (Trinidad)*, **66**, 249–55.

Fagbami, A., Ajayi, S.O. & Ali, E.M. (1985). Nutrient distribution in the basement complex soils of the tropical, dry rainforest of southwestern Nigeria. II. Micronutrients, zinc and copper. *Soil Science*, **139**, 531–7.

FAO (1966). *Agricultural Development in Nigeria, 1965–1980*, pp. 392–400. Rome: Food and Agriculture Organization.

FAO (1970). Key to soil units for the soil map of the world. AGL:SM/70–2, WS/A7/460. Rome: Food and Agriculture Organization, 16 pp.

FAO (1974). *Shifting Cultivation and Soil Conservation in Africa.* FAO Soils Bulletin No. 24. Rome: Food and Agriculture Organization, 248 pp.

FAO (1976). *Production Yearbook, 1975*, vol. 29. Rome: Food and Agriculture Organization.

FAO (1981). *Production Yearbook, 1980*, vol. 34. Rome: Food and Agriculture Organization.

FAO (1990). Interim Report on Forest Resources, Assessment 1990 Project. Document COFO-90/8(a). Committee on Forestry. Tenth Session, Rome, 24–28 September 1990. Rome: Food and Agriculture Organization.

FAO (1991). *Protein Quality Evaluation.* FAO Food and Nutrition Paper 51. Rome: Food and Agriculture Organization, 66 pp.

FAO (1992). *Production Yearbook, 1991*, vol 45. Rome: Food and Agriculture Organization.

Fawusi, M.O.A. & Agboola, A.A. (1980). Soil moisture requirements for germination of sorghum, millet, tomato and *Celosia. Agronomy Journal*, **72**, 353–7.

Fehr, W.R. & Caviness, C.E. (1977). Stages of soybean development. *Iowa Agricultural Experiment Station Special Report*, No. 80.

Fenn, L.B., Taylor, R.M. & Horst, G.L. (1987). *Phaseolus vulgaris* growth in ammonium-based nutrient solution with variable calcium. *Agronomy Journal*, **79**, 89–91.

Ferguson, T.V. & Haynes, P.H. (1970). The response of yams (*Dioscorea* spp.) to nitrogen, phosphorus, potassium, and organic fertilizers. In *Second International Symposium on Tropical Root and Tuber Crops*, vol. 1, ed. D.L. Plucknett, pp. 93–6. Honolulu, Hawaii: University Press of Hawaii.

Fernandez-Caldas, E. & Garcia, V. (1970). Contribution a l'étude de la fertilité des sols de bananeraies de l'île de Tenerife. *Fruits*, **25**, 175–85.

Ferraris, R. (1973). Pearl millet (*Pennisetum typhoides*). *Commonwealth Bureau of Pastures and Field Crops Review Series*, No. 1/1973. Farnham Royal: Commonwealth Agricultural Bureaux, 70 pp.

Ferraris, R., Norman, M.J.T. & Andrews, A.C. (1973). Adaptation of pearl millet (*Pennisetum typhoides*) to coastal New South Wales. I. Preliminary evaluation. *Australian Journal of Experimental Agriculture and Animal Husbandry*, **13**, 685–91.

Fischer, K.S. & Palmer, A.F.E. (1980). Yield efficiency in tropical maize. In *Symposium on Potential Productivity of Field Crops Under Different Environments*, 22–26 September 1980. Los Baños, Philippines: International Rice Research Institute.

Fischer, K.S. & Wilson, G.L. (1971). Studies of grain production in *Sorghum bicolor* (L. Moench): I. *Australian Journal of Agricultural Research*, **22**, 33–7.

Fischer, K.S. & Wilson, G.L. (1975a). Studies of grain production in *Sorghum bicolor* (L. Moench): V. *Australian Journal of Agricultural Research*, **26**, 31–41.

Fischer, K.S. & Wilson, G.L. (1975b). Studies of grain production in *Sorghum bicolor* (L. Moench): III. *Australian Journal of Agricultural Research*, **26**, 11–23.

Fischer, R.A. (1980). Influence of water stress on crop yield in semi-arid regions. In *Adaptation of Plants to Water and High Temperature Stress*, ed. N.C. Turner & P.J. Kramer, pp. 323–40. New York: Wiley.

Fist, A.J., Smith, F.W. & Edwards, D.G. (1987). External phosphorus requirement of five tropical grain legumes grown in flowing-culture solution. *Plant and Soil*, **99**, 75–84.

Fitzpatrick, E.A. & Nix, H.A. (1970). The climatic factor in Australian grassland ecology. In *Australian Grasslands*, ed. R.M. Moore, pp. 3–26. Canberra: Australian National University Press.

Fleming, H.S. & Rogers, D.J. (1970). A classification of *Manihot esculenta* Crantz. using the information carrying content of a character as a measure of its classificatory rank. In *Tropical Root and Tuber Crops Tomorrow*, vol. 1, ed. D.L. Plucknett, pp. 66–71. Honolulu, Hawaii: University Press of Hawaii.

Flinn, J.C. & Hoyoux, J.M. (1976). Le bananier plantain en Afrique. *Fruits*, **31**, 520–30.

Flores, C.I., Clark, R.B. & Gourley, L.M. (1988). Growth and yield traits of sorghum grown on acid soil at varied aluminium saturations. *Plant and Soil*, **106**, 49–57.

Flores, C.I., Clark, R.B. & Gourley, L.M. (1991). Genotypic variation of pearl millet for growth and yield on acid soil. *Field Crops Research*, **26**, 347–54.

Floresca. E.T. (1968). Cultural methods for soybean growth in Maahas clay with special reference to establishment, weed control and fertilization. MSc thesis, University of the Philippines, College of Agriculture, Los Baños.

Folquer, F. (1974). Varietal efficiency in the spring production of sweet potato seeds (*Ipomoea batatas* (L.) Lam.). *Revista Agronomica del Noroeste Argentino*, **11**, 193–225. (Abstract in *Field Crop Abstracts* (1976), **29**, 881.)

Fontaine, S., Delvaux, B., Duffey, J.E. & Herbillon, A.J. (1989). Potassium exchange behaviour in Caribbean volcanic ash soils under banana cultivation. *Plant and Soil*, **120**, 283–90.

Forsythe, W.M., Victor, A. & Gomez, M. (1979). Flooding tolerance and surface drainage requirements of *Phaseolus vulgaris* L. In *Soil Physical Properties and Crop Production in the Tropics*, ed. R. Lal & D.J. Greenland, pp. 205–25. Chichester: Wiley.

Fortanier, E.J. (1957). De Bienvloeding van de Bloei by *Arachis hypogaea* L. *Mendelingen Landbouwhogeschool, Wageningen*, **57**, 1–116.

Foster, H.L. (1970). Liming continuously cultivated soil in Uganda. East Africa. *East African Agriculture and Forestry Journal*, **36**, 58–69.

Fox, R.H., Talleyrand, H. & Scott, T.W. (1975). Effect of N fertilizer on yield and N content of cassava, Llanera cultivar. *University of Puerto Rico Journal of Agriculture*, **59**, 115–24.

Fox, R.L. (1974). Examples of anion and cation adsorption by soils of tropical America. *Tropical Agriculture (Trinidad)*, **51**, 200–10.

Fox, R.L. (1978). Studies on phosphorus nutrition in the tropics. In *Mineral Nutrition of Legumes in Tropical and Subtropical Soils*, ed. C.S. Andrew & E.J. Kamprath, pp. 169–87. Melbourne: Commonwealth Scientific and Industrial Research Organization.

Fox, R.L. (1981). Soils with variable charge: agronomic and fertility aspects. In *Soils with Variable Charge*, ed. B.K.G. Theng, pp. 195–224. Lower Hutt, New Zealand: New Zealand Society of Soil Science.

Fox, R.L. & Kamprath, E.J. (1970). Phosphate sorption isotherms for evaluating the phosphate requirements of soils. *Soil Science Society of America Proceedings*, **34**, 902–7.

Fox, R.L. & Searle, P.G.E. (1978). Phosphate adsorption by soils of the tropics. In *Diversity of Soils in the Tropics*, ed. J.J. Nicholaides & L.D. Swindale, pp. 97–119. Madison, Wisconsin: American Society of Agronomy.

Francis, C.A. (1972). Photoperiod sensitivity and adaptation in maize. In *Proceedings of the 27th Annual Corn and Sorghum Research Conference*, pp. 119–31. American Seed Trade Association.

Francis, C.A. (ed.) (1986). *Multiple Cropping Systems.* New York: Macmillan.

Francis, C.A. & Sanders, J.H. (1978). Economic analysis of bean and maize systems: monoculture versus associated cropping. *Field Crops Research*, 1, 319–35.

Francis, C.A., Flor, C.A. & Temple, S.R. (1976). Adapting varieties for intercropped systems in the tropics. In *Multiple Cropping*, pp. 235–53. Madison, Wisconsin: American Society of Agronomy.

Francis, C.A., Flor, C.A. & Prager, M. (1978a). Effects of bean association on yields and yield components of maize. *Crop Science*, 18, 760–4.

Francis, C.A., Prager, M. & Laing, D.R. (1978b). Genotype × environment interactions in climbing bean cultivars in monoculture and associated with maize. *Crop Science*, 18, 242–6.

Francis, C.A., Prager, M. & Tejado, G. (1982). Effects of relative planting dates in bean (*Phaseolus vulgaris* L.) and maize (*Zea mays* L.) intercropping patterns. *Field Crops Research*, 5, 45–54.

Franco, A.A. (1977a). Nutritional restraints for tropical grain legume symbiosis. In *Exploiting the Legume–Rhizobium Symbiosis in Tropical Agriculture*, ed. J.M. Vincent, A.S. Whitney & J. Bose, pp. 237–52. Honolulu, Hawaii: University of Hawaii Press.

Franco, A.A. (1977b). Micronutrient requirements of legume–*Rhizobium* symbiosis in the tropics. In *Limitations and Potentials for Biological Nitrogen Fixation in the Tropics*, ed. J. Dobereiner, R.H. Burris & A. Hollaender, pp. 161–71. New York: Plenum Press.

Franco, A.A. & Day, J.M. (1980). Effects of lime and molybdenum on nodulation and nitrogen fixation of *Phaseolus vulgaris* L. in acid soils of Brazil. *Turrialba*, 30, 99–105.

Francois, L.E., Donovon, T. & Maas, E.V. (1984). Salinity effects on seed yield, growth, and germination of grain sorghum. *Agronomy Journal*, 76, 741–4.

French, B. & Bridle, C. (1978). *Food Crops of Papua New Guinea.* Madang, Papua New Guinea: Kristen Press, 55 pp.

Friedrich, J.W., Schrader, L.E. & Nordheim, E.V. (1979). N deprivation in maize during grain filling. I. Accumulation of dry matter, nitrate-N, and sulfate-S. *Agronomy Journal*, 71, 461–5.

Fuhrmann, J. & Wollum, A.G. (1989). Symbiotic interactions between soybean and competing strains of *Bradyrhizobium japonicum*. *Plant and Soil*, 119, 139–45.

Fussell, L.K. (1992). Semi-arid cereal and grazing systems of West Africa. In *Field Crop Ecosystems*, ed. C.J. Pearson, pp. 485–578. Amsterdam: Elsevier.

Fussell, L.K. & Dwarte, D.M. (1980). Structural changes of the grain associated with black region formation in *Pennisetum americanum*. *Journal of Experimental Botany*, 31, 645–54.

Fussell, L.K. & Pearson, C.J. (1978). Course of grain development and its relationship to black region appearance in *Pennisetum americanum*. *Field Crops Research*, 1, 21–31.

Fussell, L.K. & Pearson, C.J. (1980). Effects of grain development and thermal history on grain maturation and seed vigour of *Pennisetum americanum*. *Journal of Experimental Botany*, 31, 635–43.

Fussell, L.K., Pearson, C.J. & Norman, M.J.T. (1980). Effect of temperature during various growth stages on grain development and yield of *Pennisetum americanum*. *Journal of Experimental Botany*, 31, 621–33.

Galinat, W.C. (1977). The origin of corn. In *Corn and Corn Improvement*, ed. G.F. Sprague, pp. 1–47. Madison, Wisconsin: American Society of Agronomy.

Galinat, W.C. (1983). The origin of maize as shown by key morphological traits of its ancestor, teosinte. *Maydica*, **28**, 121–38.

Ganry, J. (1973). Etude du developpement du systeme foliaire du bananier en fonction de la temperature. *Fruits*, **28**, 499–576.

Ganry, J. & Meyer, J.P. (1975). Recherche d'une loi d'action de la temperature sur la croissance des fruits du bananier. *Fruits*, **30**, 375–92.

Gaur, Y.D., Sen, A.N. & Subba Rao, N.S. (1974). Problem regarding groundnut (*Arachis hypogaea* L.) inoculation in tropics with special reference to India. *Indian National Science Academy Proceedings*, **40**, 562–70.

Gauthier, D., Diem, H.G., Dommergues, Y.R. & Ganry, F. (1985). Assessment of N_2 fixation by *Casuarina equisetifolia* inoculated with *Frankia* ORSO21001 using ^{15}N methods. *Soil Biology and Biochemistry*, **17**, 375–9.

Gay, S., Egli, D.B. & Reicosky, D.A. (1980). Physiological aspects of yield improvement in soybeans. *Agronomy Journal*, **72**, 387–91.

George, T., Ladha, J.K., Buresh, R.J. & Garrity, S.P. (1992). Managing native and legume-fixed nitrogen in lowland rice-based cropping systems. *Plant and Soil*, **141**, 69–91.

Gepts, P. & Debouck, D. (1991). Origin, domestication, and evolution of the common bean (*Phaseolus vulgaris* L.). In *Common Beans: Research for Crop Improvement*, ed. A. van Schoonhoven & O. Voysest, pp. 7–53. Wallingford, UK: CAB International/Centro Internacional de Agricultura Tropical.

Gepts, P., Osborn, T.C., Rashka, K. & Bliss, F.A. (1986). Phaseolin-protein variability in wild forms and landraces of the common bean (*Phaseolus vulgaris*): evidence for multiple centers of domestication. *Economic Botany*, **40**, 451–68.

Gerakis, P.A. & Tsangarakis, C.Z. (1969). Effect of the preceding crop and agronomic practice on sorghum (*Sorghum bicolor* L. Moench.) and groundnuts (*Arachis hypogaea* L.) in the Central Sudan. *Agronomy Journal*, **61**, 681–3.

Ghavami, M. (1974). Irrigation of Valery bananas in Honduras. *Tropical Agriculture (Trinidad)*, **51**, 443–6.

Ghosh, S.P., Nair, G.M., Pillai, N.G., Ramanujam, T., Mohankumar, B. & Lakshmi, K.R., (1987). Growth, productivity and nutrient uptake by cassava in association with four perennial species. *Tropical Agriculture (Trinidad)*, **64**, 233–6.

Ghuman, B.S. & Lal, R. (1989). Soil temperature effects of biomass burning in windrows after clearing a tropical rainforest. *Field Crops Research*, **22**, 1–10.

Ghuman, B.S., Lal, R. & Shearer, W. (1991). Land clearing and use in the humid Nigerian tropics. I. Soil physical properties. *Soil Science Society of America Proceedings*, **55**, 178–83.

Gibbons, R.W. (1979). Groundnut improvement research technology for the semi-arid tropics. In *International Symposium on Development and Transfer of Technology for Rainfed Agriculture and the SAT Farmer, August 1979*. Hyderabad, India: International Crops Research Institute for the Semi-Arid Tropics, 23 pp.

Gibbons, R.W. (1980). Adaptation and utilization of groundnuts in different environments and farming systems. In *Advances in Legume Science*, ed. R.J. Summerfield & A.H. Bunting, pp. 483–93. Kew: Royal Botanic Gardens.

Gibbons, R.W., Bunting, A.H. & Smartt, J. (1972). The classification of varieties of groundnut (*Arachis hypogaea*). *Euphytica*, **21**, 78–85.

Gillman, G.P. (1973). Studies on some deep sandy soils in Cape York Peninsula, North Queensland. III. Losses of applied phosphorus and sulphur. *Australian Journal of Experimental Agriculture and Animal Husbandry*, **13**, 418–22.

Gillman, G.P. (1981). Effects of pH and ionic strength on the cation exchange capacity of soils with variable charge. *Australian Journal of Soil Research*, **19**, 93–6.

Gillman, G.P. (1984). Nutrient availability in acid soils of the tropics following clearing and cultivation. In *Proceedings of the International Workshop on Soils: Research to Resolve Selected Problems of Soils in the Tropics, Townsville, Queensland, 12–16 September 1983*, ed. E.T. Craswell & R.F. Isbell, pp. 39–43. Australian Centre for International Agricultural Research, Proceedings Series No. 2.

Gillman, G.P. & Fox, R.L. (1980). Increases in cation exchange capacities of variable charge soils following superphosphate additions. *Soil Science Society of America Journal*, **44**, 934–8.

Giri, G. & De, R. (1980). Effect of preceding grain legumes on growth and nitrogen uptake of dryland pearl millet. *Plant and Soil*, **56**, 459–64.

Gitte, R.R., Rai, P.V. & Patil, R.B. (1978). Chemotaxis of *Rhizobium* sp. towards root exudates of *Cicer arietinum* L. *Plant and Soil*, **50**, 553–66.

Godefroy, J., Lassoudière, A., Lossois, P. & Penel, J.P. (1978). Action du chaulage sur les caracteristiques physico-chemiques et la productivité d'un sol tourbeux en culture bananiere. *Fruits*, **33**, 77–90.

Godefroy, J., Roose, E.J. & Muller, M. (1975). Estimation des partes par les eaux de ruissellement et de drainage des éléments fertilisants dans un sol de bananeraie du sud la Cote d'Ivoire. *Fruits*, **30**, 223–35.

Godfrey-Sam-Aggrey, W. (1978). Effects of plant population on sole-crop cassava in Sierra Leone. *Experimental Agriculture*, **14**, 239–44.

Godon, P. (1985). Précédent cultural et erosion sous culture de manioc. *L'Agronomie Tropicale*, **40**, 217–22.

Goedert, W.J., Corey, R.B. & Syers, J.K. (1975). Lime effects on potassium equilibrium in soils of Rio Grande do Sul, Brasil. *Soil Science*, **120**, 107–11.

Goel, N. & Varshney, K.A. (1987). Note on seed germination and early seedling growth of two chickpea varieties under saline conditions. *Legume Research*, **10**, 34–6.

Goldman, I.L., Carter, T.E. & Patterson, R.P. (1989). Differential genotype response to drought stress and subsoil aluminum in soybean. *Crop Science*, **29**, 330–4.

Goldsworthy, P.R. (1970a). The canopy structure of tall and short sorghum. *Journal of Agricultural Science*, **75**, 123–31.

Goldsworthy, P.R. (1970b). The growth and yield of tall and short sorghums in Nigeria. *Journal of Agricultural Science*, **75**, 109–22.

Goldsworthy, P.R. (1970c). The sources of assimilate for grain development in tall and short sorghum. *Journal of Agricultural Science*, **75**, 523–31.

Goldsworthy, P.R. & Fisher, N.M. (eds.) (1984). *The Physiology of Tropical Field Crops*. Chichester: Wiley.

Gollifer, D.E. (1971). A cultivar trial with yams *Dioscorea alata* L. in the British Solomon Islands. *Papua New Guinea Agricultural Journal*, **25**, 25–30.

Gollifer, D.E. (1972). Effect of applications of potassium on annual crops grown on soils of the Dala Series in Malaita, British Solomon Islands. *Tropical Agriculture (Trinidad)*, **49**, 261–8.

Gollifer, D.E. (1973). Staking trials with sweet potatoes. *Tropical Agriculture (Trinidad)*, **50**, 279–85.

Gomez, A.A. & Zandstra, H.G. (1977). An analysis of the role of legumes in multiple cropping systems. In *Exploiting the Legume–Rhizobium Symbiosis in Tropical Agriculture*, pp. 81–95. Honolulu, Hawaii: University of Hawaii Press.

Gonzalez-Erico, E., Kamprath, E.J., Naderman, G.C. & Soares, W.V. (1979). Effect of depth of lime incorporation on the growth of corn on an Oxisol of Central Brazil. *Soil Science Society of America Journal*, **43**, 1155–8.

Goodbody, S. & Humphreys, G.S. (1986). Soil chemical status and the prediction of sweet potato yields. *Tropical Agriculture (Trinidad)*, **63**, 209–11.

Gooding, H.J. (1964). Some aspects of the methods and results of sweet potato selection. *Empire Journal of Experimental Agriculture*, **32**, 279–89.

Gorman, C. (1977). A priori models and Thai prehistory: a reconsideration of the beginnings of agriculture in southeastern Asia. In *Origins of Agriculture*, ed. C.A. Reed, pp. 321–55. The Hague: Mouton.

Goto, K. & Yamamoto, T. (1972). Studies on cool injury in bean plants: III. *Research Bulletin of Hokkaido National Agricultural Experiment Station*, **100**, 14–19.

Graham, P.H. (1981). Some problems of nodulation and symbiotic nitrogen fixation in *Phaseolus vulgaris* L.: a review. *Field Crops Research*, **4**, 93–112.

Graham, P.H. & Halliday, J. (1977). Inoculation and nitrogen fixation in the genus *Phaseolus*. In *Exploiting the Legume–Rhizobium Symbiosis in Tropical Agriculture*, ed. J.M. Vincent, A.S. Whitney & J. Bose, pp. 313–14. Honolulu, Hawaii: University of Hawaii Press.

Graham, P.H. & Rosas, J.C. (1977). Growth and development of indeterminate bush and climbing cultivars of *Phaseolus vulgaris* L. inoculated with *Rhizobium*. *Journal of Agricultural Science*, **88**, 503–8.

Graham, P.H. & Rosas, J.C. (1978). Nodule development and nitrogen fixation in cultivars of *Phaseolus vulgaris* L. as influenced by planting density. *Journal of Agricultural Science*, **90**, 311–17.

Graham, P.H., Viteri, S.E., Mackie, F., Vargas, A.T. & Palacios, A. (1982). Variation in acid soil tolerance among strains of *Rhizobium phaseoli*. *Field Crops Research*, **5**, 121–8.

Graham, R.A. (1986). Effects of soil ameliorants on lime-induced chlorosis, growth and nodulation in groundnuts (*Arachis hypogaea* L.). *Tropical Agriculture (Trinidad)*, **63**, 61–2.

Graham, R.A. & Donawa, A.L. (1982). Greenhouse and field evaluation of *Rhizobium* strains nodulating groundnut (*Arachis hypogaea* L.). *Tropical Agriculture (Trinidad)*, **59**, 254–6.

Grant, R.F., Jackson, B.S., Kiniry, J.R., & Arkin, G.F. (1989). Water deficit timing effects on yield components in maize. *Agronomy Journal*, **81**, 61–5.

Grattan, S.R. & Maas, E.V. (1988). Effect of salinity on phosphate accumulation and injury in soybean. II. Role of substrate Cl and Na. *Plant and Soil*, **109**, 65–71.

Gregory, P.J. (1979). Uptake of N, P and K by irrigated and unirrigated pearl millet (*Pennisetum typhoides*). *Experimental Agriculture*, **15**, 217–23.

Gregory, P.J. & Reddy, M.S. (1982). Root growth in an intercrop of pearl millet/groundnut. *Field Crops Research*, **5**, 241–52.

Gregory, P.J. & Squire, G.R. (1979). Irrigation effects on roots and shoots of pearl millet (*Pennisetum typhoides*). *Experimental Agriculture*, **15**, 161–8.

Gregory, W.C. & Gregory, M.P. (1976). Groundnut: *Arachis hypogaea* (Leguminosae Papilionatae). In *Evolution of Crop Plants*, ed. N.W. Simmonds, pp. 151–4. London: Longman.

Gregory, W.C., Gregory, M.P., Krapovickas, A., Smith, B.W. & Yarbrough, J.A. (1973). Structures and genetic resources of peanuts. In *Peanuts: Culture and Uses*, pp. 47–133. Stillwater, Oklahoma: American Peanut Research and Education Association.

Gregory, W.C., Krapovickas, A. & Gregory, M.P. (1980). Structure, variation, evolution and classification in *Arachis*. In *Advances in Legume Science*, ed. R.J. Summerfield & A.H. Bunting, pp. 469–81. Kew: Royal Botanic Gardens.

Grieve, C.M. & Maas, E.V. (1988). Differential effects of sodium/calcium ratios on sorghum genotypes. *Crop Science*, **28**, 659–65.

Griffith, J.F. (1972). *Climates of Africa*. Amsterdam: Elsevier, 604 pp.

Grigg, D.B. (1974). *The Agricultural Systems of the World, An Evolutionary Approach*. Cambridge: Cambridge University Press, 358 pp.

Grove, T.L. (1979). Nitrogen fertility in Oxisols and Ultisols of the humid tropics. *Cornell International Agriculture Bulletin*, No. 36, 28 pp.

Grove, T.L., Ritchey, K.D. & Naderman, G.C. (1980). Nitrogen fertilization of maize on an Oxisol of the Cerrado of Brazil. *Agronomy Journal*, **72**, 261–5.

Grundon, N.J., Edwards, D.G., Takkar, P.N., Asher, C.J. & Clark, R.B. (1987). *Nutritional Disorders of Grain Sorghum*. Australian Centre for International Agricultural Research, Monograph No. 2, 99 pp.

Gupta, R.K. & Agrawal, G.G. (1976). Consumptive use of water by gram and linseed. *Indian Journal of Agricultural Sciences*, **47**, 22–6.

Gupta, S.K. & Sharma, S.K. (1990). Response of crops to high exchangeable sodium percentage. *Irrigation Science*, **11**, 173–9.

Gurnah, A.M. (1974). Effects of spacing, sett weight and fertilizers on yield and yield components in yams. *Experimental Agriculture*, **10**, 17–22.

Gutierrez, U., Infante, M. & Pinchinat, A. (1975). Situacian del cultivo de frijol en America Latina. *Boletin Informe*. Cali, Colombia: Centro Internacional de Agricultura Tropical.

Habish, H.A. & Ishag, H.M. (1974). Nodulation of legumes in the Sudan. III. Responses of haricot bean to inoculation. *Experimental Agriculture*, **10**, 45–50.

Hack, H.R.B. (1970). Emergence of crops in clay soils of the Central Sudan Rainlands, in relation to soil water and air-filled pore space. *Experimental Agriculture*, **6**, 287–302.

Hackett, C. (1991). *Plantgro: A Software Package for Coarse Prediction of Plant Growth*. Melbourne: Commonwealth Scientific and Industrial Research Organization, 242 pp.

Hadad, M.A., Loynachan, T.E., Musa, M.M. & Mukhtar, N.O. (1986). Inoculation of groundnut (peanut) in Sudan. *Soil Science*, **141**, 155–62.

Hadas, S. & Russo, D. (1974*a*). Water uptake by seeds as affected by water stress, capillary conductivity and seed soil water contact: I. *Agronomy Journal*, **66**, 64–7.

Hadas, S. & Russo. D. (1974*b*). Water uptake by seeds as affected by water stress, capillary conductivity and seed soil water contact: II. *Agronomy Journal*, **66**, 647–52.

Haen, H. de & Runge-Metzger, A. (1989). Improvements in efficiency and sustainability of traditional land use systems through learning from farmers' practice. *Quarterly Journal of International Agriculture*, **28**, 326–50.

Hahn, S.K. & Hozyo, Y. (1984). Sweet potato. In *The Physiology of Tropical Field Crops*. ed. P.R. Goldsworthy & N.M. Fisher, pp. 551–8. Chichester: Wiley.

Hahn, S.K., Terry, E.R., Leuschner, K., Akobundu, I.O., Okali, C. & Lal, R. (1979). Cassava improvement in Africa. *Field Crops Research*, **2**, 193–226.

Hall, A.J., Ginzo, H.D., Lemcoff, J.H. & Soriano, A. (1980). Influence of drought during pollen-shedding on flowering, growth and yield of maize. *Zeitschrift für Acker- und Pflanzenbau*, **149**, 287–98.

Hall, A.J., Lemcoff, J.W. & Trapani, N. (1981). Water stress before and during flowering in maize and its effects on yield, its components, and their determinants. *Maydica*, **26**, 19–38.

Hall, A.J., Vilella, F., Trapani, N. & Chimenti, C. (1982). The effects of water stress, its timing, and genotype on the dynamics of anthesis and on pollen production in maize. *Field Crops Research*, **5**, 349–63.

Hall, N.S., Chandler, W.F., Van Bavel, C.H.M., Reid, P.H. & Anderson, J.H. (1953). A tracer technique to measure growth and activity of plant root systems. *North Carolina Agricultural Experiment Station Bulletin*, No. 101.

Hamel, C. & Smith, D.L. (1991). Interspecific N-transfer and plant development in a mycorrhizal field-grown mixture. *Soil Biology and Biochemistry* **23**, 661–5.

Hammons, R.O. (1973). Early history and origin of the peanut. In *Peanuts: Culture and Uses*, pp. 17–45. Stillwater, Oklahoma: American Peanut Research and Education Association.

Hamon, P. & Toure, B. (1990). Characterization of traditional yam varieties belonging to the *Dioscorea cayenensis–rotundata* complex by the isozymic patterns. *Euphytica*, **46**, 101–7.

Hanawalt, R.B. (1969). Environmental factors affecting the sorption of atmospheric ammonia. *Soil Science Society of America Proceedings*, **33**, 231–3.

Hanfei, D. (1992). Upland rice systems. In *Field Crop Ecosystems*, ed. C.J. Pearson, pp. 183–204. Amsterdam: Elsevier.

Hanway, J.J. (1963). Growth stages of corn. *Agronomy Journal*, **55**, 487–92.

Hanyu, J., Uchijima, T. & Sugawara, S. (1966). Studies on the agroclimatological method for expressing the paddy rice products: I. *Bulletin of Tohoku Agricultural Experiment Station*, **34**, 27–36.

Haque, I. & Walker, W.M. (1980). Effect of nitrogen sources on cassava yields in Sierra Leone. *Communications in Soil Science and Plant Analysis*, **11**, 1167–73.

Haque, I., Walker, W.M. & Funnah, S.M. (1980). Effects of phosphorus and zinc on soybean in Sierra Leone. *Communications in Soil Science and Plant Analysis*, **11**, 1029–40.

Harlan, J.R. (1971). Agricultural origins: centers and non-centers. *Science*, **154**, 468–74.

Harlan, J.R. (1977). The origins of cereal agriculture in the Old World. In *Origins of Agriculture*, ed. C.A. Reed, pp. 357–83. The Hague: Mouton.

Harlan, J.R. (1986). Plant domestication: diffuse origins and diffusion. In *The Origin and Domestication of Cultivated Plants*, ed. C. Barigozzi, pp. 21–34. Amsterdam: Elsevier.

Harlan, J.R. & Stemler, A.D.B. (1976). The races of sorghum in Africa. In *Origins of African Plant Domestication*, ed. J.R. Harlan, J.M.J. de Wet & A.D.B. Stemler, pp. 465–78. The Hague: Mouton.

Harlan, J.R. & Wet, J.M.J. de (1972). A simplified classification of cultivated sorghum. *Crop Science*, **12**, 172–6.

Harper, R.S. (1973). Cassava growing in Thailand. *World Crops*, **25**, 94–7.

Hartley, A.C., Aland, F.P. & Searle, P.G.E. (1967). *The Balima-Tiauru Area, New Britain*. Soil Survey Report, No. 1. Port Moresby: Papua New Guinea Department of Agriculture, Stock and Fisheries, 170 pp.

Hartzog, D. & Adams, F. (1973). Fertilizer, gypsum and lime experiments with peanuts in Alabama. *Alabama Agricultural Experimental Station Bulletin*, No. 448.

Harwood, R.R. (1975). Farmer-oriented research aimed at crop intensification. In *Proceedings of the Cropping System Workshop*, pp. 12–32. Los Baños, Philippines: International Rice Research Institute.

Harwood, R.R. & Price, E.C. (1976). Multiple cropping in tropical Asia. In *Multiple Cropping*, pp. 11–40. Madison, Wisconsin: American Society of Agronomy.

Haseba, T. & Ho, D. (1975). Studies of transportation in relation to environment: VII. *Journal of Agricultural Meteorology (Tokyo)*, **30**, 173–82.

Hawkes, J.G. (1989). The domestication of roots and tubers in the Americas. In *Foraging and Farming: The Evolution of Plant Exploitation*, ed. D.R. Harris & G.C. Hillman, pp. 481–99. One World Archaeology, No. 13. London: Unwin Hyman.

Hawkins, R.C. & Cooper, P.J.M. (1981). Growth, development and grain yield of maize. *Experimental Agriculture*, **17**, 203–7.

Haynes, R.J. (1984). Lime and phosphate in soil–plant systems. *Advances in Agronomy*, **37**, 249–315.

Haynes, P.H. & Wholey, D.W. (1972). Variability in commercial sweet potatoes (*Ipomoea batatas* (L.) Lam.) in Trinidad. *Experimental Agriculture*, **7**, 27–32.

Hedge, D. M. & Srinivas, K. (1989). Effect of soil matric potential and nitrogen on growth, yield, nutrient uptake and water use of banana. *Agricultural Water Management*, **16**, 109–17.

Hedge, D.M., Raghunatha, G. & Narayanswamy, H. (1975). Changes in seed characters with maturity in gram (*Cicer arietinum* L.). *Agricultural Research Journal of Kerala*, **13**, 165–8.

Heenan, D.P. (1973). Preliminary observations on the growth and production of bananas in the northern district of Papua New Guinea. *Papua New Guinea Agricultural Journal*, **24**, 145–55.

Heenan, D.P. & Campbell, L.C. (1980). Growth, yield components and seed composition of two soybean cultivars as affected by manganese supply. *Australian Journal of Agricultural Research*, **31**, 471–6.

Heiser, C.B. (1979). Origins of some cultivated New World plants. *Annual Review of Ecology and Systematics*, **10**, 309–26.

Helal, H.M. (1990). Varietal differences in root phosphatase activity as related to the utilization of organic phosphates. *Plant and Soil*, **123**, 161–3.

Hemsath, D.L. & Mazurak, A.P. (1974). Seedling growth of sorghum in clay–sand mixtures at various compactions and water contents. *Soil Science Society of America Proceedings*, **38**, 387–90.

Henmi, T. & Wada, K. (1976). Morphology and composition of allophane. *American Mineralogist*, **61**, 370–90.

Henzell, E.F. (1988). The role of biological nitrogen fixation in solving problems in tropical agriculture. *Plant and Soil*, **108**, 15–21.

Hernandez, Te.P., Hernandez, Ta.P. & Miller, F.M. (1964). Frequency of somatic mutations in several sweet potato varieties. *Proceedings of the American Society for Horticultural Science*, **85**, 430–3.

Herrera, W.A.T. & Zandstra, H.G. (1979). The response of some major upland crops to excessive soil moisture. In *Tenth Annual Scientific Meeting, Crop Science Society of the Philippines*. Laguna, Philippines: University of the Philippines.

Herrero, M.P. & Johnson, R.R. (1981). Drought stress and its effects on maize reproductive systems. *Crop Science*, **21**, 105–10.

Herridge, D.F. & Betts, J.H. (1988). Field evaluation of soybean genotypes selected for enhanced capacity to nodulate and fix nitrogen in the presence of nitrate. *Plant and Soil*, **110**, 129–35.

Hesketh, J.D. (1963). Limitations to photosynthesis responsible for differences among species. *Crop Science*, **3**, 493–6.

Hesketh, J.D. & Musgrave, R.B. (1962). Photosynthesis under field conditions. IV. Light studies with individual corn leaves. *Crop Science*, **2**, 311–15.

Hesketh, J.D., Ogren, W.L., Hageman, M.E. & Peters, D.B. (1981). Correlations among leaf CO_2-exchange rates, areas and enzyme activities among soybean cultivars. *Photosynthesis Research*, **2**, 21–30.

Hill, W.A., Dodo, H., Hahn, S.K., Mulongoy, K. & Adeyeye, S.O. (1990). Sweet potato root and biomass production with and without nitrogen fertilization. *Agronomy Journal*, **82**, 1120–2.

HilleRisLambers, D.H. (1977). Results of 1975 deep-water rice flowering date survey. In *Deep-Water Rice*, pp. 193–204. Los Baños, Philippines: International Rice Research Institute.

Hiltbold, A.E., Patterson, R.M. & Reed, R.B. (1985). Soil populations of *Rhizobium japonicum* in a cotton–corn–soybean rotation. *Soil Science Society of America Journal*, **49**, 343–8.

Hinton, P. (1978). Declining production among sedentary swidden cultivators: the case of the Pwo Karen. In *Farmers in the Forest*, ed. P. Kunstadter. E.C. Chapman & S. Sabhasri, pp. 185–98. Honolulu, Hawaii: University of Hawaii Press.

Hipsley, E.H. & Kirk, N.E. (1955). Studies of dietary intake and the expenditure of energy by New Guineans. *South Pacific Commission Technical Paper*, No. 147. Mimeo, 158 pp.

Ho, P.-T. (1977). The indigenous origins of Chinese agriculture. In *Origins of Agriculture*, ed. C.A. Reed, pp. 413–84. The Hague: Mouton.

Hobman, F.R. (1985). Evaluation of groundnuts (*Arachis hypogaea* L.) for the wet tropical coast of Queensland, Australia. *Tropical Agriculture (Trinidad)*, **62**, 217–21.

Hodgson, A.S., Holland, J.F. & Rayner, P. (1989). Effects of field slope and duration of furrow irrigation on growth and yield of six grain-legumes on a waterlogging-prone Vertisol. *Field Crops Research*, **22**, 165–80.

Hofstra, G. & Hesketh, J.D. (1969). Effects of temperature on the gas exchange of leaves in the light and dark. *Planta*, **85**, 228–37.

Hofstra, G. & Nelson, C.D. (1969). The translocation of photosynthetically assimilated ^{14}C in corn. *Canadian Journal of Botany*, **47**, 1435–42.

Holder, G.D. & Gumbs, F.A. (1983a). Effects of irrigation on the growth and yield of banana. *Tropical Agriculture (Trinidad)*, **60**, 25–30.

Holder, G.D. & Gumbs, F.A. (1983b). Effects of waterlogging on the growth and yield of banana. *Tropical Agriculture (Trinidad)*, **60**, 111–16.

Holder, G.D. & Gumbs, F.A. (1983c). Effects of nitrogen and irrigation on the growth and yield of banana. *Tropical Agriculture (Trinidad)*, **60**, 179–83.

Hommertzheim, D.L. (1979). Analytical description of a soybean canopy. *Agronomy Journal*, **71**, 405–9.

Horton, D.E. (1988). World patterns and trends in sweet potato production. *Tropical Agriculture (Trinidad)*, **65**, 268–70.

Houghton, R.A. (1990). The global effects of tropical deforestation. *Environmental Science and Technology*, **24**, 414–22.

Howeler, R.H. (1980). Soil-related cultural practices for cassava. In *Cassava Cultural Practices*, ed. E.J. Weber, J.C. Toro & M. Graham, pp. 59–69. Ottawa, Ontario: International Development Research Centre.

Howeler, R.H. (1991). Long-term effect of cassava cultivation on soil productivity. *Field Crops Research*, **26**, 1–18.

Howeler, R.H., Cadavid, L.F. & Calvo, F.A. (1977). The interaction of lime with minor elements and phosphorus in cassava production. In *Fourth Symposium of the International Society of Tropical Root Crops*, pp. 113–17. Ottawa, Ontario: International Development Research Centre.

Howeler, R.H., Edwards, D.G. & Asher, C.J. (1982). The effect of soil sterilization and mycorrhizal inoculation on the growth, nutrient uptake and critical phosphorus concentration of cassava. In *Fifth International Symposium on Tropical Root and Tuber Crops*, pp. 519–37. Los Baños, Philippines.

Howeler, R.H., Sieverding, E. & Saif, S. (1987). Practical aspects of mycorrhizal technology in some tropical crops and pastures. *Plant and Soil*, **100**, 249–83.

Howell, R.W. (1963). Physiology of the soybean. In *The Soybean*, ed. A.G. Norman, pp. 37–115. New York: Academic Press.

Howle, D.S. & Caviness, C.E. (1988). Influence of cultivar and seed characteristics on vertical weight displacement by soybean seedlings. *Crop Science*, **28**, 321–4.

Hsiao, T.C., O'Toole, J.C. & Tomar, V.S. (1980). Water stress as a constraint to crop production in the tropics. In *Priorities for Alleviating Soil-Related Constraints to Food Production in the Tropics*, pp. 339–70. Los Baños, Philippines: International Rice Research Institute.

Hsu, F.C. (1979). A developmental analysis of seed size in common bean. *Crop Science*, **19**, 226–30.

Huang, H.C. & Cheng, D.C. (1960). Runoff and soil erosion on cultivated slopes: II. *Taiwan Forestry Research Institute Bulletin*, No. 68.

Huang, H.C., Chang, L.J. & Cheng, D.C. (1958). Runoff and erosion on cultivated slopes: I. *Taiwan Forestry Research Institute Bulletin*, No. 56.

Huda, A.K.S. & Virmani, S.M. (1987). Agroclimatic environment of chickpea and pigeonpea. In *ICRISAT Annual Report, 1987*, pp. 13–16. Patancheru, India: International Crops Research Institute for the Semi-Arid Tropics.

Hudson, N. (1971). *Soil Conservation*. Ithaca, New York: Cornell University Press, 320 pp.

Huett, D.O. (1975). A study of factors contributing to variability in the yield and quality of the sweet potato (*Ipomoea batatas* (L.) Lam.). MScAgr thesis, University of Sydney, Australia, 197 pp.

Huett, D.O. & O'Neill, G.H. (1976). Growth and development of short and long season sweet potatoes in sub-tropical Australia. *Experimental Agriculture*, **12**, 385–94.

Hughes, J.D. & Searle, P.G.E. (1964). Observations on the residual value of accumulated phosphorus in a red loam. *Australian Journal of Agricultural Research*, **15**, 377–83.

Huke, R. (1976). Geography and climate of rice. In *Climate and Rice*, pp. 31–50. Los Baños, Philippines: International Rice Research Institute.

Hulugalle, N.R.. & Lal, R. (1986). Root growth in a compacted gravelly alfisol as affected by rotation with a woody perennial. *Field Crops Research*, **13**, 33–44.

Hume, D.J. & Jackson, A.K.H. (1981). Pod formation in soybeans at low temperatures. *Crop Science*, **21**, 933–7.

Hunt, L.A., Wholey, D.W. & Cock, J.H. (1977). Growth physiology of cassava (*Manihot esculenta* Crantz.). *Field Crop Abstracts*, **30**, 77–91.

Hunter, J.R. & Erickson, A.E. (1952). Relation of seed germination to soil moisture tension. *Agronomy Journal*, **44**, 107–9.

Hutchinson, M., Nix, H.A. & McMahon, J.P. (1992). Climate constraints on cropping systems. In *Field Crop Ecosystems*, ed. C.J. Pearson, pp. 37–58. Amsterdam: Elsevier.

Hymowitz, T. (1970). On the domestication of the soybean. *Economic Botany*, **24**, 408–21.

ICAR (1961). *Handbook of Agriculture*, ed. K. Sawhney & J.R. Daji. New Delhi: Indian Council of Agricultural Research.

ICAR (1972). *Cropping Patterns in India*. New Delhi: Indian Council of Agricultural Research, 621 pp.

ICRISAT (1974). International Workshop on Farming Systems. Hyderabad: International Crops Research Institute for the Semi-Arid Tropics. Mimeo, 548 pp.

ICRISAT (1976). *Annual Report, 1975–76*. Patancheru, India: International Crops Research Institute for the Semi-Arid Tropics, 234 pp.

ICRISAT (1978). *Annual Report, 1977–78*. Patancheru, India: International Crops Research Institute for the Semi-Arid Tropics, 295 pp.

ICRISAT (1979). *Annual Report, 1978–79*. Patancheru, India: International Crops Research Institute for the Semi-Arid Tropics, 288 pp.

ICRISAT (1980). *Annual Report, 1979–80*. Patancheru, India: International Crops Research Institute for the Semi-Arid Tropics, 288 pp.

ICRISAT (1981). *Annual Report, 1980–81*. Patancheru, India: International Crops Research Institute for the Semi-Arid Tropics, 304 pp.

ICRISAT (1985). *Annual Report, 1984*. Patancheru, India: International Crops Research Institute for the Semi-Arid Tropics, 376 pp.

ICRISAT (1986). *Annual Report, 1985*. Patancheru, India: International Crops Research Institute for the Semi-Arid Tropics, 379 pp.

ICRISAT (1987). *Annual Report, 1986*. Patancheru, India: International Crops Research Institute for the Semi-Arid Tropics, 367 pp.

ICRISAT (1988). *Annual Report, 1987*. Patancheru, India: International Crops Research Institute for the Semi-Arid Tropics, 390 pp.

ICRISAT (1989). *Soil, Crop, and Water Management Systems for Rainfed Agriculture in the Sudano-Sahelian Zone*. Proceedings of an International Workshop, 11–16 January 1987, ICRISAT Sahelian Centre, Niamey, Niger. Patancheru, India: International Crops Research Institute for the Semi-Arid Tropics, 385 pp.

ICRISAT (1990). *Annual Report, 1989*. Patancheru, India: International Crops Research Institute for the Semi-Arid Tropics, 355 pp.

IDRC (1980). *Cassava Research Project Progress Report IX, Faculty of Agriculture, Brawijaya University, Indonesia*. Ottawa, Ontario: International Development Research Centre, 94 pp.

Igwilo, N. (1989). Response of yam cultivars to staking and fertilizer application. *Tropical Agriculture (Trinidad)*, **66**, 38–42.

IITA (1980). *Annual Report for 1979*. Ibadan, Nigeria: International Institute of Tropical Agriculture, 152 pp.

IITA (1990). *Cassava in Tropical Africa: A Reference Manual*. Ibadan, Nigeria: International Institute of Tropical Agriculture, 176 pp.

Ike, I.F. (1986). Effects of soil moisture stress on the growth and yield of Spanish variety peanut. *Plant and Soil*, **96**, 297–8.

Ikehashi, H. (1977). New procedures for breeding photoperiod-sensitive deep-water rice with rapid generation advance. In *Deep-water Rice*, pp. 45–54. Los Baños, Philippines: International Rice Research Institute.

Ilag, L.L., Rosales, A.M., Elazegui, R.A. & Mew, T.W. (1987). Changes in the population of infective endomycorrhizal fungi in a rice-based cropping system. *Plant and Soil*, **103**, 67–73.

Iltis, H.H. (1983). From teosinte to maize: the catastrophic sexual transmutation. *Science*, **222**, 886–94.

Iltis, H.H. & Doebley, J. (1980). Taxonomy of *Zea* (Gramineae). II. Subspecific categories in the *Zea mays* complex and a generic synopsis. *American Journal of Botany*, **67**, 994–1004.

Inanaga, S., Utunomiya, M., Horiguchi, T. & Nishihara, T. (1990). Behaviour of fertilizer-N absorbed through root and fruit in peanut. *Plant and Soil*, **122**, 85–9.

Inforzato, R. & Tella, R. de (1960). Sistema radicular do amendoim. *Bragantia*, **19**, 119–23.

Inouye, C. (1953). Influence of temperature on the germination of seeds: IX. *Proceedings of the Crop Science Society of Japan*, **21**, 276–7.

International Seed Testing Association (1976). International. Rules for seed testing rules, 1976. *Seed Science and Technology*, **4**, 3–49.

IPA (1975). *Fertilizer Use and Protein Production*. Proceedings of the 11th Colloquium, Ronne-Bornholm, Denmark, pp. 179–91. Berne, Switzerland: International Potash Institute.

IRAT (1972). *Rapport Annuel d'Activité*, 1971–72. Bambey, Senegal: Institut de Recherches Agronomiques Tropicales.

Iremiren, G.O. (1989). Response of maize to trash burning and nitrogen fertilizer in a newly opened secondary forest. *Journal of Agricultural Science*, **113**, 207–10.

Irikura, Y., Cock, J.H. & Kawano, K. (1979). The physiological basis of genotype–temperature interaction in cassava. *Field Crops Research*, **2**, 227–39.

Irizarry, H., Silva, S. & Vincente-Chandler, J. (1980). Effect of water table level on yield and root system of plantains. *University of Puerto Rico Journal of Agriculture*, **64**, 33–6.

Irizarry, H., Vincente-Chandler, J. & Silva, S. (1981). Root distribution of plantains growing on five soil types. *University of Puerto Rico Journal of Agriculture*, **65**, 29–34.

IRRI (1969). *Annual Report for 1969*. Los Baños, Philippines: International Rice Research Institute.

IRRI (1972). *Rice Breeding*. Los Baños, Philippines: International Rice Research Institute, 738 pp.

IRRI (1975). *Changes in Rice Farming in Selected Areas of Asia*. Los Baños, Philippines: International Rice Research Institute, 377 pp.

IRRI (1976). *Annual Report for 1975*. Los Baños, Philippines: International Rice Research Institute, 418 pp.

IRRI (1977a). *Cropping Systems Research and Development for the Asian Rice Farmer*. Los Baños, Philippines: International Rice Research Institute, 454 pp.

IRRI (1977c). *Annual Report for 1976*. Los Baños, Philippines: International Rice Research Institute, 418 pp.

IRRI (1977b). *Proceedings, 1976 Deep-Water Rice Workshop*. Los Baños, Philippines: International Rice Research Institute, 238 pp.

IRRI (1978a). *Annual Report for 1978*. Los Baños, Philippines: International Rice Research Institute.

IRRI (1978b). *The IRRI Reporter, 1978*. Los Baños, Philippines: International Rice Research Institute.

IRRI (1979a). *Nitrogen and Rice*. Los Baños, Philippines: International Rice Research Institute, 499 pp.

IRRI (1979b). *Annual Report for 1978*. Los Baños, Philippines: International Rice Research Institute, 478 pp.

IRRI (1979c). *Rainfed Lowland Rice: Selected Papers from the 1978 International Rice Research Conference*. Los Baños, Philippines: International Rice Research Institute, 341 pp.

IRRI (1984). *Cropping Systems in Asia: On-farm Research and Management*. Los Baños, Philippines: International Rice Research Institute, 196 pp.

IRRI (1985a). *Soil Physics and Rice*. Los Baños, Philippines: International Rice Research Institute, 430 pp.

IRRI (1985b). *Wetland Soils: Characterization, Classification and Utilization*. Proceedings of an International Workshop, 26 March – 5 April 1984. Los Baños, Philippines: International Rice Research Institute, 558 pp.

IRRI (1988). *Annual Report for 1987*. Los Baños, Philippines: International Rice Research Institute, 640 pp.

IRRI (1989). *Annual Report for 1988*. Los Baños, Philippines: International Rice Research Institute, 646 pp.

Irvine, F.R. (1969). *West African Crops*. Oxford: Oxford University Press, 272 pp.

Irving, H. (1956). Fertilizer experiments with yams in eastern Nigeria, 1947–51. *Tropical Agriculture (Trinidad)*, **33**, 67–78.

Isbell, R.F. (1983). Soil classification problems in the tropics and subtropics. In *The USDA Soil Taxonomy in Relation to Some Soils of Eastern Queenland*, ed. A.W. Moore, pp. 17–26. CSIRO Division of Soils Divisional Report, No. 84. Melbourne: Commonwealth Scientific and Industrial Organization.

Isbell, R.F. (1987). Pedological research in relation to soil fertility. In *Management of Acid Tropical Soils for Sustainable Agriculture*, ed. P.A. Sanchez, E.R. Stoner & P. Pushparajah, pp. 131–46.. International Board for Soil Research and Management proceedings held in Bangkok, Thailand, 24 April – 3 May 1985.

Ishag, H.M. & Agoub, A.T. (1974). Effect of sowing date and soil type on yield, yield components and survival of dry beans (*Phaseolus vulgaris* L.). *Journal of Agricultural Science*, **82**, 343–7.

Islam, A.K.M.S. (1981). Effect of soil pH on yield and mineral nutrition of ginger (*Zingiber officinale* Roscoe). PhD thesis, Department of Agriculture, University of Queensland, St Lucia.

Islam, A.K.M.S., Asher, C.J. & Edwards, D.G. (1987). Response of plants to calcium concentration in flowing solution culture with choride or sulphate as the counter ion. *Plant and Soil*, **98**, 377–95.

Islam, R. (1978). The role of nitrogen fixation in food legume production. In *Food Legume Improvement and Development*, ed. G.C. Hawtin & G.J. Chancellor, pp. 166–9. Ottawa, Ontario: International Development Research Centre.

Ismunadji, M., Blair, G.J., Momuat, E. & Sudjadi, M. (1983). Sulphur in the agriculture of Indonesia. In *Sulphur in South East Asian and South Pacific Agriculture*, ed. G.J. Balir & A.R. Till, pp. 165–79. Canberra, ACT: Australian International Development Assistance Bureau.

Isoi, T. & Yoshida, S. (1991). Low nitrogen fixation of common bean (*Phaseolus vulgaris*). *Soil Science and Plant Nutrition*, **37**, 559–63.

Jackson, G. (1965). *Cattle, Coffee and Land among the Wain*. New Guinea Research Unit Bulletin, No. 8. Canberra, ACT: Australian National University, 69 pp.

Jacobs, B.C. & Pearson, C.J. (1991). Potential yield of maize, determined by rates of growth and development of ears. *Field Crops Research*, **27**, 281–98.

Jain, S.V. & Mathur, C.M. (1961). Efficiency of different nitrogenous fertilizers for bajra production in desert soils of Rajasthan. *Indian Journal of Agronomy*, **53**, 185–98.

Janssen, B.H. & Weert, R. van der (1977). The influence of fertilizers, soil organic matter and soil compaction on maize yields on the Surinam 'Zanderij' soils. *Plant and Soil*, **46**, 445–58.

Jarret, R.L., Gawel, N. & Whittemore, A. (1992). Phylogenetic relationships of the sweetpotato (*Ipomoea batatas* (L.) Lam.). *Journal of the American Society for Horticultural Science*, **117**, 633–7.

Jayaweera, G.R. & Mikkelsen, D.S. (1991). Assessment of ammonia volatilisation from flooded soil systems. *Advances in Agronomy*, **45**, 303–56.

JCRR (1977). *Abstracts on Soil Conservation Research in Taiwan*, vol. 1, 1958–76. Taipei, Taiwan: Joint Commission on Rural Reconstruction.

Jennings, D.L. (1976). Cassava: *Manihot esculenta* (Euphorbiaceae). In *Evolution of Crop Plants*, ed. N.W. Simmonds, pp. 81–4. London: Longman.

Jintakanon, S., Edwards, D.G. & Asher, C.J. (1982). An anomalous, high external phosphorus requirement for young cassava plants in solution culture. In *Fifth International Symposium on Tropical Root and Tuber Crops*, pp. 507–18. Los Baños, Philippines.

Johannessen, C.L. & Parker, A.J. (1989). Maize ears sculptured in 12th and 13th century A.D. India as indicators of pre-Columbian diffusion. *Economic Botany*, **43**, 164–80.

Johnson, D.R. & Luedders, V.D. (1974). Effect of planted seed size on emergence and yield of soybeans (*Glycine max* (L.) Merr.). *Agronomy Journal*, **66**, 117–18.

Johnson, D.R. & Tanner, J.W. (1972). Calculation of the rate and duration of grain filling in corn (*Zea mays* L.). *Crop Science*, **12**, 485–6.

Johnson, J.F., Voorhees, W.B., Nelson, W.W. & Randall, G.W. (1990). Soybean growth and yield as affected by surface and subsoil compaction. *Agronomy Journal*, **82**, 973–9.

Jones, J.P. & Fox, R.L. (1978). Phosphorus nutrition of plants influenced by manganese and aluminium uptake from an Oxisol. *Soil Science*, **126**, 230–6.

Jones, M.J. & Bromfield, A.R. (1970). Nitrogen in the rainfall at Samaru, Nigeria. *Nature (London)*, **227**, 86.

Jones, M.J. & Wild, A. (1975). Soils of the West African Savanna. *Commonwealth Bureau of Soils Technical Communication*, No. 55, 246 pp.

Jones, P.G. & Laing, D.R. (1978). Simulation of the phenology of soybeans. *Agricultural Systems*, 3, 295–311.

Joshi, N.L. (1987). Seedling emergence and yield of pearl millet on naturally crusted arid soils in relation to sowing and cultural methods. *Soil and Tillage Research*, 10, 103–12.

Joshi, P.K., Kulkarni, J.H. & Bhatt, D.M. (1990). Interaction between strains of *Bradyrizobium* and groundnut (*Arachis hypogaea* L.) cultivars. *Tropical Agriculture (Trinidad)*, 67, 115–18.

Juma, N.G. & Tabatabai, M.A. (1988). Hydrolysis of organic phosphates by corn and soybean roots. *Plant and Soil*, 107, 31–8.

Juo, A.S.R. (1989). New farming systems development in the wetter tropics. *Experimental Agriculture*, 25, 145–63.

Juo, A.S.R. & Ezumah, H.C. (1992). Mixed root-crop systems in wet Sub-Saharan Africa. In *Field Crop Ecosystems*, ed. C.J. Pearson, pp. 243–58. Amsterdam: Elsevier.

Juo, A.S.R. & Fox, R.L. (1977). Phosphate sorption of some benchmark soils in West Africa. *Soil Science*, 124, 370–6.

Juo, A.S.R. & Maduakor, H.O. (1974). Phosphate sorption of some Nigerian soils and its effect on cation exchange capacity. *Communications in Soil Science and Plant Analysis*, 5, 479–97.

Kadem, S.S., Kachnave, K.G., Chavan, J.K. & Salunkhe, D.K. (1977). Effect of nitrogen, *Rhizobium* inoculation and simazine on yield and quality of Bengal gram (*Cicer arietinum* L.). *Plant and Soil*, 47, 279–81.

Kaigama, B.K., Teare, I.D., Stone, L.R. & Powers, W.L. (1977). Root and top growth of irrigated and non-irrigated grain sorghum. *Crop Science*, 17, 555–9.

Kailasanthan, K., Rao, G.G. & Sinha, S.K. (1976). Effect of temperature on the partitioning of seed reserves in cowpea and sorghum. *Indian Journal of Plant Physiology*, 19, 171–9.

Kamprath, E.J. (1984). Crop responses to lime on soils in the tropics. In *Soil Acidity and Liming*, 2nd edn. American Society of Agronomy Monograph, No. 12, 380 pp.

Kandiah, A. (1979). Influence of soil properties and crop cover on the erodibility of soils. In *Soil Physical Properties and Crop Production in the Tropics*, ed. R. Lal & D.J. Greenland, pp. 475–88. Chichester: Wiley.

Kang, B.T. (1975). Effects of inoculation and nitrogen fertilizer on soybean in Western Nigeria. *Experimental Agriculture*, 11, 23–31.

Kang, B.T. (1978). Effect of some biological factors on soil variability in the tropics. III. Effect of *Macrotermes* mounds. *Plant and Soil*, 50, 241–51.

Kang, B.T. & Moormann, F.R. (1977). Effect of some biological factors on soil variability in the tropics. I. Effect of preclearing of vegetation. *Plant and Soil*, 47, 441–9.

Kang, B.T. & Wilson, J.E. (1981). Effect of mound size and fertilizer on White Guinea Yam (*Dioscorea rotundata*) in southern Nigeria. *Plant and Soil*, 61, 319–27.

Kang, B.T., Nangju, D. & Ayanaba, A. (1977). Effects of fertilizer use on cowpea and soybean nodulation and nitrogen fixation in the tropics. In *Biological Nitrogen Fixation in Farming Systems of the Tropics*, ed. A. Ayanaba & P.J. Dart, pp. 205–16. Chichester: Wiley.

Kanwar, J.S., Kampen, J. & Virmani, S.M. (1982). Management of Vertisols for maximising crop production: ICRISAT experience. In *Vertisols and Rice Soils of the Tropics*, pp. 94–118. New Delhi, India: Twelfth International Congress of Soil Science.

Kaplan, L. (1965). Archaeology and domestication in American *Phaseolus* beans. *Economic Botany*, **19**, 358–68.

Kaplan, L., Lynch, T.F. & Smith, C.E. (1973). Early cultivated beans (*Phaseolus vulgaris*) from an intermontane Peruvian valley. *Science*, **179**, 76–7.

Karikari, S.K. (1981). Intercropping of plantains, cocoyams and cassava. In *Tropical Root Crops: Research Strategies for the 1980s*, ed. E.R. Terry, K.A. Oduro & F. Caveness, pp. 120–3. Ottawa, Ontario: International Development Research Centre.

Kaspar, T.C., Zahler, J.B. & Timmons, D.R. (1989). Soybean response to phosphorus and potassium fertilizers as affected by soil drying. *Soil Science Society of America Journal*, **53**, 1448–54.

Kassam, A.H. (1976). *Crops of the West African Semi-Arid Tropics*. Hyderabad, India: International Crops Research Institute for the Semi-Arid Tropics. Mimeo, 154 pp.

Kassam, A.H. & Andrews, D.J. (1975). Effect of sowing date on growth, development and yield of photosensitive sorghum at Samaru, northern Nigeria. *Experimental Agriculture*, **11**, 227–40.

Kassam, A.H. & Kowal, J.M. (1975). Water use, energy balance and growth of Gero millet at Samaru, Northern Nigeria. *Agricultural Meteorology*, **15**, 333–42.

Kassam, A.H., Kowal, J.M., Dagg, M. & Harrison, M.N. (1975). Maize in West Africa and its potential in the Savanna areas. *World Crops*, **27**, 75–8.

Katiyar, R.P. (1980). Developmental changes in leaf area index and other growth parameters in chickpea. *Indian Journal of Agricultural Sciences*, **50**, 684–91.

Kawaguchi, K. & Kyuma, K. (1974). Paddy soils in tropical Asia. II. Description of material characteristics. *Southeast Asia Studies*, **12**, 177–92.

Kawano, K., Dasa, P., Amaya, A., Rios, M. & Concalves, W.M.F. (1978). Evaluation of cassava germplasm for productivity. *Crop Science*, **18**, 377–80.

Kay, D.E. (1973). *Root Crops*. London: Tropical Products Institute, 245 pp.

Kay, D.E. (1979). *Food Legumes*. London: Tropical Products Institute, 435 pp.

Kayode, G.O. (1985). Effects of NPK fertilizer on tuber yield, starch content and dry matter accumulation of white Guinea yam (*Dioscorea rotundata*) in a forest Alfisol of South Western Nigeria. *Experimental Agriculture*, **21**, 389–93.

Kayode, G.O. (1986). Further studies on the response of maize to K fertilizer in the tropics. *Journal of Agricultural Science*, **106**, 141–7.

Kayode, G.O. (1987). Potassium requirement of groundnut (*Arachis hypogaea*) in the lowland tropics. *Journal of Agricultural Science*, **108**, 643–7.

Keating, B.A. & Evenson, J.P. (1979). Effect of soil temperature on sprouting and sprout elongation of stem cuttings of cassava (*Manihot esculenta* Crantz.). *Field Crops Research*, **2**, 241–51.

Keating, B.A., Evenson, J.P. & Edwards, D.G. (1982a). Effect of pre-harvest fertilization of cassava (*Manihot esculenta* Crantz.) prior to cutting for planting material on subsequent establishment and root yield. In *Fifth International Symposium on Tropical Root and Tuber Crops*, pp. 301–6.

Keating, B.A., Evenson, J.P. & Fukai, S. (1982b). Environmental effects on growth and development of cassava (*Manihot esculenta* Crantz.). I. Crop development. *Field Crops Research*, **5**, 271–81.

Keating, B.A., Evenson. J.P. & Fukai, S. (1982c). Environmental effects on growth and development of cassava (*Manihot esculenta* Crantz.). II. Crop growth rate and biomass yield. *Field Crops Research*, **5**, 283–92.

Keen, F.G.B. (1978). Ecological relationships in a Hmong (Meo) economy. In *Farmers in the Forest*, ed. P. Kunstadter, E.C. Chapman & S. Sabhasri, pp. 210–21. Honolulu, Hawaii: University of Hawaii Press.

Kehr, A.E., Ting, Y.C. & Miller, J.C. (1953). Induction of flowering in the Jersey type sweet potato. *Proceedings of the American Society for Horticultural Science*, **62**, 437–40.

Kerven, G.L., Asher, C.J., Edwards, D.G. & Ostatek-Boczynski, Z. (1991). Sterile solution culture techniques for aluminium toxicity studies involving organic acids. *Jounal of Plant Nutrition*, **14**, 975–85.

Keyser, H.H., Bohlool, B.B., Hu, T.S. & Weber, D.F. (1982). Fast-growing rhizobia isolated from root nodules of soybean. *Science*, **215**, 1631–2.

Khan, A.A., Thakur, R., HilleRisLambers, D. & Seshu, D.V. (1987). Relationship of ethylene production to elongation in deepwater rice. *Crop Science*, **27**, 1188–96.

Khan, A.G. (1972). The effect of vesicular-arbuscular mycorrhizal associations on growth of cereals. I. Effects on maize growth. *New Phytologist*, **71**, 613–19.

Kimber, A.J. (1972). The sweet potato in subsistence agriculture. *Papua New Guinea Agricultural Journal*, **23**, 80–100.

Kirda, C., Danso, S.K.A., & Zapata, F. (1989). Temporal water stress effects on modulation, nitrogen accumulation and growth of soybean. *Plant and Soil*, **120**, 49–55.

Kirkham, M.B. (1988). Hydraulic resistance of two sorghums varying in drought resistance. *Plant and Soil*, **105**, 19–24.

Klinkenberg, K. & Higgins, G.M. (1968). An outline of Northern Nigerian soils. *Nigerian Journal of Science*, **2**, 91–111.

Knapp, R. (1966). Effect of various temperatures on the germination of tropical and subtropical plants. *Angewandte Botanik*, **32**, 230–41.

Knight, P.T. (1971). *Brazilian Agricultural Technology and Trade*. New York: Praeger, 223 pp.

Koli, S.E. (1973). The response of yam (*D. rotundata*) to fertilizer applications in Northern Ghana. *Journal of Agricultural Science*, **80**, 245–9.

Kondo, M., Kobayashi, M. & Takahashi, E. (1989). Effect of *Azolla* and its utilization in rice culture in Niger. *Plant and Soil*, **120**, 165–89.

Kordan, H.A. (1972). Rice seedlings germinated in water with normal and impeded environmental gas exchange. *Journal of Applied Ecology*, **9**, 527–33.

Kowal, J.M. & Kassam, A.H. (1973). Water use, energy balance and growth of maize at Samaru, northern Nigeria. *Agricultural Meteorology*, **12**, 391–406.

Kowal, J.M. & Kassam, A.H. (1976). Energy load and instantaneous intensity of rainstorms at Samaru, Northern Nigeria. *Tropical Agriculture (Trinidad)*, **53**, 185–98.

Kowal, J.M. & Kassam, H. (1978). *Agricultural Ecology of Savannah*. Oxford: Clarendon Press, 403 pp.

Kpeglo, K.D., Obigbesan, G.O. & Wilson, J.E. (1981). Yield and shelf-life of white yam as influenced by fertilizer. In *Tropical Root Crops: Research Strategies for the 1980s*, ed. E.R. Terry, K.A. Oduro & F. Caveness, pp. 198–202. Ottawa, Ontario: International Development Research Centre.

Krantz, B.A. *et al.* (1974). Cropping patterns for increasing and stabilising agricultural production in the semi-arid tropics. In *International Workshop on Farming Systems*, November 1974, pp. 217–48. Hyderabad, India: International Crops Research Institute for the Semi-Arid Tropics.

Krantz, B.A., Kampen, J. & Russell, M.B. (1978). Soil management differences of Alfisols and Vertisols in the semi-arid tropics. In *Diversity of Soils in the Tropics*, ed. J.J. Nicholaides & L.D. Swindale, pp. 77–95. American Society of Agronomy Special Publication, No. 34.

Krapovickas, A. (1968). Origen, variabilidad y diffusion del mani (*Arachis hypogaea*). In *Actas y Memorias XXXVII Congreso International Americanistas*, vol. 2, pp. 517–34.

Krapovickas, A. (1973). Evolution of the genus *Arachis*. In *Agricultural Genetics*. ed. R. Moav, pp. 135–51. Jerusalem: National Council for Research and Development.

Kretchmer, P.J., Ozbun, J.L., Kapan, S.L., Laing, D.R. & Wallace, D.H. (1977). Red and far-red light effects on climbing in *Phaseolus vulgaris* L. *Crop Science*, **17**, 797–9.

Kucey, R.M.N. & Janzen, H.H. (1987). Effects of VAM and reduced nutrient availability on growth and phosphorus and micronutrient uptake of wheat and field beans under greenhouse conditions. *Plant and Soil*, **104**, 71–8.

Kueneman, E.A. & Camacho, L. (1987). Production and goals for expansion of soybeans in Latin America. In *Soybeans for the Tropics: Research, Production and Utilization*, ed. S.R. Singh, K.O. Rachie & K.E. Dashiell, pp. 125–34. Chichester: Wiley.

Kumar, C.R.M. & Hrishi, N. (1979). Intercropping systems with cassava in Kerala State, India. In *Intercropping with Cassava*, ed. E. Weber, B. Nestel & M. Campbell, pp. 31–4. Ottawa, Ontario: International Development Research Centre.

Kumar, J., Gowda, C.L., Saxena, N.P., Sethi, S.C. & Singh, U. (1980). Effect of salinity on the seed size and germinability of chickpea and protein content. *International Chickpea Newsletter*, **3**, 10.

Kung, P. (1969). Multiple cropping in Taiwan. *World Crops*, **21**, 128–30.

Kung, P. (1971). *Irrigation Agronomy in Monsoon Asia*. FAO-AGPC Miscellaneous Paper, No. 2. Rome: Food and Agriculture Organization, 106 pp.

Kunstadter, P., Chapman, E.C. & Sanga Sabhasri (eds.) (1978). *Farmers in the Forest*. Honolulu, Hawaii: University of Hawaii Press, 402 pp.

Kurian, T. (1976). The effect of supplemental irrigation with sea water on growth and chemical composition of pearl millet. *Pflanzenphysiologie*, **79**, 377–83.

Kvien, C.S., Branch, W.D., Sumner, M.E. & Csinos, A.S. (1988). Pod characteristics influencing calcium concentrations in the seed and hull of peanut. *Crop Science*, **28**, 666–71.

Ladizinsky, G. & Adler, A. (1976). The origin of chickpea, *Cicer arietinum* L. *Euphytica*, **25**, 211–17.

La Favre, A.K., Sinclair, M.J., La Favre, J.S. & Eaglesham, A.R.J. (1991). *Bradyrhizobium japonicum* native to tropical soils: novel sources of strains for inoculants for US-type soya bean. *Tropical Agriculture (Trinidad)*, **68**, 243–9.

Lahav, E. & Turner, D.W. (1983). *Fertilizing for High Yield Bananas*. Berne, Switzerland: International Potash Institute, Bulletin 7, 62 pp.

Laing, D.R., Jones, P.G. & Davis, J.H.C. (1984). Common bean (*Phaseolus vulgaris* L.). In *The Physiology of Field Crops*, ed. P.R. Goldsworthy & N.M. Fisher, pp. 305–52. Chichester: Wiley.

Lal, R. (1976a). *Soil Erosion on Alfisols in Western Nigeria.* IITA Monograph, No. 1. Ibadan, Nigeria: International Institute of Tropical Agriculture, 126 pp.

Lal, R. (1976b). Soil erosion on Alfisols in Western Nigeria. IV. Nutrient element losses in runoff and eroded sediments. *Geoderma,* **16**, 403–17.

Lal, R. (1976c). Soil erosion on Alfisols in Western Nigeria. V. Changes in physical properties and the response of crops. *Geoderma,* **16**, 419–31.

Lal, R. (1979a). Physical properties and moisture retention characteristics of some Nigerian soils. *Geoderma,* **21**, 209–23.

Lal, R. (1979b). Physical characteristics of soils of the tropics: determination and management. In *Soil Physical Properties and Crop Production in the Tropics,* ed. R. Lal & D.J. Greenland, pp. 7–46. Chichester: Wiley.

Lal, R. (1979c). Modification of soil fertility characteristics by management of soil physical properties. In *Soil Physical Properties and Crop Production in the Tropics,* ed. R. Lal & D.J. Greenland, pp. 397–405. Chichester: Wiley.

Lal, R. (1981a). Deforestation of tropical rainforest and hydrological problems. In *Tropical Agricultural Hydrology,* ed. R. Lal & E.W. Russell, pp. 131–40. Chichester: Wiley.

Lal, R. (1981b). Effects of soil moisture and bulk density on growth and development of two cassava cultivars. In *Tropical Root Crops: Research Strategies for the 1980s,* ed. E.R. Terry, K.A. Oduro & F. Caveness, pp. 104–10. Ottawa, Ontario: International Development Research Centre.

Lal, R. (1984). Soil erosion from tropical arable lands and its control. *Advances in Agronomy,* **37**, 183–248.

Lal, R. (1985). Tillage in lowland rice-based cropping systems. In *Soil Physics and Rice,* pp. 284–307. Los Baños: International Rice Research Institute.

Lal, R., Kang, B.T., Moormann, F.R., Juo, A.S.R. & Moomaw, J.C. (1975). Soil management problems and possible solutions in Western Nigeria. In *Soil Management in Tropical America,* ed. E. Bornemisza & A. Alvarado, pp. 372–408. Raleigh, North Carolina: North Carolina State University Press.

Lambert, D.H., Baker, D.E. & Cole, H. (1979). The role of mycorrhizae in the interactions of phosphorus with Zn, Cu and other elements. *Soil Science Society of America Journal,* **43**, 976–80.

Landrau, P. & Samuels, G. (1951). The effect of fertilizers on the yield and quality of sweet potatoes. *University of Puerto Rico Journal of Agriculture,* **35**, 71–8.

Lanzar, E.A., Paris, Q. & Williams, W.A. (1981). A dynamic model for technical and economic analysis of fertilizer recommendations. *Agronomy Journal,* **73**, 733–7.

Lathwell, D.J. (1979a). Crop response to liming of Ultisols and Oxisols. *Cornell International Agriculture Bulletin,* No. 35, 36 pp.

Lathwell, D.J. (1979b). Phosphorus response on Oxisols and Ultisols. *Cornell International Agriculture Bulletin,* No. 33, 40 pp.

Lathwell, D.J. (1990). Legume green manures: principles for management based on recent research. *TropSoils Bulletin No. 90–01,* 30 pp.

Launders, T.E. (1971). The effects of early season soil temperatures on emergence of summer crops in the north-western plains of New South Wales. *Australian Journal of Experimental Agriculture and Animal Husbandry,* **11**, 39–44.

Lauter, D.J., Munns, D.N. & Clarkin, K.L. (1981). Salt response of chickpea as influenced by N supply. *Agronomy Journal,* 73, 961–6.

Lavy, T.L. & Eastin, J.D. (1969). Effect of soil depth and plant age on phosphorus-32 uptake by corn and sorghum. *Agronomy Journal*, **61**, 677–80.

Lawan, M., Barnett, F.L., Khaleeq, B. & Vanderlip, R.L. (1985). Seed density and seed size of pearl millet as related to field emergence and several seed and seedling traits. *Agronomy Journal*, **77**, 567–71.

Lawn, R.J. & Byth, D.E. (1973). Response of soybeans to planting date in South-Eastern Queensland. I. Influence of photoperiod and temperature on phasic developmental patterns. *Australian Journal of Agricultural Research*, **24**, 67–80.

Lawn, R.J. & Byth, D.E. (1974). Responses of soybeans to planting date in South-Eastern Queensland. II. Vegetative and reproductive development. *Australian Journal of Agricultural Research*, **25**, 723–37.

Lawn, R.J. & Byth, D.E. (1979). Soybean. In *Australian Field Crops*, vol. 2, ed. J.V. Lovett & A. Lazenby, pp. 198–231. Sydney: Angus & Robertson.

Lea, D.A.M. (1966). Yam growing in the Maprik area. *Papua and New Guinea Agricultural Journal*, **18**, 5–16.

Leakey, C.L.A. (1972). The effect of plant population and fertility level on yield and its components in two determinate cultivars of *Phaseolus vulgaris* (L.) Savi. *Journal of Agricultural Science*, **79**, 259–67.

Leamy, M.L., Smith. G.D., Colment-Daage, F. & Otowa, M. (1981). The morphological characteristics of Andisols. In *Soils with Variable Charge*, ed. B.K.G. Theng, pp. 17–34. Lower Hutt, New Zealand: New Zealand Society of Soil Science.

Lee, D., Han, X.G. & Jordan, C.F. (1990). Soil phosphorus fractions, aluminium, and water retention as affected by microbial activity in an Ultisol. *Plant and Soil*, **121**, 125–36.

Lee, T.A., Ketring, D.L. & Powell, R.D. (1972). Flowering and growth response of peanut plants (*Arachis hypogaea* L. var. Starr) at two levels of relative humidity. *Plant Physiology*, **49**, 190–3.

Leggett, J.E. & Egli, D.B. (1980). Cation nutrition and ion balance. In *World Soybean Research Conference*, vol. II, ed. F.T. Corbin, pp. 19–34. Boulder, Colorado: Westview Press.

Leihner, D.E. (1982). *Management and Evaluation of Cassava Intercropping Systems.* Cassava Newsletter, No. 11. Cali, Colombia: Centro Internacional de Agricultura Tropical.

Leon, C.A. & Medina, C.J. (1977). *Diagnosis and Recovery of Saline and Sodic Soils for Growing Beans.* Cali, Colombia: Centro Internacional de Agricultura Tropical, 25 pp.

Leong, S.K. & Ong, C.K. (1983). The influence of temperature and soil water deficit on the development and morphology of groundnut (*Arachis hypogaea* L.). *Journal of Experimental Botany*, **34**, 1551–61.

Lepsch, I.F. & Buol, S.W. (1974). Investigations in an Oxisol-Ultisol toposequence in São Paulo State, Brazil. *Soil Science Society of America Proceedings*, **38**, 491–7.

Lepsch, I.F., Buol, S.W. & Daniels, R.B. (1977). Soil landscape relationships in the Occidental Plateau of São Paulo State, Brazil. I. Geomorphic surfaces and soil mapping units. *Soil Science Society of America Proceedings*, **41**, 104–9.

Lim, C.K., Chin, Y.K. & Bolle-Jones, E.W. (1973). Crop indications of nutrient status of peat soil. *Malaysian Agricultural Journal*, **49**, 198–207.

Lins, I.D.G. & Cox, F.R. (1989). Effect of extractant and selected soil properties on predicting the optimum phosphorus fertiliser rate for growing soybeans under field conditions. *Communications in Soil Science and Plant Analysis*, **20**, 319–33.

Litzenberger, S.C. (1973). The improvement of food legumes as a contribution to improved human nutrition. In *Potentials of Field Beans and Other Food Legumes in Latin America*, pp. 3–16. Cali, Colombia: Centro Internacional de Agricultura Tropical.

Locke, M.A. & Hons, F.M. (1988). Tillage effect on seasonal accumulation of labeled fertilizer nitrogen in sorghum. *Crop Science*, **28**, 694–700.

Lockwood, J.G. (1974). *World Climatology: An Environmental Approach*. London: Edward Arnold, 330 pp.

Loftis, S.G. & Kurtz, E.B. (1980). Field studies of inorganic nitrogen added to semi-arid soils by rainfall and blue-green algae. *Soil Science*, **129**, 150–5.

Logsdon, S.D., Reneau, R.B. & Parker, J.C. (1987). Corn seedling root growth as influenced by soil physical properties. *Agronomy Journal*, **79**, 221–4.

Lombin, G. (1983a). Evaluating the micronutrient fertility of Nigeria's semiarid savanna soils. I. Copper and manganese. *Soil Science*, **135**, 377–83.

Lombin, G. (1983b). Evaluating the micronutrient fertility of Nigeria's semiarid savanna soils. II. Zinc. *Soil Science*, **136**, 41–7.

Lombin, L.G. (1981). Continuous cultivation and soil productivity in the semi-arid savannah: the influence of crop rotation. *Agronomy Journal*, **73**, 357–63.

Londono, R.N. de (1977). *Factors Limiting to Bean Production on Farms in Colombia*. Cali, Colombia: Centro Internacional de Agricultura Tropical, 28 pp.

Lopes, E.S. (1977). Ecology of legume–*Rhizobium* symbiosis. In *Limitations and Potentials for Biological Nitrogen Fixation in the Tropics*, ed. J. Dobereiner, R.H. Burris & A. Hollaender, pp. 173–90. New York: Plenum Press.

Lopez, A.S. & Cox, F.R. (1977). A survey of fertility status of surface soils under Cerrado vegetation in Brazil. *Soil Science Society of America Journal*, **41**, 742–7.

Lorenzi, J.O., Normanha, E.S. & Conceicao, A.J. de (1980). Cassava production and planting systems in Brazil. In *Cassava Cultural Practices*, ed. E.J. Weber, J.C. Toro & M. Graham, pp. 38–43. Ottawa, Ontario: International Development Research Centre.

Lowe, S.B. & Wilson, L.A. (1974a). Comparative analysis of tuber development in six sweet potato (*Ipomoea batatas* (L.) Lam.) cultivars: I. *Annals of Botany*, **38**, 307–17.

Lowe, S.B. & Wilson, L.A. (1974b). Comparative analysis of tuber development in six sweet potato (*Ipomoea batatas* (L.) Lam.) cultivars: II. *Annals of Botany*, **38**, 319–26.

Lowe, S.B. & Wilson, L.A. (1975). Yield and yield components of six sweet potato (*Ipomoea batatas*) cultivars: I. *Experimental Agriculture*, **11**, 39–48.

Lozano, J.C., Cock, J.H. & Castano, J. (1978). New developments in cassava storage. In *Proceedings of the Cassava Protection Workshop, Cali, Colombia, 1977*, ed. T. Brekelbaum, A. Bellotti & J.C. Lozano, pp. 135–41. Cali, Colombia: Centro Internacional de Agricultura Tropical.

Lucas, E.O. (1986). The effect of density and nitrogen fertilizer on the growth and yield of maize (*Zea mays* L.) in Nigeria. *Journal of Agricultural Science*, **107**, 573–8.

Lugo, A.E. & Sanchez, M.J. (1986). Land use and organic carbon content of some subtropical soils. *Plant and Soil*, **96**, 185–96.

Lugo-Lopez, M.A., Badillo-Feliciano, J. & Calduch, L. (1977). Response of native white beans *Phaseolus vulgaris* to various N levels in an Oxisol. *University of Puerto Rico Journal of Agriculture*, **61**, 438–42.

Lynch, J., Lauchli, A. & Epstein, E. (1991). Vegetative growth of the common bean in response to phosphorus nutrition. *Crop Science*, **31**, 380–7.

Ma, J. & Takahashi, E. (1991). Availability of rice straw Si to rice plants. *Soil Science and Plant Nutrition*, **37**, 111–16.

Madsen, H.B. & Holst, K.A. (1990). Mapping of irrigation need based on computerised soil and climatic data. *Agricultural Water Management*, **17**, 391–407.

Maduakor, H.O., Lal, R. & Opara-Nadi, O.A. (1984). Effects of methods of seedbed preparation and mulching on the growth and yield of white yam (*Dioscorea rotundata*) on an ultisol in south-east Nigeria. *Field Crops Research*, **9**, 119–30.

Maesen, L.J.G. van der (1972). *Cicer. A Monograph on the Genus with Special Reference to the Chickpea (Cicer arietinum), its Ecology and Cultivation.* Mededelingen Landbouwhogeschool, Wageningen, Publication No. 72/10. Wageningen, The Netherlands: Veenman & Zonen, 342 pp.

Maesen, L.J.G. van der (1984). Taxonomy, distribution and evolution of chickpea and its wild relatives. In *Genetic Resources and Their Exploitation: Chickpea, Faba Beans and Lentils*, ed. J.R. Witcombe & W. Erskine, pp. 95–104. The Hague: Martinus Nijhoff/Junk.

Magalhaes, A.C., Montojos, J.C. & Miyasaka, S. (1971). Effect of dry organic matter on growth and yield of beans (*Phaseolus vulgaris* L.). *Experimental Agriculture*, **7**, 137–43.

Magoon, M.L., Krishnan, R. & Bai, K.V. (1969). Morphology of the pachytene chromosomes and meiosis in *Manihot esculenta*. *Cytologia*, **34**, 612–24.

Mahalakshmi, V. & Bidinger, F.R. (1985). Water stress and time of floral initiation in pearl millet. *Journal of Agricultural Science*, **105**, 437–45.

Mahalakshmi, V., Bidinger, F.R. & Raju, D.S. (1991). Effect of drought stress during grain filling in near-isogenic tall and dwarf hybrids of pearl millet (*Pennisetum glaucum*). *Journal of Agricultural Science*, **116**, 67–72.

Mahler, R.L. & Wollum, A.G. (1981). The influence of soil water potential and soil texture on the survival of *Rhizobium japonicum* and *Rhizobium leguminosarum* isolates in soil. *Soil Science Society of America Journal*, **45**, 761–6.

Mahon, J.D., Lowe, S.B. & Hunt, L.A. (1976). Photosynthesis and assimilate distribution in relation to yield of cassava grown in controlled environments. *Canadian Journal of Botany*, **54**, 1322–31.

Major, D.J., Johnson, D.R., Tanner, J.W. & Anderson, I.C. (1975). Effects of daylength and temperature on soybean development. *Crop Science*, **15**, 174–9.

Malima, V.F. (1976). Banana and plantain growing in Tanzania. *Fruits*, **31**, 651–4.

Mallik, M.A.B. & Tesfai, K. (1988). Allelopathic effect of common weeds on soybean growth and soybean–*Bradyrhizobium* symbiosis. *Plant and Soil*, **112**, 177–82.

Manchanda, H.R. & Sharma, S.K. (1989). Tolerance of chloride and sulphate salinity in chickpea (*Cicer arietinum*). *Journal of Agricultural Science*, **113**, 407–10.

Mangelsdorf, P.C., MacNeish, R.S. & Galinat, W.C. (1967). Prehistoric wild and cultivated maize. In *The Prehistory of the Tehuacan Valley*, ed. D.S. Byers, pp. 178–200. Austin, Texas: University of Texas Press.

Mann, J.D. & Jaworski, E.G. (1970). Comparison of stresses which may limit soybean yields. *Crop Science*, **10**, 620–4.

Mansfield, J.E. (1979). Land capability for annual rainfed arable crops in Northern Nigeria based on soil physical limitations. In *Soil Physical Properties and Crop Production in the Tropics*, ed. R. Lal & D.J. Greenland, pp. 407–26. Chichester: Wiley.

Manshard, W. (1974). *Tropical Agriculture*. London: Longman, 226 pp.

Marchal, J. & Mallessard, R. (1979). Comparaison des immobilisations minérales de quatre cultivars de bananiers à fruits pour cuisson et de deux 'Cavendish'. *Fruits*, **34**, 373–92.

Marsh, K.B., Tillman, R.W. & Syers, J.K. (1987). Charge relationships of sulphate sorption by soils. *Soil Science Society of America Proceedings*, **51**, 318–23.

Martin, F.W. (1972). Yam production methods. US *Department of Agriculture, Production Research Report*, No. 147, 17 pp.

Martin, F.W. (1985). Differences among sweet potatoes in response to shading. *Tropical Agriculture (Trinidad)*, **62**, 161–3.

Martin, F.W. & Jones, A. (1972). The species of *Ipomoea* closely related to the sweet potato. *Economic Botany*, **26**, 201–15.

Martin, G. (1959). The decalcification of soils in Niari: effect of calcareous amendments. *Oleagineaux*, **14**, 213–20.

Martin, J.H., Leonard, W.L. & Stamp, D.L. (1976). *Principles of Field Crop Production.* New York: Macmillan, 1118 pp.

Martin-Prevel, P. (1967). Etude dynamique des éléments minéreaux dans la nutrition d'une plante cultivée: le bananier. *Bulletin de la Societé Française de Physiologie Végétale*, **13**, 3–17.

Martin-Prevel, P. (1978). Effects of magnesium and potassium nutrition on phosphorus uptake and redistribution in a cultivated plant, *Musa* sp. In *Plant Nutrition*, ed. A.R. Ferguson, R.L. Bieleski & I.B. Ferguson, pp. 329–38. Wellington, New Zealand: New Zealand Department of Scientific and Industrial Research.

Martin-Prevel, P. (1980*a*). La nutrition minérale du bananier dans le monde: I. *Fruits*, **35**, 503–18.

Martin-Prevel, P. (1980*b*). La nutrition minérale du bananier dans le monde: II. *Fruits*, **35**, 583–93.

Martin-Prevel, P. & Montagut, G. (1966). Essais sol-plante sur bananiers. *VIII.* Dynamique de l'azote dans la croissance et le développement du végétal. *Fruits*, **21**, 283–94.

Martini, J.A. & Luzuriaga, C. (1989). Classification and productivity of six Costa Rican Andepts. *Soil Science*, **147**, 326–38.

Martini, J.A., Kochhann, R.A., Siqueira, O.J. & Borkert, C.M. (1974). Response of soybeans to liming as related to soil acidity, Al and Mn toxicities and P in some Oxisols of Brazil. *Soil Science Society of America Proceedings*, **38**, 616–20.

Mason, S.C., Leihner, D.E. & Vorst, J.J. (1986). Cassava–cowpea and cassava–peanut intercropping I. Yield and land use efficiency. *Agronomy Journal*, **78**, 43–6.

Mason, W.K., Rowse, H.R., Bennie, A.T.P., Kasper, T.C. & Taylor, H.M. (1982). Responses of soybeans to two row spacings and two soil water levels: II. *Field Crops Research*, **5**, 15–29.

Mass, E.V. & Hoffman, G.J. (1977). Crop salt tolerance: current assessment. *Journal of Irrigation Drainage Division, Proceedings of the American Society of Civil Engineers*, **103**, 115–34.

Matheny, T.A. & Hunt, P.G. (1983). Effects of irrigation on accumulation of soil and symbiotically fixed N by soybean grown on a Norfolk loamy sand. *Agronomy Journal*, **75**, 719–22.

McClelland, T.B. (1928). Studies on the photoperiodism of some economic plants. *Journal of Agricultural Research*, **37**, 603–28.

McClure, J.W. & Harvey, C. (1962). Use of radiophosphorus in measuring root growth of sorghums. *Agronomy Journal*, **54**, 457–9.

McGinnis, J.T. & Golley, F.B. (1967). *Atlantic Pacific Interocean Canal. Phase 1, Bioenvironmental and Radiological Safety Feasibility Studies.* Columbus, Ohio: Batelle Memorial Institute.

McIntosh, J.L. & Effendi, S. (1979). Soil fertility implications of cropping patterns and practices for cassava. In *Intercropping with Cassava*, ed. E. Weber, B. Nestel & M. Campbell, pp. 77–85. Ottawa, Ontario: International Development Research Centre.

McIntosh, J.L., Effendi, S. & Syarifuddin, A. (1977). Testing cropping patterns for upland conditions. In *Cropping Systems Research and Development for the Asian Rice Farmer*, pp. 201–21. Los Baños, Philippines: International Rice Research Institute.

McLeod, D.E. (1966). Light relations and photosynthesis within plant communities. In *Proceedings of the 9th International Grassland Congress, São Paulo*, pp. 511–17. São Paulo Departamento da Producao Animal da Secretaria da Agricultura do Estadode de São Paulo.

McNeil, D.L., Croft, L. & Sandhu, T.S. (1981). Response of chickpeas to inoculation with *Rhizobium* in Hawaii. *International Chickpea Newsletter*, 4, 25–6.

Medeiros, R.B., Saibro, J.C. de & Jacques, A.V. (1978). Efeito do nitrogenio e da populacao de plantas no rendimento e qualidade do milheto (*Pennisetum americanum* Schum.). *Revista da Sociedade Brasileira de Zootecnia*, 7, 276–85.

Mehta, B.V. & Shah, C.C. (1958). Accumulation and movement of minerals in 'bajiri' (*Pennisetum typhoideum*). *Indian Journal of Agricultural Science*, 28, 583–95.

Meiners, J.P. & Elden, T.C. (1980). Resistance to insects and diseases in *Phaseolus*. In *Advances in Legume Science*, ed. R.J. Summerfield & A.H. Bunting, pp. 359–64. Kew: Royal Botanic Gardens.

Meisner, C.A. & Gross, H.D. (1980). Some guidelines for the evaluation of the need for and response to inoculation of tropical legumes. *North Carolina Agricultural Research Service Technical Bulletin*, No. 265, 59 pp.

Melillo, J.M., Aber, J.D., Linkins, A.E., Ricca, A., Fry, B. & Nadelhoffer, K.J. (1989). Carbon and nitrogen dynamics along the decay continuum: plant litter to soil organic matter. *Plant and Soil*, 115, 189–98.

Mendel, K. & Kirkby, E.A. (1978). *Principles of Plant Nutrition.* Berne, Switzerland: International Potash Institute.

Menendez, T. & Shepherd, K. (1975). Breeding new bananas. *World Crops*, 27, 104–12.

Merrill, S.D. & Rawlins, S.L. (1979). Distribution and growth of sorghum roots in response to irrigation frequency. *Agronomy Journal*, 71, 738–45.

Messing, J.H.L. (1974). Long term changes in potassium, magnesium and calcium content of banana plants and soils in the Windward Islands. *Tropical Agriculture (Trinidad)*, 51, 154–60.

Meyer, W.S. & Green, G.C. (1981). Comparison of stomatal action of orange, soybean and wheat under field conditions. *Australian Journal of Plant Physiology*, 8, 65–76.

Meyer, W.S. & Mateos, L. (1990). Effects of soil type on soybean crop water use in weighing lysimeters. III. Effect of lysimeter canopy height discontinuity on evaporation. *Irrigation Science*, 11, 233–7.

Meyer, W.S., Dugas, W.A., Barrs, H.D., Smith, R.C.G. & Fleetwood, F.J. (1990). Effects of soil type on soybean crop water use in weighing lysimeters. I. Evaporation. *Irrigation Science*, 11, 69–75.

Middleton, N.J. & Thomas, D.S.G. (1992). *World Atlas of Desertification.* London: United Nations Environment Programme/Edward Arnold.

Miege, J. (1957). Influence de quelques caractères des tubercules semences sur la levée et le rendement des ignames cultivées. *Journal of Tropical Agriculture and Applied Botany,* **4**, 315–42.

Mikkelsen, D.S. (1987). Nitrogen budgets in flooded soils used for rice production. *Plant and Soil,* **100**, 71–97.

Mikkelsen, D.S., Freitas, L.M. & McClung, A.C. (1963). *Effects of Liming and Fertilizing Cotton, Corn and Soybeans on Campo Cerrado Soils, State of São Paulo, Brazil.* IRI Research Institute Bulletin, No. 29.

Mills, H.A. & McElhannon, W.S. (1982). Nitrogen uptake by sweet corn. *Horticultural Science,* **17**, 743–44.

Milthorpe, F.L. & Moorby, J. (1979). *An Introduction to Crop Physiology,* 2nd edn. Cambridge: Cambridge University Press, 244 pp.

Minchin, F.R., Summerfield, R.J., Hadley, P. & Roberts, E.H. (1980). Growth, longevity and nodulation of roots in relation to seed yield in chickpeas (*Cicer arietinum*). *Experimental Agriculture,* **16**, 241–61.

Miracle, M.P. (1965). Introduction and spread of maize in Africa. *Journal of African History,* **6**, 39–55.

Miracle, M.P. (1967). *Agriculture in the Congo Basin.* Madison, Wisconsin: University of Wisconsin Press, 355 pp.

Mitsui, S. (1954). *Inorganic Nutrition, Fertilization and Soil Amelioration for Lowland Rice.* Tokyo: Yokendo, 107 pp.

Mokwunye, U. (1979). Phosphorus needs of soils and crops of the Savanna zones of Nigeria. *Phosphorus in Agriculture,* **76**, 87–95.

Momen, N.N., Carlson, R.E., Shaw, R.H. & Arjmand, O. (1979). Moisture stress effects on yield components of two soybean cultivars. *Agronomy Journal,* **71**, 86–90.

Moncur, M.W. (1980). Floral initiation in chickpea. *International Chickpea Newsletter,* **2**, 10–11.

Money, D.C. (1978). *Climate, Soils and Vegetation.* Slough: University Tutorial Press, 272 pp.

Monnier, G. (1965). Effect of organic matter on soil structural stability. *African Soils,* **10**, 29–42.

Monteith, J.L. (1972). Solar radiation and productivity in tropical ecosystems. *Journal of Applied Ecology,* **9**, 747–66.

Monteith, J.L. (1976). *Vegetation and the Atmosphere,* vols. 1 and 2. London: Academic Press, 278 & 440 pp.

Monteith, J.L. (1980). *Microclimatology in Tropical Agriculture.* Report No. 4. Sutton Bonington: University of Nottingham School of Agriculture. Mimeo, 123 pp.

Monteith, J.L., Gregory, P.J., Marshall, B., Ong, C.K., Saffell, R.A. & Squire, G.R. (1981). Physical measurements in crop physiology. I. Growth and gas exchange. *Experimental Agriculture,* **17**, 113–26.

Montojos, J.C. & Magalhaes, A.C. (1971). Growth analysis of dry beans (*Phaseolus vulgaris* L. var. Pintado) under varying conditions of solar radiation and nitrogen application. *Plant and Soil,* **35**, 217–23.

Moomaw, J.C., Park, H.G. & Shanmugasundaram, S. (1977). Role of legumes in south and southeast Asia. In *Exploiting the Legume–Rhizobium Symbiosis in Tropical Agriculture,* pp. 155–66. Honolulu, Hawaii: University of Hawaii Press.

Moore, P.A. & Patrick, W.H. (1989*a*). Iron availability and uptake in acid sulphate soils. *Soil Science Society of America Journal*, **53**, 471–6.

Moore, P.A. & Patrick, W.H. (1989*b*). Calcium and magnesium availability and uptake by rice in acid sulfate soils. *Soil Science Society of America Journal*, **53**, 816–22.

Moore, P.A., Attanandana, T. & Patrick, W.H. (1990). Factors affecting rice growth on acid sulfate soils. *Soil Science Society of America Journal*, **54**, 1651–6.

Moormann, F.R. (1963). Acid sulphate soils (cat-clays) of the tropics. *Soil Science*, **95**, 271–5.

Moormann, F.R. & Breeman, N. van (1979). *Rice: Soil, Water, Land*. Los Baños, Philippines: International Rice Research Institute, 185 pp.

Moraghan, J.T. (1984). Differential response of five species to phosphorus fertilisers. *Communications in Soil Science and Plant Analysis*, **15**, 437–47.

Moreau, B. (1965). La croissance et le développement du bananier 'Gros Michel' en Equateur. *Fruits*, **20**, 201–20.

Moreno, R.A. & Hart, R.D. (1979). Intercropping with cassava in Central America. In *Intercropping with Cassava*, ed. E. Weber, B. Nestel & M. Campbell, pp. 17–24. Ottawa, Ontario: International Development Research Centre.

Morgan, J. (1984). Osmoregulation and water stress in higher plants. *Annual Review of Plant Physiology*, **35**, 299–319.

Morgan, W.B. (1969). Peasant agriculture in tropical Africa. In *Environment and Land Use in Africa*, ed. M.F. Thomas & G.W. Whittington, pp. 241–72. London: Methuen.

Morris, H.D. & Pierre, W.H. (1949). Minimum concentrations of manganese necessary for injury to various legumes in culture solutions. *Agronomy Journal*, **41**, 107–12.

Morris, R.A. & Zandstra, H.G. (1979). Land and climate in relation to cropping patterns. In *Rainfed Lowland Rice: Selected Papers from the 1978 International Rice Research Conference*, pp. 255–74. Los Baños, Philippines: International Rice Research Institute.

Morris, R.A., Furoc, R.E., Rajbhandari, N.K., Marqueses, E.P. & Dizon, M.A. (1989). Rice responses to waterlog-tolerant green manures. *Agronomy Journal*, **81**, 803–9.

Mortley D.G. & Hill, W.A. (1990). Sweetpotato growth and nitrogen content following nitrogen application and inoculation with *Azospirillum*. *Hortscience*, **25**, 758–9.

Mosse, B. (1981). *Vesicular-Arbuscular Mycorrhiza Research for Tropical Agriculture*. Research Bulletin, No. 194. Hawaiian Institute of Tropical Agriculture and Human Resources, University of Hawaii, 82 pp.

Muchow, R.C., Fisher, M.J., Ludlow, M.M. & Myers, R.J.K. (1980). Stomatal behaviour of kenaf and sorghum in a semiarid tropical environment: II. *Australian Journal of Plant Physiology*, **7**, 621–8.

Muchow, R.C., Robertson, M.J. & Pengelly, B.C. (1993). Radiation-use efficiency of soybean, mungbean and cowpea under different environmental conditions. *Field Crops Research*, **32**, 1–16.

Muhammed, S., Akbar, M. & Neue, H.U. (1987). Effect of Na/Ca and Na/K ratios in saline culture solution on the growth and mineral nutrition of rice (*Oryza sativa* L.). *Plant and Soil*, **104**, 57–62.

Mukasa, S.K. & Thomas, D.G. (1970). Staple food crops. In *Agriculture in Uganda*, ed. J.D. Jameson, pp. 139–53. Oxford: Oxford University Press.

Muldoon, D.K. & Pearson, C.J. (1979). The hybrid between *Pennisetum americanum* and *Pennisetum purpureum*. *Herbage Abstracts*, **49**, 189–99.

Munevar, F. & Wollum, A.G. (1981). Effect of high root temperature and *Rhizobium* strain on nodulation, nitrogen fixation and growth of soybeans. *Soil Science Society of America Journal*, **45**, 1113–20.

Munns, D.N. & Fox, R.L. (1977). Comparative lime requirements of tropical and temperate legumes. *Plant and Soil*, **46**, 533–48.

Munns, D.N., Fox, R.L. & Koch, B.L. (1977). Influence of lime on nitrogen fixation by tropical and temperate legumes. *Plant and Soil*, **46**, 591–601.

Munns, D.N., Hohenberg, J.S., Righetti, T.L. & Lauter, D.J. (1981). Soil acidity tolerance of symbiotic and nitrogen-fertilized soybeans. *Agronomy Journal*, **73**, 407–10.

Murata, Y. (1975). The effect of climatic factors and ageing on net assimilation rate of crop stands. In *Japanese International Biological Programme Synthesis: Crop Productivity and Solar Energy Utilization in Various Climates in Japan*, ed. Y. Murata, pp. 172–86. Tokyo: University of Tokyo Press.

Murata, Y. & Matsushima, S. (1975). Rice. In *Crop Physiology*, ed. L.T. Evans, pp. 73–99. Cambridge: Cambridge University Press.

Murray, D.B. (1961). Shade and fertilizer relations in the banana. *Tropical Agriculture (Trinidad)*, **38**, 123–32 .

Murtha, G.G. (1986). Soil temperature regimes: a tropical experience. In *The USDA Soil Taxonomy in Relation to Some Soils of Eastern Queensland*, ed. A.W. Moore, pp. 23–7. CSIRO Division of Soils Divisional Report, No. 84. Melbourne: Commonwealth Scientific and Industrial Research Organisation.

Murty, B.R. (1975). Biology of adaptation in chickpea. In *International Workshop on Grain Legumes*, pp. 239–52. Hyderabad, India: International Crops Research Institute for the Semi-Arid Tropics.

Murumkar, C. V. & Chavan, P.D. (1986). Influence of salt stress on biochemical processes in chickpea, *Cicer arietinum* L. *Plant and Soil*, **96**, 439–43.

Mutsaers, H.J.W. (1978). Mixed cropping experiments with maize and groundnuts. *Netherlands Journal of Agricultural Science*, **26**, 344–53.

Myers, R.J.K. (1978*a*). Nitrogen and phosphorus nutrition of dryland grain sorghum at Katherine, Northern Territory. I. Effect of rate of nitrogen fertilizer. *Australian Journal of Experimental Agriculture and Animal Husbandry*, **18**, 554–63.

Myers, R.J.K. (1978*b*). Nitrogen and phosphorus nutrition of dryland sorghum at Katherine, Northern Territory. II. Effect of rate and source of phosphorus fertilizer. *Australian Journal of Experimental Agriculture and Animal Husbandry*, **18**, 564–72.

Myers, R.J.K. (1980). The root system of a grain sorghum crop. *Field Crops Research*, **3**, 53–64.

Myers, R.J.K. & Asher, C.J. (1982). Mineral nutrition of grain sorghum: macronutrients. In *Sorghum in the Eighties*, ed. L.R. House, L.K. Mughogho & J.M. Peacock, pp. 161–77. Hyderabad, India: International Crops Research Institute for the Semi-Arid Tropics.

Nagarajrao, Y., Mallick, S. & Singh, G. (1980). Moisture depletion and root growth of different varieties of chickpea under rainfed conditions. *Indian Journal of Agronomy*, **25**, 289–93.

Nambiar, P.T.C. & Rao, B.S. (1987). Effect of sowing depth on nodulation, nitrogen fixation, root and hypocotyl growth, and yield in groundnut (*Arachis hypogaea*). *Experimental Agriculture*, **23**, 283–91.

Nambiar, P.T.C., Rego, T.J. & Rao, B.S. (1986). Comparison of the requirements and utilization of nitrogen by genotypes of sorghum (*Sorghum bicolor* (L.) Moench), and nodulating groundnut (*Arachis hypogaea* L.). *Field Crops Research*, **15**, 165–79.

Namdeo, S.I., Gupta, S.C., Jain, R.C. & Kakran, M.S. (1989). Response of chickpea genotypes to inoculation with *Rhizobium* strains under rainfed conditions. *Legume Research*, **12**, 98–100.

Nandpuri, K.S., Dhillon, R.S. & Singh, S. (1971). The influence of fertilizers and irrigation on growth and yield of sweet potatoes. *Indian Journal of Horticulture*, **28**, 139–43.

Nangju, D. (1979). Seed characters and germination in soybean. *Experimental Agriculture*, **15**, 385–92.

Nangju, D. (1980). Soybean response to indigenous rhizobia as influenced by cultivar origin. *Agronomy Journal*, **72**, 403–6.

Nayar, N.M. (1973). Origin and cytogenetics of rice. *Advances in Genetics*, **17**, 153–292.

Negeve, J.M. & Roncadori, R.W. (1985). The interaction of vesicular-arbuscular mycorrhizae and soil phosphorus fertility on growth of sweet potato (*Ipomoea batatas*). *Field Crops Research*, **12**, 181–5.

Nembozanga Sauti, R.F. (1981). Problems of cassava production in Malawi. In *Tropical Root Crops: Research Strategies for the 1980s*, ed. E.R. Terry, K.A. Oduro & F. Caveness, pp. 95–8. Ottawa, Ontario: International Development Research Centre.

Netting, R. McC. (1968). *Hill Farms of Nigeria*. Seattle, Washington: University of Washington Press, 251 pp.

Newell, R.E., Kidson, J.W., Vincent, D.G. & Boer, G.J. (1974). *The General Circulation of the Tropical Atmosphere*, vols. 1 and 2. Cambridge, Mass.: Massachusetts Institute of Technology Press, 278 & 371 pp.

Newton, K. (1960). Shifting cultivation and crop rotation in the tropics. *Papua New Guinea Agricultural Journal*, **13**, 81–118.

Newton, K. & Jamieson, G.I. (1968). Cropping and soil fertility studies at Keravat, New Britain, 1954–62. *Papua New Guinea Agricultural Journal*, **20**, 25–51.

Ngeve, J.M., Hahn, S.K. & Bouwkamp, J.C. (1992). Effects of altitude and environment on sweet potato yield in Cameroon. *Tropical Agriculture (Trinidad)*, **69**, 43–8.

Ngongi, A.G.N., Howeler, R. & MacDonald, H.A. (1977). Effects of potassium and sulphur on growth, yield and composition of cassava. In *Fourth Symposium of the International Society for Tropical Root Crops*, ed. J.H. Cock, R. MacIntyre & M. Graham, pp. 107–13. Ottawa, Ontario: International Development Research Centre.

Nicholaides, J.J. & Cox, F.R. (1970). Effect of mineral nutrition on chemical composition and early reproductive development of Virginia type peanuts (*Arachis hypogaea* L.). *Agronomy Journal*, **62**, 262–5.

Nicholaides, J.J., Bandy, D.E., Sanchez, P.A., Benites, J.R., Villachica, J.H., Couto, A.J. & Valverde, C.S. (1985). Agricultural alternatives for the Amazon Basin. *BioScience*, **35**, 279–85.

Nicou, R. & Chopart, J.L. (1979). Root growth and development in sandy and sandy clay soils of Senegal. In *Soil Physical Properties and Crop Production in the Tropics*, ed. R. Lal & D.J. Greenland, pp. 375–85. Chichester: Wiley.

Nishimoto, R.K., Fox, R.L. & Parvin, P.E. (1977). Response of vegetable crops to phosphorus concentrations in soil solution. *Journal of the American Society of Horticultural Science*, **102**, 705–9.

Nishiyama, I. (1971). Evolution and domestication of the sweet potato. *Botanical Magazine (Tokyo)*, **84**, 377–87.

Njoku, A. (1963). The propagation of yams (*Dioscorea* spp.) by vine cuttings. *Journal of the West African Science Association*, **8**, 29–32.

Njoku, E., Nwoke F.I.O., & Okonkwo, S.N.C. (1984). Pattern of growth and development in *Dioscorea rotundata* Poir. *Tropical Agriculture (Trinidad)*, **61**, 17–19.

Noble, A.D., Fey, M.V. & Sumner, M.E. (1988). Calcium–aluminium balance and the growth of soybean roots in nutrient solutions. *Soil Science Society of America Journal*, **52**, 1651–6.

Noorsyamsi, H. & Sarwani, M. (1989). Mangement of tidal swampland for food crops: southern Kalimantan experiences. *Indonesian Agricultural Research Development Journal*, **11**, 18–24.

Norman, D.W. (1972). *An Economic Study of Three Villages in Zaria Province: II. Input–output study*, vol. I (text). Samaru Miscellaneous Paper, No. 37. Samaru, Nigeria: Institute for Agricultural Research.

Norman, M.J.T. (1966). *Katherine Research Station 1956–64: A Review of Published Work*. CSIRO Division of Land Research Technical Paper, No. 28. Melbourne: Commonwealth Scientific and Industrial Research Organization.

Norman, M.J.T. (1978). Energy inputs and outputs of subsistence cropping systems in the tropics. *Agro-Ecosystems*, **4**, 355–66.

Norman, M.J.T. (1979). *Annual Cropping Systems in the Tropics*. Gainesville, Florida: University Presses of Florida, 276 pp.

Norman, M.J.T. & Begg, J.E. (1968). Bulrush millet (*Pennisetum typhoides* (Burm.) S. & H.) at Katherine, N.T.: a review. *Journal of the Australian Institute of Agricultural Science*, **34**, 59–68.

Northwood, P.J. & Macartney, J.C. (1971). The effect of different amounts of cultivation on the growth of maize on some soil types in Tanzania. *Tropical Agriculture (Trinidad)*, **48**, 25–33.

Nunozawa, F. & Tanaka, A. (1984). Comparision of nutrient management among acid soils differing in phosphorus fixation capacity and cation exchange capacity. *Soil Science and Plant Nutrition*, **30**, 51–61.

Nutman, P.S. (1976). IBP field experiments on nitrogen fixation by nodulated legumes. In *Symbiotic Nitrogen Fixation in Plants*, ed. P.S. Nutman, pp. 211–37. Cambridge: Cambridge University Press.

Nwoke, F.I.O. & Okonkwo, S.N.C. (1978). Effects of periodic removal of mother tubers on yield of *Dioscorea rotundata*. *Experimental Agriculture*, **14**, 145–50.

Nye, M.M. (1991). The mis-measure of manioc (*Manihot esculenta*, Euphorbiaceae). *Economic Botany*, **45**, 47–57.

Nye, P.H. (1954). Fertilizer responses in the Gold Coast in relation to time and method of application. *Empire Journal of Experimental Agriculture*, **22**, 101–11.

Nye, P.H. & Greenland, D.J. (1960). *The Soil Under Shifting Cultivation*. Farnham Royal: Commonwealth Agricultural Bureaux, 156 pp.

Obiefuna, J.C. (1983). Effect of propagule type and depth of planting on the yield decline of plantains (*Musa* AAB) in the rain forest belt of Nigeria. *Tropical Agriculture (Trinidad)*, **60**, 107–10.

Obigbesan, G.O. (1977a). Investigations on Nigerian root and tuber crops: response to cassava cultivars to potassium fertilizer in Western Nigeria. *Journal of Agricultural Science*, **89**, 23–7.

Obigbesan, G.O. (1977*b*). Investigations on Nigerian root and tuber crops: effect of potassium on starch yields, HCN content and nutrient uptake of cassava cultivars (*Manihot esculenta*). *Journal of Agricultural Science*, **89**, 29–34.

Obigbesan, G.O. & Agboola, A.A. (1978). Uptake and distribution of nutrients by yams (*Dioscorea* spp.) in western Nigeria. *Experimental Agriculture*, **14**, 349–55.

Obigbesan, G.O. & Fayemi, A.A.A. (1976). Investigations on Nigerian root and tuber crops: influence of nitrogen fertilization on the yield and chemical composition of two cassava cultivars. *Journal of Agricultural Science*, **86**, 401–6.

Obigbesan, G.O., Agboola, A.A. & Fayemi, A.A.A. (1977). Effect of potassium on tuber yield and nutrient uptake in yams. In *Fourth Symposium of the International Society of Tropical Root Crops*, pp. 104–7. Ottawa, Ontario: International Development Research Centre.

Odurukwe, S.O. (1986). Yam–maize intercropping investigations in Nigeria. *Tropical Agriculture (Trinidad)*, **63**, 17–21.

Ofori, C.S. (1970). Absorption and translocation of phosphate through cassava tubers (*Manihot esculenta* Crantz.). *Ghana Journal of Agricultural Science*, **3**, 203–5.

Ofori, C.S. (1973). *Use of Isotopes in Studies on the Nutrition of Groundnuts*. Part of a coordinated program on the use of isotopes and radiation in studies on physicochemical relation of soils and plants. Report, IAEA-R-729-F. Vienna: International Atomic Energy Authority, 30 pp.

Ofori, C.S. (1975). Effect of time and rate of nitrogen application on yield and fertilizer nitrogen utilization by groundnuts (*Arachis hypogaea* L.). *Ghana Journal of Agricultural Science*, **8**, 213–17.

Ohiro, A.C. & Ezumah, H.C. (1990). Tillage effects on cassava (*Manihot esculenta*) production and some soil properties. *Soil and Tillage Research*, **17**, 221–9.

Ojeniyi, S.O. (1990). Effect of bush clearing and tillage methods on soil physical and chemical properties of humid tropical Alfisols. *Soil and Tillage Research*, **15**, 269–77.

Okezie, C.E., Okonkwo, S.N.C. & Nweke, F.I. (1981). Growth pattern and growth analysis of the white Guinea yam raised from seed. In *Tropical Root Crops: Research Strategies for the 1980s*, ed. E.R. Terry, K.A. Oduro & F. Caveness, pp. 180–8. Ottawa, Ontario: International Development Research Centre.

Okigbo, B.N. (1971). Effect of planting date on the yield and general performance of cassava (*Manihot utilissima* Pohl.). *Nigerian Agricultural Journal*, **8**, 115–22.

Okigbo, B.N. (1977*a*). Role of legumes in small holdings of the humid tropics in Africa. In *Exploiting the Legume–Rhizobium Symbiosis in Tropical Agriculture*, pp. 119–54. Honolulu, Hawaii: University of Hawaii Press.

Okigbo, B.N. (1977*b*). Legumes in farming systems of the humid tropics. In *Biological Nitrogen Fixation in Farming Systems of the Humid Tropics*, ed. A. Ayanaba & P.J. Dart, pp. 61–72. New York: Wiley.

Okigbo, N.B. & Greenland, D.J. (1976). Intercropping systems in tropical Africa. In *Multiple Cropping*, pp. 63–101. Madison, Wisconsin: American Society of Agronomy.

Okoli, O.O. (1980). Dry matter accumulation and tuber sprouting in *Dioscorea* species. *Experimental Agriculture*, **16**, 161–7.

Okon, Y. & Kapulnik, Y. (1986). Development and function of *Azospirillum*-inoculated roots. *Plant and Soil*, **90**, 3–16.

Oldeman, L.R. & Suardi, D. (1977). Climatic determinants in relation to cropping patterns. In *Cropping Systems Research and Development for the Asian Rice Farmer*, pp. 61–80. Los Baños, Philippines: International Rice Research Institute.

Olsen, F.J. & Santos, G.L. (1976). Effect of nitrogen fertilization on the productivity of Sorghum Sudan grass cultivars and millet in Rio do Sul, Brazil. *Tropical Agriculture (Trinidad)*, **53**, 211–16.

Olsen, R.A., Dreier, C.A., Hoover, C.A. & Rhoades, H.F. (1962). Factors responsible for poor response of corn and grain sorghum to phosphorus fertilization. I. Phosphorus level and climatic factors. *Soil Science Society of America Proceedings*, **26**, 571–4.

Olsen, F.J., Hamilton, G. & Elkins, D.M. (1975). Effect of nitrogen on nodulation and yield of soybean. *Experimental Agriculture*, **11**, 289–94.

Olsson, K.A., Cary, P.R. & Turner, D.W. (in press). Fruit crops. In *Control of Crop Productivity*, ed. C.J. Pearson, pp. 219–38. Sydney: Academic Press.

Onwualu, A. P. & Anazodo, U.G. N. (1989). Soil compaction effects on maize production under various tillage methods in a derived savannah zone of Nigeria. *Soil and Tillage Research*, **14**, 99–114.

Onwueme, I.C. (1978). *The Tropical Tuber Crops*. Chichester: Wiley, 234 pp.

Opara-Nadi, O.A. & Lal, R. (1987). Effects of land clearing and tillage methods on soils properties and maize root growth. *Field Crops Research*, **15**, 193–206.

Opik, H. (1973). Effect of anaerobiosis on respiratory rate, cytochrome oxidase activity and mitochondrial structures in coleoptiles of rice (*Oryza sativa* L.). *Journal of Cell Science*, **12**, 725–39.

Orchard, P.W. & Jessop, R. S. (1984). The response of sorghum and sunflower to short-term waterlogging. I. Effects of stage of development and duration of waterlogging on growth and yield. *Plant and Soil*, **81**, 119–32.

Orphanos, P.I. (1977). Emergence of *Phaseolus vulgaris* seedlings from wet soil. *Journal of Horticultural Science*, **52**, 447–55.

Osa-Afiana, L.O.D. & Alexander, M. (1979). Effect of moisture on survival of rhizobium in soil. *Soil Science Society of America Journal*, **43**, 925–30.

Osafo, D.M. (1977). Effects of population density on yields of two soybean (*Glycine max*) varieties in Ghana forest zone. *Experimental Agriculture*, **13**, 235–40.

O'Toole, J.C. & Moya, T.B. (1981). Water deficits and yield in upland rice. *Field Crops Research*, **4**, 247–59.

Owen, P.C. (1971). The effects of temperature on the growth and development of rice. *Field Crops Abstracts*, **24**, 1–8.

Pacovsky, R.S., Da Silva, P., Carvalho, M.T. & Tsai, S.M. (1991). Growth and nutrient allocation in *Phaseolus vulgaris* L. colonised with endomycorrhizae or *Rhizobium*. *Plant and Soil*, **132**, 127–37.

Pal, U.R. & Saxena, M.C. (1975). Responses of soybean to symbiosis and nitrogen fertilization under humid sub-tropical conditions. *Experimental Agriculture*, **11**, 221–6.

Paliwal, K.V. & Anjaneyulu, B.S.R. (1967). Growth of wheat and gram under semi-alkali field conditions. *Madras Agricultural Journal*, **51**, 169–75.

Pallas, J.E. & Samish, Y.B. (1974). Photosynthetic response of peanuts. *Crop Science*, **14**, 478–81.

Pallas, J.E. & Stansell, J.R. (1978). Solar energy utilization of peanut under several soilwater regimes in Georgia. *Oleagineaux*, **33**, 235–8.

Pallas, J.E., Samish, Y.B. & Willmer, C.M. (1974). Endogenous rhythmic activity in photosynthesis, transpiration, dark respiration and carbon dioxide compensation point of peanut leaves. *Plant Physiology*, **53**, 907–11.

Palm, C.A. & Sanchez, P.A. (1991). Nitrogen release from the leaves of some tropical legumes as affected by their lignin and polyphenolic contents. *Soil Biology and Biochemistry*, **23**, 83–8.

Palvadi, H.K. & Lal, B. (1976). Note on the susceptibility of maize to waterlogging at different growth stages. *Pantnagar Journal of Research*, **1**, 141–2.

Papadakis, J. (1970). *Agricultural Potentialities of World Climates*. Buenos Aires: Libro Edicion Argentina, 70 pp.

Papadakis, J. (1975). *Climates of the World and their Potentialities*. Buenos Aires: Libro Edicion Argentina, 200 pp.

Pardales, J.R., Kono, Y. & Yamauchi, A. (1991). Response of the different root system components of sorghum to incidence of waterlogging. *Environmental and Experimental Botany*, **31**, 107–15.

Parfitt, R.L. & Clayden, B. (1991). Andisols: the development of a new order in soil taxonomy. *Geoderma*, **49**, 181–98.

Park, S.J. & Buttery, B.R. (1989). Identification and characterization of common bean (*Phaseolus vulgaris* L.) lines well nodulated in the presence of high nitrate. *Plant and Soil*, **119**, 237–44.

Parker, M.U. & Borthwick, H.A. (1939). Effect of variation in temperature during photoperiodic induction upon initiation of flower primordia in Biloxi soybean. *Botanical Gazette*, **101**, 145–67.

Parsons, D.J. (1970). Agricultural systems. In *Agriculture in Uganda*, 2nd edn, ed. J.D. Jameson, pp. 127–38. Oxford: Oxford University Press.

Parton, W.J., Sanford, R.L., Sanchez, P.A. & Stewart, J.W.B. (1989). Modelling soil organic matter dynamics in tropical soils. In *Dynamics of Soil Organic Matter in Tropical Ecosystems*, ed. D.C. Coleman, J.M. Oades & G. Uehara, pp. 153–71. Department of Agronomy and Soil Science, College of Tropical Agriculture and Human Resources, University of Hawaii.

Passam, H.C. (1977). Sprouting and apical dominance of yam tubers. *Tropical Science*, **19**, 29–39.

Passam, H.C., Read, S.J. & Rickard, J.E. (1978). The respiration of yam tubers and its contribution to storage losses. *Tropical Agriculture (Trinidad)*, **55**, 207–14.

Pate, J.S. & Minchin, F.R. (1980). Comparative studies of carbon and nitrogen nutrition of selected grain legumes. In *Advances in Legume Science*, ed. R.J. Summerfield & A.H. Bunting, pp. 105–14. Kew: Royal Botanic Gardens.

Patel, P.M., Wallace, A. & Wallihan, E.G. (1975). Influence of salinity and NP fertility levels on mineral content and growth of sorghum in sand culture. *Agronomy Journal*, **67**, 622–5.

Patil, B.D., Reddy, B.B. & Gill, A.S. (1978). Photoperiodism in relation to forage yield in pearl millet. *Sabrao Journal*, **10**, 126–9.

Patil, J.D. & Patil, N.D. (1981). Effect of calcium carbonate and organic matter on the growth and concentration of iron and manganese in sorghum *(Sorghum bicolor)*. *Plant and Soil*, **60**, 295–300.

Patrick, W.H. & Reddy, K.R. (1978). Chemical changes in rice soils. In *Soils and Rice* pp. 361–79. Los Baños, Philippines: International Rice Research Institute.

Peacock, J.M. & Wilson, G.L. (1984). Sorghum. In *The Physiology of Tropical Field Crops*, ed. P.R. Goldsworthy & N.M. Fisher, pp. 163–212. Chichester: Wiley.

Pearson, C.J. (1975). Thermal adaptation of *Pennisetum:* seedling development. *Australian Journal of Plant Physiology*, **2**, 413–24.

Pearson, C.J. (1984). *Pennisetum* millet. In *The Physiology of Tropical Field Crops*, ed. P.R. Goldsworthy & N.M. Fisher, pp. 281–304. Chichester: Wiley.

Pearson, C.J. (1994). *Sustainable Dryland Cropping in Relation to Soil Productivity*. Rome: Food and Agriculture Organization.

Pearson, C.J. & Jacobs, B.C. (1986). Elongation and retarded growth of rice during short-term submergence at three stages of development. *Field Crops Research*, **13**, 331–43.

Pearson, C.J., Masduki, S. & Moenandir, J. (1980). Effects of plant population and canopy manipulation on growth and yield of soybean in East Java, Indonesia. *Field Crops Research*, **3**, 337–45.

Pearson, C.J., Larson, E.M., Hesketh, J.D. & Peters, D.B. (1984). Development and source-sink effects on single leaf and canopy carbon dioxide exchange in maize. *Field Crops Research*, **9**, 391–402.

Pearson, R.W. (1975). Soil acidity and liming in the humid tropics. *Cornell International Agriculture Bulletin*, No. 30, 66 pp.

Peet, M.M., Bravo, A., Wallace. D.H. & Ozbun, J.L. (1977). Photosynthesis, stomatal resistance and enzyme activities in relation to yield of field-grown dry bean varieties. *Crop Science*, **17**, 287–93.

Peoples, M.B. & Crasswell, E.T. (1992). Biological nitrogen fixation: investments, expectations and actual contributions to agriculture. *Plant and Soil*, **141**, 13–39.

Peoples, M.B. & Herridge, D.F. (1991). Nitrogen fixation by legumes in tropical and subtropical agriculture. *Advances in Agronomy*, **44**, 155–223.

Pereira, J.A.R., Cavalcante, V.A., Baldani, J.I. & Dobereiner, J. (1988). Field inoculation of sorghum and rice with *Azospirillum* spp. and *Herbaspirillum seropedicae*. *Plant and Soil*, **110**, 269–74.

Pereira, P.A.A. & Bliss, F.A. (1989). Selection of common bean (*Phaseolus vulgaris* L.) for N_2 fixation at different levels of available phosphorus under field and environmentally-controlled conditions. *Plant and Soil*, **104**, 79–84.

Perez-Escolar, R. (1977). Effect of soil pH and related acidity factors on yields of sweet potatoes and soybean grown on typical soils of the humid tropics. *University of Puerto Rico Journal of Agriculture*, **61**, 2–9.

Perez-Escolar, R. & Lugo-Lopez, M.A. (1979). The effect of soil pH and related acidity factors on yield of plantains (cv. Maricongo) grown on Los Guineos clay, an Ultisol. *University of Puerto Rico Journal of Agriculture*, **63**, 22–6.

Perry, D.A. (1976). Seed vigour and seedling establishment. *Advances in Research and Technology of Seeds*, **2**, 62–85.

Peters, D.B. & Johnson, C. (1960). Soil moisture use by soybeans. *Agronomy Journal*, **52**, 687–9.

Phillips, L.J. (1959). *The Influence of Crop Sequence on the Yield of Peanuts, Sorghum, and Cotton at Katherine, N.T.* Division of Land Research and Regional Survey Technical Paper, No. 2. Melbourne: Commonwealth Scientific and Industrial Research Organization.

Phillips, L.J. & Norman, M.J.T. (1961). Sorghum–peanut crop sequences at Katherine, N.T. *Australian Journal of Experimental Agriculture and Animal Husbandry*, **1**, 144–9.

Phillips, L.J. & Norman M.J.T. (1962). *Fodder-Crop Cash-Crop Sequences at Katherine, N.T.* Division of Land Research and Regional Survey Technical Paper, No. 20. Melbourne: Commonwealth Scientific and Industrial Research Organization.

Phillips, L.J. & Norman, M.J.T. (1967). *A Comparison of Two Varieties of Bulrush Millet (Pennisetum typhoides) at Katherine, N.T.* Division of Land Research Technical Memorandum, No. 67/18. Melbourne: Commonwealth Scientific and Industrial Research Organization.

Phillips, P.G. (1954). The metabolic cost of common West African agricultural activity. *Journal of Tropical Medicine and Hygiene*, **57**, 12–20.

Piha, M.I. & Munns, D.N. (1987). Nitrogen fixation capacity of field grown bean (*Phaseolus vulgaris* L.) compared with other grain legumes. *Agronomy Journal*, **79**, 690–6.

Pinchinat, A.M. (1977). The role of legumes in tropical America. In *Exploiting the Legume–Rhizobium Symbiosis in Tropical Agriculture*, pp. 171–82. Honolulu, Hawaii: University Press of Hawaii.

Pinchinat, A.M., Soria, J. & Bazan, R. (1976). Multiple cropping in tropical America. In *Multiple Cropping*, pp. 51–61. Madison, Wisconsin: American Society of Agronomy.

Plucknett, D.L. (1978). Tolerance of some tropical root crops and starch-producing tree crops to suboptimal land conditions. In *Crop Tolerance to Suboptimal Land Conditions*, ed. G.A. Jung, pp. 125–14. Madison, Wisconsin: American Society of Agronomy.

Ponnamperuma, F.N. (1965). Dynamic aspects of flooded soils. In *The Mineral Nutrition of the Rice Plant*, pp. 295–328. Baltimore, Maryland: Johns Hopkins Press.

Ponnamperuma, F.N. (1972). The chemistry of submerged soils. *Advances in Agronomy*, **24**, 29–96.

Ponnamperuma, F.N. (1976). *Specific Soil Chemical Properties of Submerged Soils in Relation to Fertility*. Research Paper Series, No. 2. Los Baños, Philippines: International Rice Research Institute, 18 pp.

Ponnamperuma, F.N. & Bandyopadhya, A.K. (1980). Soil salinity as a constraint on food production in the humid tropics. In *Priorities for Alleviating Soil-Related Constraints to Food Production in the Tropics*, pp. 203–16. Los Baños, Philippines: International Rice Research Institute.

Portères, R. (1976). African cereals: *Eleusine*, Fonio, Black fonio, Teff, *Brachiaria, Paspalum, Pennisetum*, and African rice. In *Origins of African Plant Domestication*, ed. J.R. Harlan, J.M.J. de Wet & A.B.L. Stemler, pp. 409–52. The Hague: Mouton.

Porto, M.C.M., Almeida, P.A. de, Mattos, P.L.P. de & Souza, R.F. (1979). Cassava intercropping in Brazil. In *Intercropping with Cassava*, ed. E. Weber, B. Nestel & M. Campbell, pp. 25–30. Ottawa, Ontario: International Development Research Centre.

Powell, J. (1976). Ethnobotany. In *New Guinea Vegetation*, ed. K. Paijmans, pp. 106–99. Canberra: Australian National University Press.

Powers, D., Kanemasu, E.T., Singh, P. & Kreitner, G. (1980). Floral development of pearl millet (*Pennisetum americanum* (L.) K. Schum). *Field Crops Research*, **3**, 245–66.

Prabowo, A., Prastowo, B. & Wright, G.C. (1990). Growth, yield and soil water extraction of irrigated and dryland peanuts in south Sulawesi, Indonesia. *Irrigation Science*, **11**, 63–8.

Prihar, S.S., Ghildyal, B.P., Painuli, D.K. & Sur, H.S. (1985). Physical properties of mineral soils affecting rice-based cropping systems, pp. 57–70. Los Baños, Philippines: International Rice Research Institute.

Protheroe, R.M. (1972). *People and Land in Africa South of the Sahara.* Oxford: Oxford University Press, 344 pp.

Protz, R. (1981). Soil properties important for various tropical crops: Pahang Tenggara Master Planning Study. In *Soil Resource Inventories and Development Planning*, pp. 187–200. Soil Management Support Services, Technical Monograph, No. 1. United States Department of Agriculture Soil Conservation Service.

Purseglove, J.W. (1965). The spread of tropical crops. In *The Genetics of Colonizing Species*, ed. H.G. Baker & G.L. Stebbins, pp. 375–86. New York: Academic Press.

Purseglove, J.W. (1968). *Tropical Crops: Dicotyledons.* London: Longman, 719 pp.

Purseglove, J.W. (1972). *Tropical Crops: Monocotyledons.* London: Longman, 607 pp.

Quin, F.M. (1985). Report on yam research (*Dioscorea* spp.), 1981–84. East Sepik Rural Development Project, Wewak, Papua New Guinea, 90 pp.

Quinby, J.R. (1971). Time of physiologic maturity of sorghum parents and hybrids. *Sorghum Newsletter*, **14**, 98.

Rachie, K.O. (1977). The nutritional role of grain legumes in the lowland humid tropics. In *Biological Nitrogen Fixation in Farming Systems of the Tropics*, ed. A. Ayanaba & P.J. Dart, pp. 45–60. New York: Wiley.

Rachie, K.O. & Majmudar, J.V. (1980). *Pearl Millet.* University Park, Pennsylvania : Pennsylvania State University Press, 307 pp.

Rachie, K.O. & Silvestre, P. (1977). Grain legumes. In *Food Crops of the Lowland Tropics*, ed. C.L.A. Leakey & J.B. Wills, pp. 41–74. Oxford: Oxford University Press.

Rai, R. & Prasad V. (1986). Chemotaxis of *Cicer Rhizobium* strains to root exudates of chick pea (*Cicer arietinum* L.) genotypes and their interaction response on nodulation, nodulins, leghaemoglobin and grain yield in calcareous soil. *Journal of Agricultural Science*, **107**, 75–81.

Rajendrudu, G. & Williams, J.H. (1987). Effect of gypsum and drought on pod initiation and crop yield in early maturing groundnut (*Arachis hypogaea*) genotypes. *Experimental Agriculture*, **23**, 259–71.

Raju, P.S., Clark, R.B., Ellis, J.R., Duncan, R.R. & Maranville, J.W. (1990). Benefit and cost analysis and phosphorus efficiency of VA mycorrhizal fungi colonizations with sorghum (*Sorghum bicolor*) genotypes grown at varied phosphorus levels. *Plant and Soil*, **124**, 199–204.

Ram, H.Y.M., Ram, M. & Steward, F.C. (1962). Growth and development of the banana plant. *Annals of Botany*, **26**, 657–73.

Ramanujam, S. (1976). Chickpea. *Cicer arietinum* (Leguminosae–Papilionatae). In *Evolution of Crop Plants*, ed. N.W. Simmonds, pp. 157–9. London: Longman.

Rao, M.R. (1986). Cereals in multiple cropping. In *Multiple Cropping Systems*, ed. C.A. Francis, pp. 96–132. New York: Macmillan.

Rao, M.R. & Singh, M. (1990). Productivity and risk evaluation in contrasting intercropping systems. *Field Crops Research*, **23**, 279–93.

Rao, V. R., Ayanaba, A., Eaglesham, A.R.J. & Thottappilly, G. (1985). Effects of *Rhizobium* inoculation on field-grown soybeans in Western Nigeria and assessment of inoculum persistence during a two-year fallow. *Tropical Agriculture (Trinidad)*, **62**, 125–30.

Rathore, T.R., Ghildyal, B.P. & Sachan, R.S. (1981). Germination and emergence of soybean under crusted soil conditions. I. Effect of crust impedance on seedling emergence. *Plant and Soil*, **62**, 97–105.

Ratnaswamy, M.C. (1960). Studies in cereals: structure in relation to drought resistance. *Madras Agricultural Journal*, 47, 427–36.

Rawson, H.M. & Constable, G.A. (1980). Carbon production of sunflower cultivars in field and controlled environments: I. *Australian Journal of Plant Physiology*, 7, 555–7.

Rawson, H.M., Turner, N.C. & Begg, J.E. (1978). Agronomic and physiological response of soybean and sorghum crops to water deficits: IV. *Australian Journal of Plant Physiology*, 5, 195–209.

Raymond, C. (1968). Pour une meilleure connaissance de la croissance et du développement des mils *Pennisetum*. *L'Agronomie Tropicale*, 23, 844–63.

Redden, R.J., Diatloff, A. & Usher, T. (1990). Field screening accessions of *Phaseolus vulgaris* for capacity to nodulate over a range of environments. *Australian Journal of Experimental Agriculture*, 30, 265–70.

Reddy, K.R., Rao, P.S.C. & Patrick, W.H. (1980). Factors influencing oxygen consumption rates in flooded soils. *Soil Science Society of America Journal*, 44, 741–4.

Reddy, M.S. & Willey, R.W. (1980). Growth and resource use studies in an intercrop of pearl millet/groundnut. *Field Crops Research*, 4, 13–24.

Reddy, M.V., Raju, T.N. & Nene, V.L. (1987). Field screening of pigeonpea germplasm for resistance to bacterial leaf spot and stem canker. *International Pigeonpea Newsletter*, 6, 62–5.

Reicosky, D.C., Deaton, D.E. & Parsons, J.E. (1980). Canopy air temperatures and evapotranspiration from irrigated stressed soybeans. *Agricultural Meteorology*, 21, 21–36.

Reid, P.H. & Cox, F.R. (1973). Soil properties, mineral nutrition and fertilization practices. In *Peanuts: Culture and Uses*, pp. 271–97. Stillwater, Oklahoma: American Peanut Research and Education Association.

Reneau, R.B., Jones, G.D. & Friedericks, J.B. (1983). Effect of P and K on yield and chemical composition of forage sorghum. *Agronomy Journal*, 75, 5–9.

Renvoize, B.S. (1972). The area of origin of *Manihot esculenta* as a crop plant: a review of the evidence. *Economic Botany*, 26, 352–60.

Rigaud, J. (1976). Effet des nitrates sur la fixation d'azote par les nodules de haricot (*Phaseolus vulgaris* L.). *Physiologie Végétale*, 14, 297–308.

Ritchey, K.D. (1979). Potassium fertility in Oxisols and Ultisols of the humid tropics. *Cornell International Agriculture Bulletin*, No. 37, 45 pp.

Ritchie, J.T. (1973). Influence of soil water status and meteorological conditions on evaporation from a corn canopy. *Agronomy Journal*, 65, 893–7.

Roberts, E.A., Summerfield, R.J., Minchin, F.R. & Hadley, P. (1980). Phenology of chickpeas (*Cicer arietinum*) in contrasting aerial environments. *Experimental Agriculture*, 16, 343–60.

Roberts, R.H. & Struckmeyer, B.E. (1939). Further studies on the effects of temperature and other environmental factors upon the photoperiodic response of plants. *Journal of Agricultural Research*, 59, 699–709.

Robertson, W.K., Hammond, L.C., Johnson, J.T. & Boote, K.J. (1980). Effects of plant water stress on root distribution of corn, soybeans, and peanuts in sandy soil. *Agronomy Journal*, 72, 548–50.

Robinson, J.C. (1981). Studies on the phenology and production potential of Williams banana in a subtropical climate. *Subtropica*, 2, 12–16.

Robinson, J.C. & Bower, J.P. (1988). Transpiration from banana leaves in the subtropics in response to diurnal and seasonal factors and high evaporative demand. *Scientia Horticulturae*, **37**, 129–43.

Rodriguez-Gomez, M. (1980). Estudios preliminares sobre la nutrición con potasio de los bananales en América central. *Fruits*, **35**, 283–94.

Roger, P.A. & Ladha, J.K. (1992). Biological N_2 fixation in wetland rice fields: estimation and contribution to nitrogen balance. *Plant and Soil*, **141**, 41–55.

Rogers, D.J. (1965). Some botanical and ethnological considerations of *Manihot esculenta*. *Economic Botany*, **19**, 369–77.

Rogers, D.J. & Appan, S.G. (1970). Untapped genetic resources for cassava improvement. In *Tropical Root and Tuber Crops Tomorrow*, vol. 1, ed. D.L. Plucknett, pp. 72–5. Honolulu, Hawaii: University of Hawaii Press.

Rogers, H.T. (1948). Liming for peanuts in relation to exchangeable soil calcium and effect on yield, quality and uptake of calcium and potassium. *Journal of the American Society of Agronomy*, **40**, 15–31.

Roose, E.J. (1977) Application of the Universal Soil Loss Equation of Wischmeier and Smith in West Africa. In *Soil Conservation and Management in the Humid Tropics*, ed. D.J. Greenland & R. Lal, pp. 177–187. New York: Wiley.

Roy, R.N. & Wright, B.C. (1974). Sorghum growth and nutrient uptake in relation to soil fertility. II. N, P and K uptake pattern by various plant parts. *Agronomy Journal*, **66**, 5–10.

Ruthenberg, H. (1980). *Farming Systems in the Tropics*, 3rd edn. Oxford: Clarendon Press, 424 pp.

Sachay, J.E., Wallace, R.L. & Johns, M.A. (1991). Phosphate stress response in hydroponically grown maize. *Plant and Soil*, **132**, 85–90.

Sadik, S. & Okereke, O.U. (1975). Flowering, pollen grain germination, fruiting, seed germination and seedling development of white yam, *Dioscorea rotundata* Poir. *Annals of Botany*, **39**, 597–604.

Saeed, M. & Fox, R.L. (1979). Influence of phosphate fertilization on zinc adsorption in tropical soils. *Soil Science Society of America Journal*, **43**, 683–6.

Sale, P.J.M. (1975). Productivity of vegetable crops in a region of high solar input: IV. *Australian Journal of Plant Physiology*, **2**, 461–70.

Sale, P.W.G. & Campbell, L.C. (1987). Differential responses to K deficiency among soybean cultivars. *Plant and Soil*, **104**, 183–90.

Salih, F.A. (1979). Food legume research and development in the Sudan. In *Food Legume Improvement and Development*, ed. G.C. Hawtin & G.J. Chancellor, pp. 58–64. Ottawa, Ontario: International Development Research Centre.

Salinas, J.G. (1978). Differential response of some cereal and bean cultivars to Al and P stress in an Oxisol in Central Brazil. PhD thesis, North Carolina State University Department of Soil Science, Raleigh, 326 pp.

Salter, P.J. & Goode, J.E. (1967). Crop responses to water at different stages of growth. *Commonwealth Bureau of Horticultural and Plantation Crops Research Review*, No. 2.

Sanchez, P.A. (1973a). Puddling tropical rice soils. I. Growth and nutritional aspects. *Soil Science*, **115**, 149–58.

Sanchez, P.A. (1973b). Puddling tropical rice soils. II. Effects of water losses. *Soil Science*, **115**, 303–8.

Sanchez, P.A. (1973c). Nitrogen fertilization. In *A Review of Soils Research in Tropical Latin America*, ed. P.A. Sanchez, pp. 90–125. Raleigh, North Carolina: North Carolina Agricultural Experimental Station.

Sanchez, P.A. (1976). *Properties and Management of Soils in the Tropics*. New York: Wiley, 618 pp.

Sanchez, P.A. (1977). Advances in the management of Oxisols and Ultisols in tropical South America. In *Soil Environment and Fertility Management in Intensive Agriculture*, pp. 535–66. Tokyo: Society of the Science of Soil and Manure.

Sanchez, P.A. (1987). Management of acid soils in the humid tropics of Latin America. In *Management of Acid Tropical Soils for Sustainable Agriculture*, ed. P.A. Sanchez, E. Stoner & E. Pushparajah, pp. 63–107. Proceedings of the Conference of the International Board for Soil Research and Management held in Bangkok, Thailand.

Sanchez, P.A. (1989). Soils. In *Tropical Rain Forest Ecosystems*, ed. H. Lieth & M.J.A. Werger, pp. 73–88. Amsterdam: Elsevier.

Sanchez, P.A. & Benites, J.R. (1987). Low-input cropping for acid soils of the humid tropics. *Science*, **238**, 1521–7.

Sanchez, P.A. & Isbell, R.F. (1979). A comparison of the soils of tropical Latin America and tropical Australia. In *Pasture Production in Acid Soils of the Tropics*, ed. P.A. Sanchez, pp. 25–54. Cali, Colombia: Centro Internacional de Agricultura Tropical.

Sanchez, P.A. & Miller, R.H. (1986). Organic matter and soil fertility management. *Thirteenth Congress of the International Society of Soil Science Transactions*, **6**, 609–25.

Sanchez, P.A. & Salinas, J.G. (1981). Low input management technology for managing Oxisols and Ultisols in tropical America. *Advances in Agronomy*, **34**, 279–406.

Sanchez, P.A. & Uehara, G. (1980). Management considerations for acid soils with high phosphorus fixation capacity. In *The Role of Phosphorus in Agriculture*, ed. F.E. Khasawneh, E.C. Sample & E.J. Kamprath, pp. 471–514. Madison, Wisconsin: American Society of Agronomy.

Sanchez, P.A., Couto, W. & Buol, S.W. (1982). The fertility capability soil classification system: interpretation, application and modification. *Geoderma*, **27**, 283–309.

Sanchez, P.A., Villachica, J.H. & Bandy, D.E. (1983). Soil fertility dynamics after clearing a tropical rainforest in Peru. *Soil Science Society of America Journal*, **47**, 1171–8.

Sanchez, P.A., Benites, J. & Bandy, D. (1987). Low-input systems and managed fallows for acid soils in the humid tropics. In *Soil Management Under Humid Conditions in Asia (Asialand)*, pp. 353–60. Proceedings of the First Regional Seminar of the International Board for Soil Research and Management in Asia and the Pacific, Khon Kaen, Thailand, 13–20 October 1986.

Sanchez, P.A., Palm, C.A. & Smyth, T.J. (1990). Approaches to mitigate tropical deforestation by sustainable soil management practices. In *Soils on a Warmer Earth*, ed. H.W. Scharpenseel, M. Shomarker & A. Ayoub, pp. 211–20. Amsterdam: Elsevier.

Sarig, S., Blum, A. & Okon, Y. (1988). Improvements in the water status and yield of field-grown grain sorghum (*Sorghum bicolor*) by inoculation with *Azospirillum brasilense*. *Journal of Agricultural Science*, **110**, 271–7.

Sarkar, K.R., Mukherjee, B.K., Gupta, D. & Jain, H.K. (1974). Maize. In *Evolutionary Studies on World Crops*, ed. J.B. Hutchinson, pp. 121–7. Cambridge: Cambridge University Press.

Satake, T. (1969). Research on cool injury of paddy rice plant in Japan. *Japanese Agricultural Research Quarterly*, **4**, 5–10.

Sato, K. & Takahashi, M. (1971). The development of rice grains under controlled environment: I. *Tohaku Journal of Agricultural Research*, **22**, 57–68.

Savithri, K.S., Ganapathy, P.S. & Sinha, S.K. (1980). Sensitivity to low temperature in pollen germination and fruit-set in *Cicer arietinum* L. *Journal of Experimental Botany*, **31**, 475–81.

Saxena, M.C. (1979). Some agronomic and physiological aspects of the important food legume crops in West Africa. In *Food Legume Improvement and Development*, ed. G.C. Hawtin & G.J. Chancellor, pp. 155–65. Ottawa, Ontario: International Development Research Centre.

Saxena, M.C. (1980). Recent advances in chickpea agronomy: I. In *International Workshop on Chickpea Improvement*, pp. 89–96. Patancheru, India: International Crops Research Institute for the Semi-Arid Tropics.

Saxena, M.C. & Singh, K.B. (eds.) (1987). *The Chickpea*. Wallingford, UK: CAB International.

Saxena, M.C. & Yadav, D.S. (1975). Some agronomic considerations of pigeonpeas and chickpeas. In *International Workshop on Grain Legumes*, pp. 61–2. Hyderabad, India: International Crops Research Institute for the Semi-Arid Tropics.

Saxena, N.P. (1980). Pod setting in relation to temperature at Hissar. *International Chickpea Newsletter*, **2**, 11–12.

Saxena, N.P. (1987). Screening of chickpea to drought: case studies with chickpea and pigeonpea. In *Adaptation of Chickpea and Pigeonpea to Abiotic Stresses*, pp. 63–76. Patancheru, India: International Crops Research Institute for the Semi-Arid Tropics.

Saxena, N.P., Krishnamurthy, L. & Sheldrake, A.R. (1981). *Pulse Physiology Progress Report 7, 1979–80*, Part II, *Chickpea Physiology*. Patancheru, India: International Crops Research Institute for the Semi-Arid Tropics, 114 pp.

Saxena, N.P., Krishnamurthy, L. & Sheldrake, A.R. (1982). *Pulse Physiology Progress Report 9, 1980–81*, Part II, *Chickpea Physiology*. Patancheru, India: International Crops Research Institute for the Semi-Arid Tropics, 99 pp.

Schank, S.C., Smith, R.L. & Weiser, G.C. (1980). Responses of two pearl millets grown *in vitro* after inoculation with *Azospirillum brasilense*. *Soil and Crop Science Society of Florida Proceedings*, **39**, 112–15.

Schettini, T.M., Gabelman, W.H. & Gerloff, G.C. (1987). Incorporation of phosphorus efficiency from exotic germplasm into agriculturally adapted germplasm of common bean (*Phaseolus vulgaris* L.). *Plant and Soil*, **90**, 175–84.

Schmidt, F.H. & Ferguson, J.H. (1951). Rainfall types based on wet and dry period ratios for Indonesia. *Verhandel Djawatan Meteorologie dan Geofis, Jakarta*, **42**, 77–87.

Schnier, H.F., Dingkuhn, M., De Datta, S.K., Mengel, K. & Faronilo, J.E., (1990*a*). Nitrogen fertilization of direct seeded flooded vs. transplanted rice. I. Nitrogen uptake, photosynthesis, growth, and yield. *Crop Science*, **30**, 1276–84.

Schnier, H.F., Dingkuhn, M., De Datta, S.K., Mengel, K., Wijangco, E. & Javellana, C., (1990*b*). Nitrogen economy and canopy CO_2 assimilation in tropical lowland rice. *Agronomy Journal*, **82**, 451–9.

Schumacher, T.E. & Smucker, A.J.M. (1981). Mechanical impedance effects on oxygen uptake and porosity of drybean roots. *Agronomy Journal*, **73**, 51–5.

Scott, H.D., DeAngulo, J., Daniels, M.B. & Wood, L.S. (1989). Flood duration effects on soybean growth and yield. *Agronomy Journal*, **81**, 631–6.

Scott, R.M., Heyligers, P.B., McAlpine, J.R., Saunders, J.C. & Speight, J.G. (1967). *Lands of Bougainville and Buka Islands, Territory of Papua New Guinea.* Land Research Series, No. 20. Melbourne: Commonwealth Scientific and Industrial Research Organization, 184 pp.

Scott-Wendt, J., Chase, R.G. & Hossner, L.R. (1988). Soil chemical variability in sandy Ustalfs in semiarid Niger, West Africa. *Soil Science*, **145**, 414–19.

Searle, P.G.E., Yuthapong Comudon, Shedden, D.E. & Nance, R.A. (1981). Effect of maize + legume intercropping systems and fertilizer nitrogen on crop yields and residual nitrogen. *Field Crops Research*, **4**, 133–45.

Sekioka, H. (1964). The effect of some environmental factors on the translocation and storage of carbohydrate in the sweet potato, potato and sugar beet: III. *Bulletin of the Faculty of Agriculture, Kyushu University*, **21**, 131–48.

Semb, G. & Garburg, P.K. (1969). Some effects of planting date and nitrogen fertilizer in maize. *East African Agricultural and Forestry Journal*, **34**, 371–9.

Semb, G. & Robinson, J.B.D. (1969). The natural nitrogen flush in different arable soils and climates in East Africa. *East African Agricultural and Forestry Journal*, **34**, 350–70.

Senaratne, R., Amornpimol, C. & Hardarson, G. (1987). Effect of combined nitrogen on nitrogen fixation of soybean (*Glycine max*, L. Merrill.) as affected by cultivar and rhizobial strain. *Plant and Soil*, **103**, 45–50.

Sethi, S.C., Bythe, D.E., Gowda, C.L.L. & Green, J.M. (1981). Photoperiodic response and accelerated generation turnover in chickpea. *Field Crops Research*, **4**, 215–25.

Setter, T.L., Kupkanchanakul, T., Bhekasut, P., Wiengweera, A. & Greenway, H. (1987). Concentrations of CO_2 and O_2 in floodwater and in internodal lacunae of floating rice growing at 1–2 metre water depths. *Plant, Cell and Environment*, **10**, 767–76.

Shanmugasundaram, S., Kuo, G.C. & Nalampang, A. (1980). Adaptation and utilisation of soybeans in different environments and agricultural systems. In *Advances in Legume Science*, ed. R.J. Summerfield & A.H. Bunting, pp. 265–77. Kew: Royal Botanic Gardens.

Sharma, D.P. & Swarup, A. (1989). Response of pearl millet (*Pennisetum americanum*) to short-term flooding in a moderately sodic soil under field conditions. *Journal of Agricultural Science*, **113**, 331–7.

Sharma, P.K. & Datta, S.K. de (1985a). Puddling influence on soil, rice development and yield. *Soil Science Society of America Journal*, **49**, 1451–7.

Sharma, P.K. & Datta, S.K. de (1985b). Effects of puddling on soil physical properties and processes. In *Soil Physics and Rice*, pp. 217–34. Los Baños, Philippines: International Rice Research Institute.

Sharma, P.K., Datta, S.K. de & Redulla, C.A. (1989). Effect of percolation rate on nutrient kinetics and rice yield in tropical rice soils. I. Role of soil organic matter. *Plant and Soil*, **119**, 111–19.

Shaw, R.H. (1977). Climatic requirement. In *Corn and Corn Improvement*, ed. G.F. Sprague, pp. 591–623. Madison, Wisconsin: American Society of Agronomy.

Shaw, R.H. & Laing, D.R. (1966). Moisture stress and plant response. In *Plant Environment and Efficient Water Use*, ed. W.H. Pierre, D. Kirkham, J. Pesek & R.H. Shaw, pp. 73–94. Madison, Wisconsin: American Society of Agronomy/Soil Science Society of America.

Sheldrake, A.R. & Saxena, N.P. (1979). Comparison of earlier- and later-formed pods of chickpeas (*Cicer arietinum* L.). *Annals of Botany*, **43**, 467–73.

Shibles, R.M. (1980). Adaptation of soybeans to different seasonal durations. In *Advances in Legume Science*, ed. R.J. Summerfield & A.H. Bunting, pp. 279–86. Kew: Royal Botanic Gardens.

Shibles, R.M., Anderson, I.C. & Gibson, A.H. (1975). Soybean. In *Crop Physiology*, ed. L.T. Evans, pp. 151–89. Cambridge: Cambridge University Press.

Shmueli, E. (1953). Irrigation studies in the Jordan Valley: I. *Bulletin of the Research Council of Israel*, 3, 228–47.

Shriniwas (1980). Root activity and soil feeding zones of some bajra hybrids (*Pennisetum typhoides* Stapf.). *Journal of Nuclear Agriculture and Biology*, 9, 124–5.

Shukla, J., Nobre, C. & Sellers, P. (1990). Amazon deforestation and climatic change. *Science*, 247, 1322–5.

Shuman, L.M., Ramseur, E.L. & Duncan, R.R. (1990). Soil aluminium effects on the growth and aluminium concentration of sorghum. *Agronomy Journal*, 82, 313–18.

Siddique, M.A. & Goodwin, P.B. (1980). Maturation temperature influences on seed quality and resistance to mechanical injury of some snap bean genotypes. *Journal of the American Society for Horticultural Science*, 105, 235–8.

Siebert, S.F. & Lassoie, J.P. (1991). Soil erosion, water runoff and their control on steep slopes in Sumatra. *Tropical Agriculture (Trinidad)*, 68, 321–4.

Sieverding, E. & Howeler, R.H. (1985). Influence of species of VA mycorrhizal fungi on cassava yield response to phosphorus fertilization. *Plant and Soil*, 88, 213–21.

Sigafus, R.E. (1972). Peanuts. *University of Kentucky College of Agriculture, Lexington, Technical Paper*.

Silva, S. & Irizarry, H. (1981). Effect of depth to water table on yields of two cultivars of sweet potatoes. *University of Puerto Rico Journal of Agriculture*, 65, 114–17.

Silveira, P.M. de & Stone, L.F. (1979). Blanco de agua na cultura do feijao em Latossolo Vermelho-Amarelo. *Pesyuisa Agropecuaria Brasileira, Agronomia*, 14, 111–16.

Simmonds, N.W. (1972). *The Evolution of the Bananas*. London: Longman, 240 pp.

Simmonds, N.W. (1976). Bananas: *Musa* (Musaceae). In *Evolution of Crop Plants*, ed. N.W. Simmonds, pp. 211–15. London and New York: Longman.

Simpson, D. & Daft, M.J. (1990). Interactions between water-stress and different mycorrhizal inocula on plant growth and mycorrhizal development in maize and sorghum. *Plant and Soil*, 121, 179–86.

Singh, A. & Manjhi, S. (1975). Fertilizer effects on protein yield of tropical pulses. In *Fertilizer Use and Protein Production*, pp. 179–91. Berne, Switzerland: International Potash Institute.

Singh, K.B. & Auckland, A.K. (1975). Chickpea breeding at ICRISAT. In *International Workshop on Grain Legumes*, pp. 3–18. Hyderabad, India: International Crops Research Institute for the Semi-Arid Tropics.

Singh, K.N., Sharma, D.K. & Chillar, R.K. (1988). Forage yield and chemical composition of pearl-millet (*Pennisetum typhoides*) as influenced by exchangeable sodium. *Journal of Agricultural Science*, 111, 465–7.

Singh, M. & Tilak, K.V.B.R. (1989). Field response of chickpea to inoculation with *Glomus versiforme*. *Plant and Soil*, 119, 281–4.

Singh, N. & Dhalival, G.S. (1972). Effect of soil temperature on seedling emergence in different crops. *Plant and Soil*, 37, 441–4.

Singh, N.P. & Saxena, M.C. (1972). Field study on nitrogen fertilization of soybean (*Glycine max* (L.) Merr.). *Indian Journal of Agricultural Science*, 42, 1028–31.

Singh, N.P. & Saxena, M.C. (1977). Phosphorus concentration and uptake in soybean (*Glycine max* (L.) Merr.) as affected by nitrogen fertilization and inoculation. *Phosphorus in Agriculture*, **69**, 23–7.

Singh, P. (1991). Influence of water-deficits on phenology, growth and day-matter allocation in chickpea (*Cicer arietinum*). *Field Crops Research*, **28**, 1–15.

Singh, P. & Russell, M.B. (1979). Water balance and profile moisture loss patterns of an Alfisol. *Agronomy Journal*, **71**, 963–6.

Singh, R.J. & Hymowitz, T. (1987). Intersubgeneric crossability in the genus *Glycine* Willd. *Plant Breeding*, **98**, 171–3.

Singh, R.P. & Reddy, G.S. (1988). Identifying crops and cropping systems with greater production stability in water deficit environments. In *Drought Research Priorities for the Dryland Tropics*, ed. F.R. Bidinger & C. Johansen, Patancheru, India: International Crops Research Institute for the Semi-Arid Tropics.

Singh, S. & Sharma, H.C. (1980). Effect of profile soil moisture and phosphorus levels on the growth and yield and nutrient uptake by chickpea. *Indian Journal of Agricultural Sciences*, **50**, 943–7.

Singh, S.B. & Abrol, I.P. (1985). Effect of soil sodicity on the growth, yield and chemical composition of groundnut (*Arachis hypogaea* Linn.). *Plant and Soil*, **84**, 123–7.

Singh, S.R., Rachie, K.O. & Dashiell, K.E. (1987). *Soybeans for the Tropics: Research, Production and Utilization*. Chichester, Wiley, 230 pp.

Singleton, P.W., AbelMagid, H.M. & Tavares, J.W. (1985). Effect of phosphorus on the effectiveness of strains of *Rhizobium japonicum*. *Soil Science Society of Americal Journal*, **49**, 613–16.

Sinthuprama, S. (1979). Cassava and cassava-based intercrop systems in Thailand. In *Intercropping with Cassava*, ed. E. Weber, B. Nestel & M. Campbell, pp. 57–65. Ottawa, Ontario: International Development Research Centre.

Sivakumar, M.V.K., Singh, P. & Williams, J.H. (1987). Agroclimatic aspects in planning for improved productivity of Alfisols. In *Alfisols in the Semi-Arid Tropics*, pp. 15–30. Patancheru, India: International Crops Research Institute for the Semi-Arid Tropics.

Sivaprasad, B. & Sundara Sarma, K.S. (1987). Seedling emergence of chickpea (*Cicer arietinum* L.), pigeonpea (*Cajanus cajan* L.) and pearl millet (*Pennisetum typhoides* L.): effects of differential soil crusting, as induced by raindrop size, and depth of sowing. *Plant and Soil*, **104**, 263–8.

Smartt, J. (1990). *Grain Legumes: Evolution and Genetic Resources*. Cambridge: Cambridge University Press, 379 pp.

Smith, B.H. & Emerson, W.W. (1976). Exchangeable aluminium in kaolinite. *Australian Journal of Soil Research*, **14**, 45–53.

Smith, B.W. (1950). *Arachis hypogaea*: aerial flower and subterranean fruit. *American Journal of Botany*, **27**, 803–15.

Smith, B.W. (1954). *Arachis hypogaea*: reproductive efficiency. *American Journal of Botany*, **41**, 607–16.

Smith, B.W. (1956). *Arachis hypogaea*: embryogeny and the effect of peg elongation upon embryo and endosperm growth. *American Journal of Botany*, **43**, 233–40.

Smith, D.T. & Clark, N.A. (1968). Effect of soil nutrients and pH on nitrate nitrogen and growth of pearl millet (*Pennisetum typhoides* Burm. Staph. and Hubbard) and sudangrass (*Sorghum sudanense* (Piper) Staph.). *Agronomy Journal*, **60**, 38–40.

Smith, F.W. & Myers, R.J.K. (1978). Patterns of uptake and distribution of phosphorus and nitrogen in grain sorghum during its development. In *Plant Nutrition 1978*, ed. A.R. Ferguson, R.L. Bieleski & I.B. Ferguson, pp. 491–8. Wellington: New Zealand Government Printer.

Smith, R.M. & Abruña, F. (1955). Soil and water conservation research in Puerto Rico, 1938 to 1947. In *University of Puerto Rico Agricultural Experiment Station Bulletin*, No. 124, pp. 18–23.

Smith, R.S. & Ellis, M.A. (1981). Soybean nodulation as influenced by seedling vigor. *Agronomy Journal*, **72**, 605–8.

Smith, R.S., Ellis, M.A. & Smith, R.E. (1981). Effect of *Rhizobium japonicum* inoculant rates on soybean nodulation in a tropical soil. *Agronomy Journal*, **73**, 505–8.

Smoliak, S., Johnston, A. & Hanna, M.R. (1972). Germination and seedling growth of alfalfa, sainfoin and cicer milkvetch. *Canadian Journal of Plant Science*, **52**, 757–62.

Smyth, A.J. & Cravo, M.S. (1990). Critical phosphorus levels for corn and cowpea in a Brazilian Amazon Oxisol. *Agronomy Journal*, **82**, 309–12.

Smyth, A.J. & Montgomery, R.J. (1962). *Soils and Land Use in Central Western Nigeria*. Ibadan, Nigeria: Government Printer.

Smyth, T.J. & Sanchez, P.A. (1980). Effects of lime, silicate, and phosphorus applications to an Oxisol on phosphorus sorption and ion retention. *Soil Science Society of America Proceedings*, **44**, 500–5.

Snaydon, R.W. (1991). The productivity of C3 and C4 plants: a re-assessment. *Functional Ecology*, **5**, 321–30.

Snyder, G.H., Jones, D.B. & Gascho, G.J. (1986). Silicon deficiency of rice on Everglades Histosols. *Soil Science Society of America Journal*, **50**, 1259–63.

Sobulo, R.A. (1972a). Studies on white yam *(Dioscorea rotundata):* I. *Experimental Agriculture*, **8**, 99–106.

Sobulo, R.A. (1972b). Studies on white yam *(Dioscorea rotundata):* II. *Experimental Agriculture*, **8**, 107–15.

Soil Survey Staff (1975). *Soil Taxonomy*. United States Department of Agriculture Handbook, No. 436. Washington, DC: United States Government Printer, 754 pp.

Soil Survey Staff (1990). *Keys to Soil Taxonomy*, 4th edn. SMSS Technical Monograph, No. 19. Blacksburg, Virginia.

Soman, P., Stomph, T.J., Bidinger, F.R. & Fussell, L.K. (1986). Improvement in stand establishment in pearl millet. In *OALL/STRC – SAFGRAD International Drought Symposium*, Nairobi, Kenya, 21 pp.

Somasegaran, P., Hoben, H.J. & Gurgun, V. (1988). Effects of inoculation rate, rhizobial strain competition, and nitrogen fixation in chickpea. *Agronomy Journal*, **80**, 68–73.

Spaans, E.J.A., Bouma, J., Lansu, A. & Wielemaker, W.G. (1990). Measuring soil hydraulic properties after clearing of tropical rain forest in a Costa Rica soil. *Tropical Agriculture (Trinidad)*, **67**, 61–5.

Spain, J.M., Francis, C.A., Howeler, R.H. & Calvo, F. (1975). Differential species and varietal tolerance to soil acidity in tropical crops and pastures. In *Soil Management in Tropical America*, ed. E. Bornemisza & A. Alvarado, pp. 308–29. Raleigh, North Carolina: North Carolina State University.

Spear, S.N., Asher, C.J. & Edwards. D.G. (1978a). Response of cassava, sunflower and maize to potassium concentrations in solution. I. Growth and plant potassium concentration. *Field Crops Research*, **1**, 347–61.

Spear, S.N., Asher, C.J. & Edwards, D.G. (1978*b*). Response of cassava, sunflower and maize to potassium concentrations in solution. II. Potassium absorption and its relation to growth. *Field Crops Research*, 1, 363–73.

Spencer, J.E. (1966). *Shifting Cultivation in Southeastern Asia*. Berkeley & Los Angeles, California: University of California Press, 247 pp.

Sprague, G.F. & Dudley, J.W. (eds.) (1988). *Corn and Corn Improvement*, 3rd edn. Madison: American Society of Agronomy.

Sprent, J.L. (1976). Nitrogen fixation by legumes subjected to water and light stress. In *Symbiotic Nitrogen Fixation in Plants*, ed. P.S. Nutman, pp. 405–20. Cambridge University Press.

Sproat, M.N. (1968). A guide to subsistence agriculture in Micronesia. *Trust Territory of the Pacific Islands, Saipan, Mariana Islands, Agricultural Extension Bulletin*, No. 9. Mimeo, 142 pp.

Squire, G.R. (1990). *The Physiology of Tropical Crop Production*. Wallingford, UK: CAB International, 236 pp.

Ssali, H. (1977). Root activity of bananas during wet and dry seasons as measured by ^{32}P uptake. *East African Agricultural and Forestry Journal*, 42, 304–8.

Ssali, H. & Keya, S.O. (1986). The effects of phosphorus and nitrogen fertilizer level on nodulation, growth, and dinitrogen fixation of three bean cultivars. *Tropical Agriculture (Trinidad)*, 63, 105–9.

Standley, J, Thomas, G.A., Hunter, H.M., Webb, A.A. & Berthelsen, S. (1990). Decreases over seven years in subsoil nitrate in a Vertisol with grain sorghum and grass. *Plant and Soil*, 125, 1–6.

Staphorst, J.L., Strifdom, B.W. & Otto, J.F. (1975). Nitrogen-fixing ability of *Rhizobia* which nodulate groundnuts in South African soils. *Phytophylactica*, 7, 133–6.

Steele, W.M. & Mehra, K.L. (1980). Structure, evolution and adaptation to farming systems and environments in *Vigna*. In *Advances in Legume Science*, ed. R.J. Summerfield & A.H. Bunting, pp. 393–404. Kew: Royal Botanic Gardens.

Steinbauer, C.E. & Kushman, L.J. (1971). *Sweet Potato: Culture and Diseases*. US Department of Agriculture, Agricultural Research Services Handbook, No. 388.

Steponkus, P.L., Cutler, J.M. & O'Toole, J.C. (1980). Adaptation to water stress in rice. In *Adaptation of Plants to Water and High Temperature Stress*, ed. N.C. Turner & P.J. Kramer, pp. 401–18. New York: Wiley.

Stern, W.R. (1968). The influence of sowing date on the yield of grain sorghum in a short summer rainfall environment. *Australian Journal of Experimental Agriculture and Animal Husbandry*, 8, 599–605.

Stewart, W.D.P., Rowell, P., Ladha, J.K. & Sampaio, M.J.A.M. (1979). Blue-green algae (Cyanobacteria): some aspects related to their role as sources of fixed nitrogen in paddy soils. In *Nitrogen and Rice*, pp. 263–85. Los Baños, Philippines: International Rice Research Institute.

Stocking, M. & Peake, L. (1986). Crop yield and losses from the erosion of Alfisols. *Tropical Agriculture (Trinidad)*, 63, 41–5.

Stofella, R.J., Sandsted, R.F., Zobel, R.W. & Hymes, W.L. (1979). Root characteristics of black beans: II. *Crop Science*, 19, 826–30.

Stoner, E.R., Freitas, E. de, Bryant, R.B. & Lathwell, D.J, (1991). Physical constraints to root growth in savanna Oxisols. *TropSoils Bulletin*, No. 91–01, 28 pp.

Stoop, W.A. (1980). Ion adsorption meehanisms in Oxidic soils: implications for point of zero charge determinations. *Geoderma*, **23**, 303–14.

Stover, R.H. (1974). Effect of measured levels of Sigatoka disease of bananas on fruit quality and leaf senescence. *Tropical Agriculture (Trinidad)*, **51**, 531–42.

Stover, R.H. (1982). 'Valery' and 'Grand Nairn': plant foliage characteristics and a proposed banana ideotype. *Tropical Agriculture (Trinidad)*, **59**, 303–5.

Stover, R.H. & Simmonds, N.W. (1987). *Bananas*. Harlow, UK: Longman.

Sturkie, D.G. & Buchanan. G.A. (1973). Cultural practices. In *Peanuts: Culture and Uses*, pp. 299–326. Stillwater, Oklahoma: American Peanut Research and Education Association.

Subba Rao, N.S. (1976). Field response of legumes in India to inoculation and fertilizer application. In *Symbiotic Nitrogen Fixation in Plants*, ed. P.S. Nutman, pp. 255–68. Cambridge: Cambridge University Press.

Suhaeti, W. & Suhaeti, R. (1989). Farmers' capability for funding tidal swamp operation and maintenance cost. *Indonesian Agricultural Research and Development Journal*, **11**, 25–9.

Summerfield, R.J. & Bunting, A.H. (eds.) (1980). *Advances in Legume Science*. Kew: Royal Botanic Gardens.

Summerfield, R.J. & Wein, H.C. (1980). Effects of photoperiod and air temperature on growth and yield of economic legumes. In *Advances in Legume Science*, ed. R.J. Summerfield & A.H. Bunting, pp. 17–36. Kew: Royal Botanic Gardens.

Summerville, W.A.T. (1944). Studies on nutrition as qualified by development in *Musa cavendishii* Lambert. *Queensland Journal of Agricultural Science*, **1**, 1–127.

Suryawanshi, B.M. & Chaudhari, C.S. (1979). Response of gram to soil and foliar application of phosphorus. *Indian Journal of Agricultural Research*, **13**, 18–22.

Suthipradit, S., Edwards, D.G. & Asher, C.J. (1990). Effects of aluminium on tap-root elongation of soybean (*Glycine max*), cowpea (*Vigna unguiculata*) and green gram (*Vigna radiata*) grown in the presence of organic acids. *Plant and Soil*, **124**, 233–7.

Swan, D., Brown, D.M. & Coligado, M.C. (1981). Leaf emergence rates of corn (*Zea mays* L.) as affected by temperature and photoperiod. *Agricultural Meteorology*, **24**, 51–73.

Tai, E.A. (1977). Banana. In *Ecophysiology of Tropical Crops*, ed. P. de T. Alvim & T.T. Kozlowski, pp. 44–60. New York: Academic Press.

Talleyrand, H. & Lugo-Lopez, M.A. (1976). Effect of five levels and three sources of N on sweet potato yield on an Ultisol. *University of Puerto Rico Journal of Agriculture*, **60**, 9–14.

Tan, K.H. (1965). The Andosols in Indonesia. *World Soil Resources Report*, **14**, 46–9.

Tan, K.H. & Binger, A. (1986). Effect of humic acids on aluminium toxicity in corn plants. *Soil Science*, **141**, 20–5.

Tan, K.H. & Keltjens, W.G. (1990). Interaction between aluminium and phosphorus in sorghum plants. II. Studies with aluminium tolerant sorghum genotype SCO283. *Plant and Soil*, **124**, 25–32.

Tanaka, I. (1976). Climatic influence on photosynthesis and respiration of rice. In *Climate and Rice*, pp. 223–47. Los Baños, Philippines: International Rice Research Institute.

Tanguilig, V.C., Yambao, E.B., O'Toole, J.C. & de Datta, S.K. (1987). Water stress effects on leaf elongation, leaf water potential, transpiration, and nutrient uptake of rice, maize and soybean. *Plant and Soil*, **103**, 155–68.

Taylor, S.E. & Sexton, O.J. (1972). Some implications of leaf tearing in Musaceae. *Ecology*, **53**, 145–9.

Tayo, T.O. (1986). Flower and pod production at various nodes of *Phaseolus vulgaris* L. *Journal of Agricultural Science*, **107**, 29–36.

Te Krony, D.M., Egli, D.B. & Henson, G. (1981). A visual indicator of physiological maturity in soybean plants. *Agronomy Journal*, **73**, 553–6.

Tella, R. de, Canecchio, V. & Da Rocha, J.L.V. (1970). Effect of increasing rates of nitrogen, phosphorus and potassium on groundnut in podzolized soils. *Bragantia*, **29**, 199–205.

Terman, G.L. (1974). Yield nutrient concentration relationships in maize. In *Plant Analysis and Fertiliser Problems*, ed. J. Wehrmann, pp. 447–58. Hanover: German Society of Plant Nutrition.

Terra, G.J.A. (1954). Mixed-garden horticulture in Java. *Malayan Journal of Tropical Geography*, **3**, 33–43.

Terra, G.J.A. (1958). Farm systems in south-east Asia. *Netherlands Journal of Agricultural Science*, **6**, 157–64.

Theng, B.K.G. (ed.) (1981). Preface. In *Soils with Variable Charge*, pp. vi–viii. Lower Hutt, New Zealand: New Zealand Society of Soil Science.

Theodorides, T.N. & Pearson, C.J. (1981). Effect of temperature on total nitrogen distribution in *Pennisetum americanum*. *Australian Journal of Plant Physiology*, **8**, 201–10.

Thibodeau, P.S. & Jaworski, E.G. (1975). Patterns of nitrogen utilization in the soybean. *Planta*, **127**, 133–47.

Thung, M. (1991). Bean agronomy in monoculture. In *Common Beans: Research for Crop Improvement*, ed. A. van Schoonhoven & O. Voysest, pp. 737–834. Wallingford, UK: CAB International/Centro International de Agricultura Tropical.

Thung, M. & Cock, J.H. (1979). Multiple cropping cassava and field beans: status of present work at the International Centre of Tropical Agriculture (CIAT). In *Intercropping with Cassava*, ed. E. Weber, B. Nestel & M. Campbell, pp. 7–16. Ottawa, Ontario: International Development Research Centre.

Tienseemuang, S. & Ponsana, P. (1977). Hydraulic properties of soils in relation to runoff and water control. In *Soil Conservation and Management in the Humid Tropics*, ed. D.J. Greenland & R. Lal, pp. 221–8. London: Wiley.

Tollenaar, M. (1977). Sink–source relationships during reproductive development in maize: a review. *Maydica*, **22**, 49–75.

Tomar, V.S. & O'Toole, J.C. (1980). Water use in lowland rice cultivation in Asia: a review of evapotranspiration. *Agricultural Water Management*, **3**, 83–106.

Ton, C.S. & Hernandez, T.P. (1978). Wet soil stress effects on sweet potatoes. *Journal of the American Society for Horticultural Science*, **103**, 600–3.

Topark-Ngarm, B., Patcharapreecha, P., Goto, I. & Kimura, M. (1990). Studies on saline soils in Khon Kaen Region, North East Thailand. II. Seasonal changes of physical properties and chemical properties. *Soil Science and Plant Nutrition*, **36**, 289–98.

Torrey, J.G. (1982). *Casuarina*, actinorhizal nitrogen-fixer of the tropics. In *Biological Nitrogen Fixation Technology for Tropical Agriculture*, ed. P.H. Graham & S.C. Harris, pp. 427–39. Cali, Colombia: Centro Internacional de Agricultura Tropical.

Townsend, C.E. (1972). Influence of seed size and depth of planting on seedling emergence of two milkvetch species. *Agronomy Journal*, **64**, 627–30.

Trewartha, G.T. (1968). *An Introduction to Climate*, 4th edn. Tokyo: McGraw Hill/Kogakusha, 408 pp.

Troedson, R.J., Lawn, R.J. & Byth, D.E. (1981). Growth and nodulation of soybeans in high water table (HWT) culture. In *Current Perspectives in Nitrogen Fixation*, ed. A.H. Gibson & W.E. Newton. Amsterdam: Elsevier/North-Holland Biomedical Press, 534 pp.

Troedson, R.J., Lawn, R.J., Byth, D.E. & Wilson, G.L. (1989). Response of field-grown soybean to saturated soil culture. I. Patterns of biomass and nitrogen accumulation. *Field Crops Research*, **21**, 171–87.

Tsay, J.S., Fukai, S. & Wilson, G.L. (1988). Effects of relative sowing time of soybean on growth and yield of cassava in cassava/soybean intercropping. *Field Crops Research*, **19**, 227–39.

Tsuno, Y. & Fujise, K. (1963). Studies on the dry matter production of the sweet potato: II. *Crop Science Society of Japan Proceedings*, **31**, 285–8.

Tsuno, Y. & Fujise, K. (1965a). Studies on the dry matter production of sweet potato: VIII. *Crop Science Society of Japan Proceedings*, **33**, 230–5.

Tsuno, Y. & Fujise, K. (1965b). Studies on the dry matter production of sweet potato: IX. *Crop Science Society of Japan Proceedings*, **33**, 236–41.

Tu, J.C. & Tan, C.S. (1991). Effect of soil compaction on growth, yield, and root rots of white beans in clay loam and sandy loam soil. *Soil Biology and Biochemistry*, **23**, 233–8.

Turner, D.W. (1971). Effects of climate on rate of banana leaf production. *Tropical Agriculture (Trinidad)*, **48**, 283–7.

Turner, D.W. (1972a). Banana plant growth: I. *Australian Journal of Experimental Agriculture and Animal Husbandry*, **12**, 209–15.

Turner, D.W. (1972b). Banana plant growth: II. *Australian Journal of Experimental Agriculture and Animal Husbandry*, **12**, 216–24.

Turner, D.W. (1990). Modelling demand for nitrogen in the banana. *Acta Horticulturae*, **275**, 497–503.

Turner, D.W. & Barkus, B. (1973). Loss of mineral nutrients from banana pseudostems after harvest. *Tropical Agriculture (Trinidad)*, **50**, 229–33.

Turner, D.W. & Barkus, B. (1980). Plant growth and dry matter production of the 'Williams' banana in relation to supply of potassium, magnesium and manganese in sand culture. *Scientia Horticulturae*, **12**, 27–45.

Turner, D.W. & Barkus, B. (1982). Yield, chemical composition, growth and maturity of 'Williams' banana fruit in relation to supply of potassium, magnesium and manganese. *Scientia Horticulturae*, **16**, 239–52.

Turner, D.W. & Lahav, E. (1985) Temperature influences the distribution of some nutrients in young banana plants independently of dry matter. *Scientia Horticulturae*, **28**, 47–58.

Turner, N.C. (1974). Stomatal response to light and water under field conditions. *Royal Society of New Zealand Bulletin*, **12**, 423–32.

Turner, N.C. (1986a). Crop water deficits: a decade of progress. *Advances in Agronomy* **39**, 1–51.

Turner, N.C. (1986b). Adaptation to water deficits: a changing perspective. In *Plant Growth, Drought and Salinity*, ed. N.C. Turner & J.B. Passioura, pp. 175–90. Melbourne: Commonwealth Scientific and Industrial Research Organization.

Turner, N.C. & Jones, M.M. (1980). Turgor maintenance by osmotic adjustment: a review and evaluation. In *Adaptation of Plants to Water and High Temperature Stress*, ed. N.C. Turner & P.J. Kramer, pp. 87–104. New York: Wiley.

Turner, N.C., Begg, J.E., Rawson, H.M., English, S.D. & Hearn, A.B. (1978). Agronomic and physiological responses of soybean and sorghum crops to water deficits: III. *Australian Journal of Plant Physiology*, **5**, 179–94.

Turner, N.C., O'Toole, J.C., Cruz, R.T., Yambao, E.B., Ahmad, S., Namuco, O.S. & Dingkuhn, M. (1986). Responses of seven diverse rice cultivars to water deficits. II. Osmotic adjustment, leaf elasticity, leaf extension, leaf death, stomatal conductance and photosynthesis. *Field Crops Research*, **13**, 273–86.

Twyford, I.T. (1965). Soil, nutrition and fertilizers. *Annual Report, 1963/64*. WINBAN Research Scheme, St Lucia, Windward Islands.

Twyford, I.T. (1967). Banana nutrition: a review of principles and practices. *Journal of the Science of Food and Agriculture*, **18**, 177–83.

Twyford, I.T. & Walmsley, D. (1974*a*). The mineral composition of the Robusta banana plant. III. Uptake and distribution of mineral constituents. *Plant and Soil*, **41**, 471–91.

Twyford, I.T. & Walmsley, D. (1974*b*). The mineral composition of the Robusta banana plant. IV. The application of fertilizers for high yields. *Plant and Soil*, **41**, 493–508.

Uchijima, Z. (1976). Maize and rice. In *Vegetation and the Atmosphere*, vol. 2, ed. J.L. Monteith, pp. 33–64. London: Academic Press.

Uehara, G. (1977). An overview of soils of the arable tropics. In *Exploiting the Legume–Rhizobium Symbiosis in Tropical Agriculture*, ed. J.M. Vincent, A.S. Whitney & J. Bose, pp. 67–80. Honolulu, Hawaii: University of Hawaii Press.

Uexküll, H.R. von (1978). Potash and rice production in Asia. *Potash Review Cereal Crops, 41st Suite*, **8**, 1–8.

Ugent, D., Pozorski, S. & Pozorski, T. (1986). Archaeological manioc from coastal Peru. *Economic Botany*, **40**, 78–102.

Unamma, R.P.A., Akobundu, I.O. & Fayemi, A.A.A. (1981). Weed interference in white yam. In *Tropical Root Crops: Research Strategies for the 1980s*, ed. E.R. Terry, K.A. Oduro & F. Caveness, pp. 203–7. Ottawa, Ontario: International Development Research Centre.

Underwood, C.V., Taylor, H.M. & Hoveland, C.S. (1971). Soil physical factors affecting peanut pod development. *Agronomy Journal*, **63**, 953–4.

Unger, P.W. (1984). Tillage effects on surface soil physical conditions and sorghum emergence. *Soil Science Society of America Journal*, **48**, 1423–32.

Vaishya, U.K. & Gajendragadkar, G.R. (1982). Inoculation trials on chickpea in Madhya Pradesh. *International Chickpea Newsletter*, **6**, 24–5.

Vanderlip, R.L. (1972). How a sorghum plant develops. *Cooperative Extension Service, Kansas State University, Circular*, 447.

Vanderlip, R.L. & Arkin, G.F. (1977). Simulating accumulation and distribution of dry matter in grain sorghum. *Agronomy Journal*, **69**, 917–23.

VanToai, T., Fausey, N. & McDonald, M. (1988). Oxygen requirements for germination and growth of flood-susceptible and flood-tolerant corn lines. *Crop Science*, **28**, 79–83.

Vargas, A.A.T. & Graham, P.H. (1988). *Phaseolus vulgaris* cultivar and *Rhizobium* strain variation in acid-pH tolerance and nodulation under acid conditions. *Field Crops Research*, **19**, 91–101.

Varma, S.K. & Kumari, P. (1978). Nature, extent, periodicity and intensity of flower and pod shedding in gram (*Cicer arietinum* L.). *Legume Research*, **1**, 108–14.

Velagaleti, R.R. & Marsh, S. (1989). Influence of host cultivars and *Bradyrhizobium* strains on the growth and symbiotic performance of soybean under salt stress. *Plant and Soil*, **119**, 133–8.

Veltkamp, H.J. (1986). Physiological causes of yield variation in cassava, (*Manihot esculenta* Crantz). Wageningen: Agricultural University, 103 pp.

Vergara, B.S. (1976). Physiological and morphological adaptability of rice varieties to climate. In *Climate and Rice*, pp. 67–86. Los Baños, Philippines: International Rice Research Institute.

Vergara, B.S. (1992). Tropical wet rice systems. In *Field Crop Ecosystems*, ed. C. J. Pearson, pp. 167–82. Amsterdam: Elsevier.

Vergara, B.S., Puranabhavung, S. & Lilis, R. (1965). Factors determining the growth duration of rice varieties. *Phyton*, **22**, 177–85.

Vergara, B.S., Jackson, B. & Datta, S.K. de (1976). Deep water rice and its response to deep water stress. In *Climate and Rice*, pp. 301–19. Los Baños, Philippines: International Rice Research Institute.

Verteuil, J. de (1971). Cassava experiments. *Bulletin of Department of Agriculture, Trinidad and Tobago*, **16**, 18–21.

Vinall, H.N. & Reed, H.R. (1918). Effect of temperature and other meteorological factors in the growth of sorghum. *Journal of Agricultural Research*, **13**, 133–9.

Vine, H. (1968). Developments in the study of shifting agriculture in tropical Africa. In *The Soil Resources of Tropical Africa*, ed. R.P. Moss, pp. 89–119. Cambridge: Cambridge University Press.

Vine, P.N. & Ahmad, N. (1987). Yield development in cassava under different soil physical conditions. *Field Crops Research*, **17**, 175–98.

Virmani, S.M. (1975). The agricultural climate of the Hyderabad region in relation to crop planning. Hyderabad, India: International Crops Research Institute for the Semi-Arid Tropics. Mimeo, 61 pp.

Virmani, S.M. & Singh, P. (1986). Agroclimatological characteristics of the groundnut-growing regions in the semi-arid tropics. In *Agrometeorology of Groundnut*, pp. 3–46. Patancheru, India: International Crops Research Institute for the Semi-Arid Tropics.

Vishnu-Mittre (1974). Palaeobotanical evidence in India. In *Evolutionary Studies on World Crops*, ed. J.B. Hutchinson, pp. 3–30. Cambridge: Cambridge University Press.

Vishnu-Mittre (1977). Changing economy in ancient India. In *Origins of Agriculture*, ed. C.A. Reed, pp. 570–88. The Hague: Mouton.

Vitorello, V.A., Cerri, C.C., Andreux, F., Feller, C. & Victoria, R.L. (1989). Organic matter and natural carbon-13 distribution in forested and cultivated Oxisols. *Soil Science Society of America Journal*, **53**, 773–8.

Vittal, K.P.R., Vijayalakshmi, K. & Rao, U.M.B. (1990). Relationships between topsoil depth and yields of rain-fed sorghum, pearl millet, and castor bean. *Soil Science*, **150**, 476–81.

Voysest, O. & Dessert, M. (1991). Bean cultivars: classes and commercial seed types. In *Common Beans: Research for Crop Improvement*, ed. A. van Schoonhoven & O. Voysest, pp. 119–62. Wallingford, UK: CAB International/Centro Internacional de Agricultura Tropical.

Vries, C.A. de, Ferwerda, J.D. & Flack, M. (1967). Choice of food crops in relation to actual and potential production in the tropics. *Netherlands Journal of Agricultural Science*, **15**, 241–8.

Wada, K. (1981). Mineralogical characteristics of Andisols. In *Soils with Variable Charge*, ed. B.K.G. Theng, pp. 87–108. Lower Hutt, New Zealand: New Zealand Society of Soil Science.

Wade, M.K. (1978). Soil management practices for increased crop production for small farms of the Amazon jungle of Peru. PhD thesis, North Carolina State University, Raleigh.

Waggoner, P.E. (1968). Meterological data and the agricultural problem. In *Agroclimatological Methods*, pp. 25–38. Paris: United Nations Educational, Scientific and Cultural Organization.

Wahab, A.H. (1979). Performance of soybeans in Guyana's peats and peaty clays. *University of Puerto Rico Journal of Agriculture*, **63**, 443–9.

Wahab, A.H., Hassan, I. & Lugo-Lopez, M.A. (1978). Performance of twenty cassava cultivars on marginal peats and peaty clays of Guyana. *University of Puerto Rico Journal of Agriculture*, **62**, 343–54.

Waitt, A.W. (1963). Yams, *Dioscorea* species. *Field Crop Abstracts*, **16**, 145–57.

Walcott, J.J., Chauviroj, M., Chinchest, A., Chotichevy, P., Ferraris, R. & Norman, B.W. (1977). Long-term productivity of intensive rice cropping systems on the Central Plain of Thailand. *Experimental Agriculture*, **13**, 305–15.

Walker, G.K. & Hatfield, J.L. (1979). Test of the stress-degree-day concept using multiple planting dates of red kidney beans. *Agronomy Journal*, **71**, 967–71.

Walker, M.E., Morris, H.D. & Carter, R.L. (1974). The effect of rate and method of application of N, P, and K on yield, quality and chemical composition of Spanish and Runner peanuts. *Georgia Agricultural Experimental Station Research Bulletin*, No. 152.

Walker, M.E., Marchant, W.H. & Ethredge, W.J. (1975). Effects of soil pH on forage yield and chemical composition of sorghum and millet. *Agronomy Journal*, **67**, 191–3.

Wallace, D.H. (1980). Adaptation of *Phaseolus* to different environments. In *Advances in Legume Science*, ed. R.J. Summerfield & A.H. Bunting, pp. 349–57. Kew: Royal Botanic Gardens.

Wallace, D.H. & Enriquez, G.A. (1980). Daylength and temperature effects on days to flowering of early and late maturing beans (*Phaseolus vulgaris* L.). *Journal of the American Society for Horticultural Science*, **105**, 583–91.

Walmsley, D. & Twyford, I.T. (1968). The uptake of ^{32}P by the Robusta banana. *Tropical Agriculture (Trinidad)*, **45**, 223–8.

Walton, D.C., Galston, E. & Harris, M.A. (1977). The relationship between stomatal resistance and abscissic acid levels in leaves of water-stressed bean plants. *Planta*, **133**, 145–8.

Wambeke, A. van (1976). Formation, distribution and consequences of acid soils in agricultural development. In *Plant Adaptation to Mineral Stress in Problem Soils*, ed. M.J. Wright, pp. 15–24. Cornell University Agricultural Experiment Station Special Publication.

Wani, S.P., Chandrapalaih, S., Zambre, M.A. & Lee, K.K. (1988). Association between N_2-fixing bacteria and pearl millet plants: responses, mechanisms and persistence. *Plant and Soil*, **110**, 289–302.

Wardlaw, C.W., Leonard, E.R. & Barnell, H.R. (1939). Studies in tropical fruits: VII. *Annals of Botany*, **3**, 845–60.

Warkentin, B.P. & Maeda, T. (1981). Physical and mechanical characteristics of Andisols. In *Soils with Variable Charge*, ed. B.K.G. Theng, pp. 281–302. Lower Hutt, New Zealand: New Zealand Society of Soil Science.

Warner, R.M. & Fox, R.L. (1977). Nitrogen and potassium nutrition of the Giant Cavendish banana in Hawaii. *Journal of the American Society for Horticultural Science*, **102**, 739–43.

Warren-Wilson, J. (1971). Maximum potential yield. In *Transition from Extensive to Intensive Agriculture with Fertilizers*, Proceedings of the Seventh Colloquium, International Potash Institute, pp. 314–56. Berne, Switzerland: International Potash Institute.

Watanabe, I. & Liu, C.C. (1992). Improving nitrogen-fixing systems and integrating them into sustainable rice farming. *Plant and Soil*, **141**, 57–67.

Watanabe, K. & Nakayama, K. (1969). Studies on the effect of soil physical conditions on the growth and yield of sweet potato: X. *Crop Science Society of Japan Proceedings*, **39**, 446–50.

Watters, R.J. (1971). Shifting Cultivation in Latin America. FAO Forestry Development Paper, No. 17. Rome: Food and Agriculture Organization.

Wedderburn, R.N. & DeLeon, C. (eds.) (1987). *Proceedings of the Second Asian Regional Maize Workshop*. El Batan, Mexico: Centro Internacional de Mejoramiento de Maiz y Trigo.

Wein, H.C., Standsted, R.F. & Wallace, D.H. (1973). The influence of flower removal on growth and seed yield of *Phaseolus vulgaris* L. *Journal of the American Society for Horticultural Science*, **98**, 45–9.

Weiss, A., Hipp, L.E., Blad, B.L. & Steadman, J.R. (1980). Comparison of within-canopy microclimate and white mold disease *(Sclerotinia sclerotiorum)* development in dry edible beans as influenced by canopy structure and irrigation. *Agricultural Meteorology*, **22**, 11–21.

Westermann, D.T. & Crothers, S.E. (1977). Plant population effects on the seed yield components of beans. *Crop Science*, **17**, 493–6.

Westermann, D.T., Kleinkopf, G.E., Porter, L.K. & Legett, G.E. (1981). Nitrogen sources for bean seed production. *Agronomy Journal*, **73**, 660–4.

Westphal, E. (1974). *Pulses in Ethiopia: Their Taxonomy and Agricultural Significance*. Wageningen, The Netherlands: Centre for Agricultural Publishing and Documentation, 276 pp.

Westphal, E. (1975). *Agricultural Systems in Ethiopia*. Wageningen, The Netherlands: Centre for Agricultural Publishing and Documentation, 278 pp.

Wet, J.M.J. de, Bidinger, F.R. & Peacock, J.M. (1992). Pearl millet (*Pennisetum glaucum*): a cereal of the Sahel. In *Desertified Grasslands: Their Biology and Management*, ed. G.P. Chapman, pp. 259–67. London: Academic Press.

Wetselaar, R. (1961a). Nitrate distribution in tropical soils. I. Possible causes of nitrogen accumulation near the surface after a long dry period. *Plant and Soil*. **15**, 110–20.

Wetselaar, R. (1961b). Nitrate distribution in tropical soils. II. Extent of capillary accumulation of nitrate during a long dry period. *Plant and Soil*, **15**, 121–33.

Wetselaar, R. (1967a). Determination of the mineralization coefficient of soil organic nitrogen in two soils at Katherine, N.T. *Australian Journal of Experimental Agriculture and Animal Husbandry*, **7**, 266–74.

Wetselaar, R. (1967*b*). Estimation of nitrogen fixation by four legumes in a dry monsoonal area of north-western Australia. *Australian Journal of Experimental Agriculture and Animal Husbandry*, **7**, 518–22.

Wetselaar, R. (1981). Nitrogen inputs and outputs of an unfertilized paddy field. In *Terrestrial Nitrogen Cycles: Processes, Ecosystem Strategies and Management Impacts*, ed. F.E. Clark & T. Rosswall. *Ecology Bulletin*, **3**, 573–83.

Wetselaar, R. & Ganry, F. (1981). Nitrogen balance in tropical agrosystems. In *Microbiology of Tropical Soils*, ed. Y. Dommergues & H.G. Diem. The Hague: Martinus Nijhoff.

Wetselaar, R. & Norman, M.J.T. (1960). Recovery of available soil nitrogen by annual fodder crops at Katherine, Northern Territory. *Australian Journal of Agricultural Research*, **11**, 693–704.

Whitt, D.M. & Bavel, C.H.M. van (1955). Irrigation of tobacco, peanuts, and soybeans. In *Water, Yearbook of Agriculture*, pp. 376–81. Washington, DC: US Government Printing Office.

Wholey, D.W. & Cock, J.H. (1974). Onset and rate of root bulking in cassava. *Experimental Agriculture*, **10**, 193–8.

Wholey, D.W. & Haynes, P.H. (1969). The analysis of foliage characteristics of sweet potato cultivars with contrasting response to nitrogen fertilizer. In *Half-Yearly Report, Faculty of Agriculture, University of the West Indies, St Augustine, 1968–69*, pp. 54–5.

Wieneke, J. (1990). Phosphorus efficiency and phosphorus remobilisation in two sorghum (*Sorghum bicolor* (L.) Moench) cultivars. *Plant and Soil*, **123**, 139–45.

Wignarajah, K. (1990). Growth response of *Phaseolus vulgaris* to varying salinity regimes. *Environmental and Experimental Botany*, **30**, 141–7.

Wilcox, J.R. & Frankenberger, E.M. (1987). Indeterminate and determinate soybean responses to planting date. *Agronomy Journal*, **79**, 1074–8.

Wild, A. (1971). The potassium status of soils in the savannah zone of Nigeria. *Experimental Agriculture*, **7**, 257–70.

Wilkinson, G.E & Aina, P.O. (1976). Infiltration of water into two Nigerian soils under secondary forest and subsequent arable cropping. *Geoderma*, **50**, 1–9.

Willatt, S.T. & Olsson, K.A. (1982). Root distribution and water uptake by irrigated soybeans on a Duplex soil. *Australian Journal of Soil Research*, **20**, 139–46.

Willett, I.R. (1986). Phosphorus dynamics in relation to redox processes in flooded soils. In *Transactions of the Thirteenth International Conference of the Soil Science Society*, vol. 6, pp. 748–55. Hamburg: International Soil Science Society.

Willey, R.W. (1979). Intercropping: its importance and research needs. I. Competition and yield advantages. *Field Crop Abstracts*, **32**, 1–10.

Willey, R.W. & Osiru, D.S.O. (1972). Studies on mixtures of maize and beans (*Phaseolus vulgaris*) with particular reference to plant population. *Journal of Agricultural Science*, **79**, 517–29.

Williams, C.N. (1972). Growth and productivity of tapioca (*Manihot ultissima*): III. *Experimental Agriculture*, **8**, 15–23.

Williams, C.N. (1974). Growth and productivity of tapioca (*Manihot utissima*): IV. *Experimental Agriculture*, **10**, 9–16.

Williams, C.N. (1975). *The Agronomy of the Major Tropical Crops*. Kuala Lumpur: Oxford University Press, 228 pp.

Williams, C.N. & Joseph. K.T. (1970). *Climate, Soil and Crop Production in the Humid Tropics.* Kuala Lumpur: Oxford University Press, 177 pp.

Williams, J.H. (1979). The physiology of groundnuts (*Arachis hypogaea* L. cv. Egret): II. *Rhodesian Journal of Agricultural Research*, **17**, 49–55.

Williams, J.H., Wilson, J.H.H. & Bate, G.C. (1975). The growth of groundnuts (*Arachis hypogaea* L. v. Makulu Red) at three altitudes in Rhodesia. *Rhodesian Journal of Agricultural Research*, **13**, 33–43.

Williams, J.H., Dutta, M. & Nambiar, P.T.C. (1990). Light interception as a source of variation for nitrogen fixation in groundnut genotypes. *Plant and Soil*, **121**, 83–8.

Williamson, R.E. & Kriz, G.J. (1970). Response of agricultural crops to flooding, depth of water table and soil gaseous composition. *American Society of Agricultural Engineers, Transactions*, **13**, 216–20.

Williamson, R.E., Willey, C.R. & Gray, T.N. (1969). Effect of water table depth and flooding on yield of millet. *Agronomy Journal*, **61**, 310–13.

Wilson, J.W., Clowes, M. St J. & Allison, J.C.S. (1973). Growth and yield of maize at different altitudes in Rhodesia. *Annals of Applied Biology*, **73**, 77–84.

Wilson, L.A. (1977). Root crops. In *Ecophysiology of Tropical Crops*, ed. P. de T. Alvim & T.T. Kozlowski, pp. 187–236. New York: Academic Press.

Wilson, L.A. (1982). Tuberization in sweet potato (*Ipomoea batatas* (L) Lam.). In *Proceedings of the First International Sweet Potato Symposium*, pp. 79–94. Tainan, Taiwan: Asian Vegetable Research and Development Centre.

Wilson, L.A., Rankine, L.B., Ferguson, T.U., Ahmad, N., Griffith, S. & Roberts-Nkrumah, L. (1992). Mixed root-crop systems in the Caribbean. In *Field Crop Ecosystems*, ed. C.J. Pearson, pp. 205–42. Amsterdam : Elsevier.

Wischmeier, W.H. & Smith, D.D. (1958). Rainfall energy and its relationship to soil loss. *Transactions of the American Geophysical Union*, **39**, 285–91.

Wit, C.T. de (1978). *Simulation of Assimilation, Respiration and Transpiration of Crops.* Wageningen: PUDOC, 141 pp.

Wolt, J.D. & Adams, F. (1979). Critical levels of soil- and nutrient-solution calcium for vegetative growth and fruit development of Florunner peanuts. *Soil Science Society of America Journal*, **43**, 1159–65.

Wong, M.T.F., Hughes, R. & Rowell, D.L. (1990). Retarded leaching of nitrate in acid soils from the tropics: measurement of the effective anion exchange capacity. *Journal of Soil Science*, **41**, 655–63.

Wood, I.M.W. (1968a). The effect of temperature at early flowering on the growth and development of peanuts (*Arachis hypogaea*). *Australian Journal of Agricultural Research*, **19**, 241–51.

Wood, I.M.W. (1968b). The effects of seed size, seed treatment, method of shelling and organic matter on the establishment of peanuts (*Arachis hypogaea*) at Katherine, N.T. *Australian Journal of Experimental Agriculture and Animal Husbandry*, **8**, 81–4.

Woolley, J. & Davis, J.H.C. (1991). The agronomy of intercropping with beans. In *Common Beans: Research for Crop Improvement*, ed. A. van Schoonhoven & O. Voysest, pp. 707–35. Wallingford, UK: CAB International/Centro Internacional de Agriculture Tropical.

Woolley, J., Ildefonso, R.L., Portes e Castro, T. de A. & Voss, J. (1991). Bean cropping systems in the tropics and subtropics and their determinants. In *Common Beans: Research for Crop Improvement*, ed. A. van Schoonhoven & O. Voysest, pp. 679–705. Wallingford, UK: CAB International/Centro Internacional de Agricultura Tropical.

Wright, G.C. (1989). Effect of pod zone moisture content on reproductive growth in three cultivars of peanut (*Arachis hypogaea*). *Plant and Soil*, **116**, 111–14.

Wright, G.C. & Smith, C.J. (1987). Soybean root distribution under wet soil culture on a red-brown earth. *Plant and Soil*, **103**, 129–33.

Wu, M.-H., Lee, S.-T. & Chiang, M.-H. (1979). Effects of submerged paddy soil on the life of soybean rhizobia. *Journal of the Agricultural Association, China (New Series)*, **64**, 13–17.

Wynne, J.C., Elkan, G.H., Meisner, C.M., Schneeweis, T.J. & Ligon, J.M. (1980). Greenhouse evaluations of strains of *Rhizobium* for peanuts. *Agronomy Journal*, **72**, 645–9.

Yadav, J.S.P. & Gupta, A.R. (1984). *Usar Bhumi Ka Sudhar* (In Hindi). New Delhi: Indian Council of Agricultural Research.

Yamakawa, H. (1962). Studies on the ecological variations of the growth of rice plants caused by shifting of cultivation season in warm region of Japan. *Agricultural Bulletin, Saga University*, **14**, 23–159.

Yamauchi, M. & Winslow, M.D. (1989). Effect of silica on yield of upland rice in the humid tropics. *Plant and Soil*, **113**, 265–9.

Yayock, J.Y. (1979). Effects of variety and spacing on growth, development and dry matter distribution in groundnut (*Arachis hypogaea* L.) at two locations in Nigeria. *Experimental Agriculture*, **15**, 339–51.

Yen, D.E. (1974). *The Sweet Potato and Oceania: An Essay in Ethnobotany*. Honolulu, Hawaii: Bishop Museum Press, 389 pp.

Yen, D.E. (1976). Sweet potato. *Ipomoea batatas* (Convolvulaceae). In *Evolution of Crop Plants*, ed. N.W. Simmonds, pp. 52–5. London: Longman.

Yong, C.K. (1961). Effects of thermoperiodism on tuber formation in *Ipomoea batatas* under controlled conditions. *Plant Physiology*, **36**, 380–4.

Yoshida, S. (1968). Varietal differences. In *IRRI Annual Report 1968*, pp. 26–8. Los Baños, Philippines: International Rice Research Institute.

Yoshida, S. (1973). Effects of temperature on growth of the rice plant (*Oryza sativa L.*) in a controlled environment. *Soil Science and Plant Nutrition (Tokyo)*, **19**, 299–310.

Yoshida, S. (1977). Rice. In *Ecophysiology of Tropical Crops*, ed. P. de T. Alvim & T. Kozlowski, pp. 57–88. New York: Academic Press.

Yoshida, S. & Parao, F.T. (1976). Climatic influence on yield and yield components of lowland rice in the tropics. In *Climate and Rice*, pp. 471–94. Los Baños, Philippines: International Rice Research Institute.

Yost, R.S. & Fox, R.L. (1979). Contribution of mycorrhizae to P nutrition of crops growing on an Oxisol. *Agronomy Journal*, **71**, 903–8.

Yost, R.S., Kamprath, E.J., Naderman, G.C. & Lobato, E. (1981). Residual effects of phosphorus applications on a high phosphorus adsorbing Oxisol of Central Brazil. *Soil Science Society of America Journal*, **45**, 540–3.

Zaag, P. van der (1979). The phosphorus requirements of root crops. PhD thesis, Department of Agronomy and Soil Science, University of Hawaii, 140 pp.

Zaag, P. van der & Fox, R.L. (1981). Field production of yams (*Dioscorea alata*) from stem cuttings. *Tropical Agriculture (Trinidad)*, **58**, 143–5.

Zaag, P. van der, Fox, R.L., Pena, R.S. de la & Yost, R.S. (1979). P nutrition of cassava including mycorrhizal effects on P, K, S, Zn and Ca uptake. *Field Crops Research*, **2**, 253–63.

Zaag, P. van der, Fox, R.L., Kwakye, P.K. & Obigbesan, G.O. (1980). The phosphorus requirements of yams (*Dioscorea* spp.). *Tropical Agriculture (Trinidad)*, **57**, 97–106.

Zandstra, H.G. (1979). Cassava intercropping research: agroclimatic and biological interactions. In *Intercropping with Cassava*, ed. E. Weber, B. Nestel & M. Campbell, pp. 67–75. Ottawa, Ontario: International Development Research Centre.

Zehni, M.S., Saad, F.S. & Morgan, D.C. (1970). Photoperiod and flower bud development in *Phaseolus vulgaris*. *Nature (London)*, **227**, 628–9.

Zhailibaev, K. & Khosenov, E.K. (1966). On the formation and productivity of the photosynthetic apparatus of soybean under varying irrigation conditions. *Soviet Plant Physiology*, **13**, 146–9.

Zinke, P.J., Stangenberger, A.G., Post, W.M., Emanuel, W.R. & Olson, J.S. (1984). *Worldwide Organic Soil Carbon and Nitrogen Data*. Oak Ridge National Laboratory, Environmental Sciences Division Publication No. 2212. US Department of Energy.

Index